Fundamentals of Photonic Crystal Guiding

If you're looking to understand photonic crystals, this systematic, rigorous, and pedagogical introduction is a must. Here you'll find intuitive analytical and semi-analytical models applied to complex and practically relevant photonic crystal structures. You will also be shown how to use various analytical methods borrowed from quantum mechanics, such as perturbation theory, asymptotic analysis, and group theory, to investigate many of the limiting properties of photonic crystals, which are otherwise difficult to rationalize using only numerical simulations.

An introductory review of nonlinear guiding in photonic lattices is also presented, as are the fabrication and application of photonic crystals. In addition, end-of-chapter exercise problems with detailed analytical and numerical solutions allow you to monitor your understanding of the material presented. This accessible text is ideal for researchers and graduate students studying photonic crystals in departments of electrical engineering, physics, applied physics, and mathematics.

Maksim Skorobogatiy is Professor and Canada Research Chair in Photonic Crystals at the Department of Engineering Physics in École Polytechnique de Montréal, Canada. In 2005 he was awarded a fellowship from the Japanese Society for Promotion of Science, and he is a member of the Optical Society of America.

Jianke Yang is Professor of Applied Mathematics at the University of Vermont, USA. He is a member of the Optical Society of America and the Society of Industrial and Applied Mathematics.

Fundamentals of Photonic Crystal Guiding

MAKSIM SKOROBOGATIY[1]

JIANKE YANG[2]

École Polytechnique de Montréal, Canada[1]
University of Vermont, USA[2]

CAMBRIDGE UNIVERSITY PRESS
Cambridge, New York, Melbourne, Madrid, Cape Town, Singapore, São Paulo, Delhi

Cambridge University Press
The Edinburgh Building, Cambridge CB2 8RU, UK

Published in the United States of America by Cambridge University Press, New York

www.cambridge.org
Information on this title: www.cambridge.org/9780521513289

First published 2009

Printed in the United Kingdom at the University Press, Cambridge

A catalog record for this publication is available from the British Library

Library of Congress Cataloging in Publication data

Skorobogatiy, Maksim, 1974–
Fundamentals of photonic crystal guiding / by Maksim Skorobogatiy and Jianke Yang.
    p.   cm.
Includes index.
ISBN 978-0-521-51328-9
1. Photonic crystals.   I. Yang, Jianke.   II. Title.
QD924.S56   2008

621.36 – dc22   2008033576

ISBN 978-0-521-51328-9 hardback

M. Skorobogatiy dedicates this book to his family. He thanks his parents
Alexander and Tetyana for never-ceasing support, encouragement, and
participation in all his endeavors. He also thanks his wife Olga,
his children, Alexander junior and Anastasia, and
his parents for their unconditional love.

J. Yang dedicates this book to his family.

Contents

Preface

The field of photonic crystals (aka periodic photonic structures) is experiencing an unprecedented growth due to the dramatic ways in which such structures can control, modify, and harvest the flow of light.

The idea of writing this book came to M. Skorobogatiy when he was developing an introductory course on photonic crystals at the Ecole Polytechnique de Montréal/ University of Montréal. The field of photonic crystals, being heavily dependent on numerical simulations, is somewhat challenging to introduce without sacrificing the qualitative understanding of the underlying physics. On the other hand, exactly solvable models, where the relation between physics and quantitative results is most transparent, only exist for photonic crystals of trivial geometries. The challenge, therefore, was to develop a presentational approach that would maximally use intuitive analytical and semi-analytical models, while applying them to complex and practically relevant photonic crystal structures.

We would like to note that the main purpose of this book is not to present the latest advancements in the field of photonic crystals, but rather to give a systematic, logical, and pedagogical introduction to this vibrant field. The text is largely aimed at students and researchers who want to acquire a rigorous, while intuitive, mathematical introduction into the subject of guided modes in photonic crystals and photonic crystal waveguides. The text, therefore, favors analysis of analytically or semi-analytically solvable problems over pure numerical modeling. We believe that this is a more didactical approach when trying to introduce a novice into a new field. To further stimulate understanding of the book content, we suggest many exercise problems of physical relevance that can be solved analytically.

In the course of the book we extensively use the analogy between the Hamiltonian formulation of Maxwell's equations and the Hamiltonian formulation of quantum mechanics. We present both frequency and propagation-constant based Hamiltonian formulations of Maxwell's equations. The latter is particularly useful for analyzing photonic crystal-based linear and nonlinear waveguides and fibers. This approach allows us to use a well-developed machinery of quantum mechanical semi-analytical methods, such as perturbation theory, asymptotic analysis, and group theory, to investigate many of the limiting properties of photonic crystals, which are otherwise difficult to investigate based only on numerical simulations.

M. Skorobogatiy has contributed Chapters 2, 3, 4, 5, and 6 of this book, and J. Yang has contributed Chapter 8. Chapters 1 and 7 were co-authored by both authors.

Acknowledgements

M. Skorobogatiy would like to thank his graduate and postgraduate program mentors, Professor J. D. Joannopoulos and Professor Y. Fink from MIT, for introducing him into the field of photonic crystals. He is grateful to Professor M. Koshiba and Professor K. Saitoh for hosting him at Hokkaido University in 2005 and for having many exciting discussions in the area of photonic crystal fibers. M. Skorobogatiy acknowledges the Canada Research Chair program for making this book possible by reducing his teaching load.

J. Yang thanks the funding support of the US Air Force Office of Scientific Research, which made many results of this book possible. He also thanks the Zhou Pei-Yuan Center for Applied Mathematics at Tsinghua University (China) for hospitality during his visit, where portions of this book were written. Both authors are grateful to their graduate and postgraduate students for their comments and help, while this book was in preparation. Especially, J. Yang likes to thank Dr. Jiandong Wang, whose help was essential for his book writing.

1   Introduction

When thinking about traditional optical materials one invokes a notion of homogeneous media, where imperfections or variations in the material properties are minimal on the length scale of the wavelength of light λ (Fig. 1.1 (a)). Although built from discrete scatterers, such as atoms, material domains, etc., the optical response of discrete materials is typically "homogenized" or "averaged out" as long as scatterer sizes are significantly smaller than the wavelength of propagating light. Optical properties of such homogeneous isotropic materials can be simply characterized by the complex dielectric constant ε. Electromagnetic radiation of frequency ω in such a medium propagates in the form of plane waves $\mathbf{E}, \mathbf{H} \sim e^{i(\mathbf{k}\cdot\mathbf{r}-\omega t)}$ with the vectors of electric field $\mathbf{E}(\mathbf{r}, t)$, magnetic field $\mathbf{H}(\mathbf{r}, t)$, and a wave vector $\mathbf{k}$ forming an orthogonal triplet. In such materials, the dispersion relation connecting wave vector and frequency is given by $\varepsilon\omega^2 = c^2\mathbf{k}^2$, where c is the speed of light. In the case of a complex-valued dielectric constant ε, one typically considers frequency to be purely real, while allowing the wave vector to be complex. In this case, the complex dielectric constant defines an electromagnetic wave decaying in space, $|\mathbf{E}|, |\mathbf{H}| \sim e^{-\mathrm{Im}(\mathbf{k})\cdot\mathbf{r}}$, thus accounting for various radiation loss mechanisms, such as material absorption, radiation scattering, etc.

Another common scattering regime is a regime of geometrical optics. In this case, radiation is incoherently scattered by the structural features with sizes considerably larger than the wavelength of light λ (Fig. 1.1 (b)). Light scattering in the regime of geometrical optics can be quantified by the method of ray tracing. There, the rays are propagating through the piecewise homogeneous media, while experiencing partial reflections on the structure interfaces. At any spatial point, the net light intensity is computed by incoherent addition of the individual ray intensities.

The regime of operation of photonic crystals (PhC) falls in between the two limiting cases presented above. This is because a typical feature size in a photonic crystal structure is comparable to the wavelength of propagating light λ (Fig. 1.1(c)). Moreover, when scatterers are positioned in a periodic array (hence the name photonic crystals), coherent addition of scattered fields is possible, thus leading to an unprecedented flexibility in changing the dispersion relation and density distribution of electromagnetic states.

Originally, photonic crystals were introduced in the context of controlling spontaneous emission of atoms. [1,2] It was then immediately suggested that in many respects the behavior of light in periodic dielectrics is similar to the behavior of electrons in the periodic potential of a solid-state crystal, [3,4] and, therefore, one can manipulate the flow of light in photonic crystal circuits in a similar manner as one can manipulate

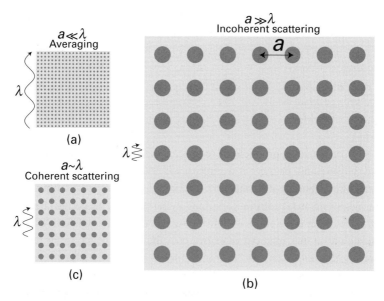

Figure 1.1 Scattering regimes. (a) Homogeneous media, averaging over individual scatterers. (b) Incoherent scattering in geometrical optics. (c) Coherent scattering from the structure with characteristic feature size comparable to the wavelength of light.

the flow of electrons in solid-state circuits. Since then, photonic crystals became a very dynamic research field, with many novel applications and fabrication methods discovered regularly. The existing body of work on photonic crystals is so vast that to do proper justice to all the great ideas is impossible in the context of a short review chapter. For a comprehensive introduction, we therefore refer the reader to several recent review articles and references thereof, which we summarize briefly in the remainder of this section.

1.1   Fabrication of photonic crystals

It is interesting to note that photonic crystals, despite their structural complexity, have many analogs in nature. Biologically occurring photonic crystals are discussed, for example, in [5] where the authors present an overview of natural photonic crystals and discuss their characteristic geometries, sizes, material composition, and bandgap structure. The authors conclude that a low-refractive-index contrast is a universal feature of biological photonic crystals, compared with artificial ones. Natural photonic crystals are also encountered in nonorganic compounds, such as opals.

Recently, artificial opals fabricated by self-assembly of colloidal particles became a popular research topic, mostly owing to the relatively simple and cost-effective methodology of producing large samples of periodic photonic structures. In [6,7], the authors survey some of the techniques used for promoting self-organization of the colloidal spheres into 2D and 3D ordered lattices, highlighting applications of these artificial crystals as photonic bandgap reflectors, as physical masks in lithographic patterning, and as diffractive elements for optical sensors. Furthermore, by infiltration of the artificial opals with high-refractive-index materials, and after removal of the original spheres, one can create "true" 3D photonic crystals with complete photonic bandgaps in all directions. Although the self-assembly approach is simple and cost effective, random lattice defects, such as dislocations and missing particles, limit the maximal size of a defect-free photonic crystal. Moreover, the lattice symmetries resulting from colloidal assembly are limited.

An alternative "mass production" technique developed for fabrication of the uniform 2D and 3D photonic crystals is a multi-beam interference lithography, also known as holographic lithography. In this technique, the periodic intensity pattern caused by several interfering beams is transferred to a photosensitive polymer substrate through a single exposure. In [8,9], the authors review the relationship between beam geometry and the symmetry of the interference patterns, the lithographic process, and various types of photoresist systems.

Just like the introduction of dopants into semiconductor crystals, to provide advanced functionalities to photonic crystals one has to introduce controlled defects into the photonic crystal's otherwise perfectly periodic lattice. While self-assembly of colloidal crystals and holography provides an efficient way of fabricating large-scale 2D and 3D photonic crystals, these methods have to be supplemented with other methods for defect incorporation. In [10] the authors cover recent advances in the fabrication of defects within photonic crystal lattices, including, among others, micromanipulation, direct writing, and multistep procedures involving the combination of 3D PhC fabrication methods and some types of 2D lithography. In particular, the micromanipulation approach uses nanopositioner robots and proceeds through serial assembly of nanoscale building blocks. Although this method has the potential of building nearly arbitrary 3D structures it suffers from the serial nature of the nanomanipulation process. Another methodology is the direct writing of photonic crystals in photosensitive materials. This method is interesting as it is relatively fast, easily parallelizable, and it treats on an equal footing the writing of periodic structures and the incorporation of nonperiodic defects. One of the implementations of a direct writing method is based on two-photon polymerization (TPP). This utilizes a nonlinear multiphoton excitation process to polymerize a photosensitive monomer in a very small volume (potentially subwavelength) around the focal point. This technique has been successfully tested in writing embedded features within holographic and self-assembled photonic crystals. Finally, several multistep methodologies have been developed for the incorporation of point, linear, and planar defects within self-assembled photonic crystals. These approaches generally involve three steps. The first step consists of growing a self-assembled colloidal crystal with flat surface termination. The second step involves deposition of an intermediate layer, followed by some kind of 2D lithographic patterning of the layer. Finally, self-assembly growth of the overlaying part of a photonic crystal concludes the process.

To date, probably the most studied process for photonic crystal device fabrication is based on 2D patterning of planar dielectric slabs using electron beam lithography. State-of-the-art lithography in combination with electrochemical or reactive-ion etching allows fabrication of low-imperfection 2D photonic crystals in various high-refractive-index material combinations, including silicon–air and III–V semiconductors. [10,11,12] The introduction of defects into these structures is carried out at the same time and on the same footing as periodic structure formation, resulting in excellent alignment between the defects and the surrounding lattice. By removing a single feature (a cylinder, for example) or a line of features from the perfectly periodic lattice, resonators (point defects) or waveguides (linear defects) can be introduced. Finally, the generalization of planar lithographical techniques to create 3D photonic crystals using multistep processes, such as wafer fusion and layer-by-layer photolithography, was recently reviewed in [10,13].

Despite their great potential, most photonic crystal-based devises are still in the research phase. Some of the few existing commercial applications of photonic crystals beyond 1D planar interference filters include glass and polymer based photonic crystal fibers (PCF). [14,15,16] Similar to conventional optical fibers, PCFs are slender cylinders with sizes ranging from below 100 μm to over 1 mm in diameter. In its cross-section, as the name suggests, the PCF features a 2D periodic pattern of holes or a 1D periodic pattern of concentric layers, as well as a localized defect serving as a fiber core. Lengthwise, PCFs can be of several meters to several kilometers. Photonic crystal fiber fabrication differs greatly from that of conventional photonic crystals as it typically proceeds through heating and elongation of a macroscale fiber preform. As a result, PCF fabrication is ideally suited for mass production, resulting in kilometers of high-quality, uniform fiber per single preform. Commercial applications of PCFs, among others, include supercontinuum generation for spectroscopic applications, [17,18] and photonic crystal fibers for radiation guidance in hollow, gas-filled, and subwavelength porous cores [19,20,21,22,23,24] thus enabling low-loss, low-nonlinearity, high-power delivery anywhere in the visible, IR, and even THZ spectral ranges.

1.2   Application of photonic crystals

In this section, we review some of the unique properties of photonic crystals that make them desirable for various light managing applications. As in the case of the overview of fabrication techniques presented earlier, we refer the reader to the recent review papers [5,6,7,8,9,10,11,12,13,14] for a more thorough introduction to possible PhC applications.

1.2.1   Photonic crystals as low-loss mirrors: photonic bandgap effects

The majority of applications of photonic crystals utilize the phenomenon of photonic bandgap (PBG). Photonic bandgap is defined as a frequency region characterized by zero density of the electromagnetic states. For the frequencies within PBG, the propagation of electromagnetic waves inside of a photonic crystal is, therefore, suppressed. The existence of PBGs opens the road to the design of efficient low-loss dielectric reflectors that can confine radiation in channels (waveguides) or localized defects (resonators) with sizes comparable to the wavelength of light.

In general, confining light presents a difficult task, as electromagnetic radiation propagates unimpeded in free space. Therefore, confining light is fundamentally different
from confining direct electric currents for which free space acts as an isolator. Consider,
for example, optical fiber, which is the optical analog of electric wire. Light in an optical
fiber is confined inside the optically dense core, like the electric current, which is confined inside a wire core of low ohmic resistance. When optical fiber is bent, substantial
radiation leakage from the fiber core into the fiber cladding is observed, as escaping
radiation propagates freely through the cladding region. A similar effect is not observed
in bent metallic wires, as free space surrounding the wire acts as an efficient current
isolator.

To date, the most studied natural materials known to expel and reflect electromagnetic waves efficiently are metals. In the microwave regime, $\lambda \sim 1$ cm, most metals
exhibit high reflectivity and low absorption loss per single reflection, which makes them
ideal candidates for confining and managing electromagnetic radiation. For example,
aluminium, which is one of the least expensive materials used to make mirrors, exhibits
almost perfect reflectivity in the microwave region with $\sim 0.01\%$ absorption-induced
loss per single reflection. However, in the visible, $\lambda \sim 400$ nm–800 nm, and infrared
(IR), $\lambda \sim 800$ nm–20 μm, the same material exhibits ~ 10–20% loss per reflection, thus,
considerably limiting its utility for managing light in these spectral regions.

Before the invention of all-dielectric photonic crystals, building compact light circuits
in the IR spectral region was problematic because of the lack of low-loss, highly reflecting materials. As most dielectric materials in the IR have considerably lower losses than
metals, all-dielectric photonic crystals have a potential to considerably outperform metal
reflectors in terms of losses. Together with the ability of designing spectral position of the
photonic bandgaps by means of varying the geometry, photonic crystals offer unprecedented flexibility in realizing low-loss, highly reflective medium for the applications of
light guiding almost anywhere in the visible and IR regions.

Although interpretation of photonic crystals as efficient reflectors was known from
their discovery, it was later established that this analogy can be advanced further, and the
concept of omnidirectional reflectivity was suggested. By definition, omnidirectional
reflectors exhibit almost perfect reflection in the vicinity of a designable wavelength
for all angles of incidence and all polarizations (Fig. 1.2 (a)). Although much prior
work has been done on dielectric multilayers, design criteria for the omnidirectional 1D
periodic multilayer reflectors were first presented in [25]. This idea was later extended to
waveguides, where the core can be of any material including lossless vacuum, while light
is confined in the core by an omnidirectional reflector in the planar (Fig. 1.2 (c)) [26,27]
or cylindrical geometry (Bragg fiber in Fig. 1.2 (d)). [19,20] Quasi-1D PhC reflectors
in the form of a transverse concentric Bragg stack led to implementation of hollow
waveguides, where most of the transmitted power is concentrated in the gas-filled hollow
core, thus reducing propagation losses and nonlinearities. This enabled high-power laser
guidance at various wavelengths in the mid-IR. [21]

It was later demonstrated theoretically [28] that, although helpful, omnidirectional
reflectivity is not necessary for opening sizable bandgaps as well as for enabling efficient guidance of all polarizations in low refractive index cores. This was also verified

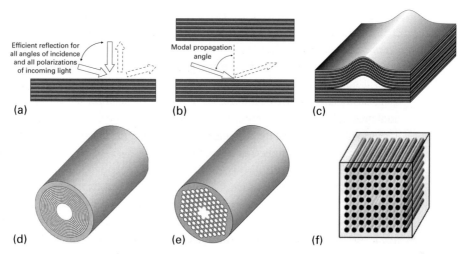

Figure 1.2 Photonic crystal waveguides and fibers guiding in the lower refractive index cores.
(a) Omnidirectional reflectors. For any polarization and angle of incidence of the incoming
light, radiation is completely reflected in a certain frequency window, also known as a bandgap.
(b) Light can be trapped inside a hollow core when completely surrounded by a photonic crystal
reflector. (c, d, e) Practical implementations of the hollow core waveguides in the form of an
integrated 1D-periodic waveguide (c), Bragg fiber featuring concentric quasi-1D periodic
reflector (d), and photonic bandgap fiber featuring 2D photonic crystal reflector (e). (f) Optically
induced reconfigurable photonic lattice with a single site defect inside of a photorefractive
crystal.

experimentally in various low index contrast photonic crystal and Bragg fibers. [29,30,31]
This is related to the fact that modal propagation inside a photonic crystal waveguide,
especially one having a large core diameter, can be described as a sequence of consec-
utive bounces of the trapped radiation at the core–reflector interface (Fig. 1.2 (b)). The
angle of incidence of such radiation onto a reflector can be typically characterized by the
effective modal propagation angle. Therefore, to enable guidance in the low refractive
index core, it is only necessary to design a photonic crystal reflector that is efficient
for a particular modal propagation angle. Finally, we note that an alternative way of
confining radiation in the low refractive index core is by using 2D photonic crystal
reflectors made of a periodic array of air holes embedded into the fiber cladding material
(Fig. 1.2 (e)). [22,23] Such fibers are typically referred to as photonic bandgap fibers
and are made of silica glass. Although periodic cladding in such fibers exhibits relatively high index contrast, the corresponding photonic crystal reflectors are, however,
not omnidirectional.

Another PCF-like system with even weaker refractive-index contrasts is a photorefractive crystal imprinted with an optically induced photonic lattice (Fig. 1.2 (f)). The
photonic lattice in such a system can be created by interference of several plane waves,
[32] while localized defects in a photonic lattice can be created using amplitude masks.
[33] Owing to the crystal's photosensitivity, the externally induced photonic lattice creates a periodic refractive-index variation on the order of 10^{-3} inside the crystal. Even

for such a small index contrast, linear bandgap guidance in the low-refractive-index core is possible. [33] The significant advantage of optically induced photonic lattices is that the experimental configuration of the photonic lattice is dynamically adjustable, thus allowing real-time control of the lattice spacing and amount of index variation. Therefore, it provides a convenient test bed for the study of bandgap guidance in the weak index-contrast systems. Another useful property of these photonic lattices is that nonlinear effects of light appear at very low intensities (on the order of 1 mW/cm^2). In addition, the nonlinearity can change from focusing to defocusing by simply reversing the DC electric field applied to a crystal. The nonlinearity gives rise to stationary solitons propagating along the direction of lattice uniformity. These solitons, when launched along a nonnormal direction, can also navigate through the lattice with little energy loss, and can, in principle, enable light-routing applications in lattice networks. [34,35]

Up till now we have described the use of photonic crystal reflectors to confine light inside low-refractive-index core waveguides and fibers (Figs. 1.2 (c),(d),(e)), where radiation guidance is along the waveguide axis. A related application of photonic crystal waveguides is their use as interconnect elements in integrated optical circuitry. In principle, by bending the waveguide, one can steer radiation away from its original propagation direction. However, when decreasing the bending radius, radiation leakage from such waveguides increases dramatically. Enhanced radiation leakage occurs because by bending a waveguide, one also perturbs the geometry and, hence, the reflection properties of a photonic crystal reflector, rendering it ineffective when the perturbation is strong. If the photonic crystal waveguides are to be used in integrated optical circuits it is, therefore, necessary to implement compact light steering elements such as bends, T and Y junctions, etc., with as low radiation loss as possible (Fig. 1.3 (a)). In [36] the authors proposed that such light-steering elements can be introduced as defects in otherwise perfectly periodic 2D photonic crystals featuring omnidirectional bandgaps (for at least one of the light polarizations). As there is no physical bending present in such systems, the omnidirectional reflector structure remains unperturbed and, thus, suppresses any radiation that is not propagating along the direction of a defect (Fig. 1.3 (b)). As a result, all the radiation coming into the steering element can be either propagated along or reflected back with no radiation loss into the cladding.

Moreover, by introduction of a topology-optimized resonator into the structure of a steering element one can design narrow [37] and wide bandwidth [38] steering elements exhibiting almost perfect transmission (Fig. 1.3 (c)), no crosstalk and no back reflection. As sizes of the individual features in photonic crystals are comparable to the wavelength of guided light, the above-mentioned approach allows design of compact steering elements with sizes on the order of the light wavelength.

Another prominent application of photonic crystals is their use as confining media for the photonic microcavities. Microcavities are essential components for various integrated optical devices, such as wavelength filters, add–drop multiplexers, and lasers. When a uniform dielectric region is surrounded by a photonic crystal reflector featuring omnidirectional reflectivity (complete bandgap), such a structure can serve as a high-quality resonator able to trap and store light. When looking at the density of electromagnetic states of a photonic crystal featuring a point defect (resonator), a localized resonator state

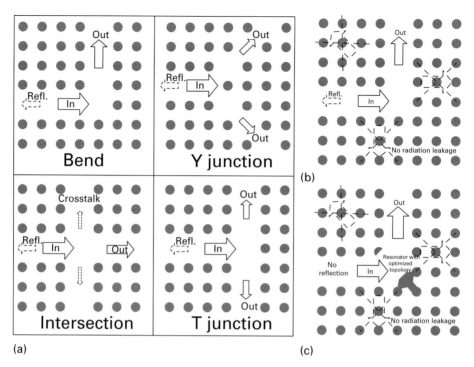

Figure 1.3 Compact light-steering elements based on photonic crystals for integrated optical circuits. (a) Various types of light steering elements such as bends, waveguide intersections, Y and T junctions. (b) Light launched into a steering element (a waveguide bend in this figure) surrounded by an omnidirectional photonic crystal reflector can only propagate along the direction of a defect. Radiation propagation outward from the defect is suppressed by the reflector bandgap. In the unoptimized defect structure, back reflection can be substantial. (c) By introduction of a topology-optimized resonator into the defect structure one can obtain narrow and wide bandwidth steering elements exhibiting almost perfect transmission and no crosstalk (for waveguide intersections).

manifests itself as a delta function positioned inside the photonic bandgap of a surrounding photonic crystal [4]. Various geometries of the microcavities have been explored over the years with the goal of increasing the quality factor of a cavity, while reducing the microcavity volume. Two main cavity geometries can be distinguished as those based on concentric Bragg reflectors [39] (Fig. 1.4 (a)), and those based on localized defects in slab photonic crystals [40,41] (Fig. 1.4 (b)) (here we omit discussion of ring resonators, which do not operate using bandgap effects).

Microcavities of both types can be used as laser cavities if a gain media is placed inside, leading to a possibility of compact lasers with designable direction of emission that can be easily integrated with other planar PhC components. [39,42,43] When the cavity is filled with a nonlinear material all-optical devices become possible where light routing and information transmission is enabled by the same beam. [44]

To excite a microcavity one typically uses the mode of a bus waveguide weakly coupled to a cavity state. When the microcavity is used for filtering, a mix of several wavelengths

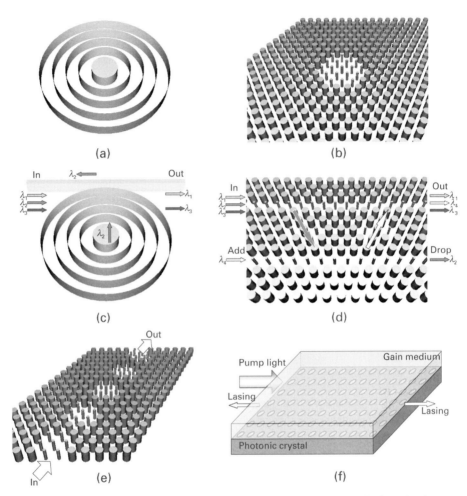

Figure 1.4 Schematics of various PhC components and devices. (a) Microcavity in a circular Bragg reflector geometry. (b) Microcavity as a point defect in a photonic crystal lattice. (c) Optical narrow band rejection filter comprising a bus waveguide and a microcavity. (d) All optical add–drop multiplexer comprising two bus waveguides, one drop resonator and one add resonator. (e) Coupled resonator optical waveguide for slow light applications. (f) Bloch-mode laser comprising a planar waveguide made of a gain media and covering a 2D PhC. The periodic structure of the PhC provides feedback for lasing.

is launched into a bus waveguide; a wavelength resonant with that of a cavity state (drop wavelength) is then reflected back into the input arm of a filter (wavelength λ_2 in Fig. 1.4 (c)), while the rest of the wavelengths are passed through. Optical coupling of the cavity to free space (vertical radiation loss) can also be used to partially extract the drop wavelength into the out-of-plane direction. [45] A complete add–drop multiplexer can be designed using two bus waveguides and a sequence of resonators of certain symmetry. [46] In such a multiplexer (Fig. 1.4 (d)), a mix of wavelengths is launched into the input arm of a device; a specific wavelength (λ_2 in Fig. 1.4 (d)) can then be dropped into the lower right arm of the device with the aid of a first resonator without any back reflection into the input or add arms of the device. A second resonator can then be used to add a different wavelength (λ_4 in Fig. 1.4 (d)) into the output arm of the device, again without any back reflection into the input or add arms of the device.

Finally, when placing several microcavities in a row (Fig. 1.4 (e)), and assuming weak interaction between individual cavities, one can realize a so-called coupled resonator optical waveguide (CROW). [47,48] The CROW guides light from one end of the chain to the other by photon tunneling between adjacent resonators; therefore, in the limit of weak coupling one expects considerable slowing down of the modes propagating in such waveguides. Coupling of the individual resonator modes creates flat bands within the photonic gap of a surrounding photonic crystal, each of them centered on a mode of the isolated microcavity. Varying the number of rows between resonators, one can adjust the coupling strength, and, as a consequence, the group velocity of guided modes. Potential applications of slow light in CROW structures include compact optical delay lines, optical memory, and devices based on enhanced nonlinear interactions.

1.2.2   Photonic crystals for out-of-bandgap operation

Another way of using PhCs is in the frequency regime outside of the photonic bandgap. In this case transmission through the bulk of a crystal is allowed, however, owing to coherent reflection from the periodic structure of a crystal, propagation of radiation will be considerably modified from the propagation in a uniform dielectric material. For example, for the frequencies right below or above the bandgap edge the group velocity of propagating light is typically greatly reduced. This, in turn, leads to a substantial increase in the propagation time through the PhC device enhancing material–radiation interaction, which is of prime interest for lasing, nonlinear, and sensor applications.

Lasers that use PhC lattices as a whole and do not require microcavities are known as Bloch mode lasers or band-edge lasers (Fig. 1.4 (f)). Such structures use the radiative losses of modes to extract light from the active region. Lasing typically occurs for slow Bloch modes characterized by wavevectors in the vicinity of the high symmetry points of a first Brillouin zone, where light strongly interacts with the active medium. [49,50] In their spirit, Bloch mode lasers can be considered as 2D analog of the widespread distributed feedback lasers.

Finally, PhCs, being geometrically anisotropic, can also possess highly anisotropic dispersion. This property enables such PhC applications as superprisms and supercollimators. Particularly, PhCs can be designed so that the direction of radiation propagation through such a device is highly sensitive to the frequency and initial launch direction. It is then possible to design superprisms [51,52] – devices that exhibit very large refraction, such that a mix of wavelengths traveling in a single direction is split into several monochromatic waves propagating at considerably different angles. Using highly anisotropic energy flow of Bloch waves, one can also design such devices as supercollimators, [53,54] in which a beam of light self-collimates and does not spread when propagating through a photonic crystal.

References

[1] V. P. Bykov. Spontaneous emission in a periodic structure, *Sov. Phys. JETP* **35** (1972), 269–273.

[2] E. Yablonovitch. Inhibited spontaneous emission in solid-state physics and electronics, *Phys. Rev. Lett.* **58** (1987), 2059–2062.

[3] S. John. Strong localization of photons in certain disordered dielectric superlattices, *Phys. Rev. Lett.* **58** (1987), 2486–2489.

[4] E. Yablonovitch, T. J. Gmitter, R. D. Meade, *et al.* Donor and acceptor modes in photonic band-structure, *Phys. Rev. Lett.* **67** (1991), 3380.

[5] V. L. Welch and J.-P. Vigneron. Beyond butterflies – the diversity of biological photonic crystals, *Opt. Quant. Electr.* **39** (2007), 295–303.

[6] Y. Xia, B. Gates, Y. Yin, and Y. Lu. Monodispersed colloidal spheres: old materials with new applications, *Adv. Mater.* **12** (2000), 693–713.

[7] C. Lopez. Materials aspects of photonic crystals, *Adv. Mater.* **15** (2003), 1679–1704.

[8] M. J. Escuti and G. P. Crawford. Holographic photonic crystals, *Opt. Eng.* **43** (2004), 1973–1987.

[9] J. H. Moon, J. Ford, and S. Yang. Fabricating three-dimensional polymeric photonic structures by multi-beam interference lithography, *Polym. Adv. Technol.* **17** (2006), 83–93.

[10] P. V. Braun, S. A. Rinne, and F. G. Santamaría. Introducing defects in 3D photonic crystals: state of the art, *Adv. Mater.* **18** (2006), 2665–2678.

[11] D. W. Prather, S. Shi, J. Murakowski, *et al.* Photonic crystal structures and applications: perspective, overview, and development, *IEEE J. Sel. Topics Quant. Electr.* **12** (2006), 1416–1437.

[12] H. Benisty, J. M. Lourtioz, A. Chelnokov, S. Combrié, and X. Checoury. Recent advances toward optical devices in semiconductor-based photonic crystals, *Proc. IEEE* **94** (2006), 997–1023.

[13] S. Noda, M. Imada, M. Okano, *et al.* Semiconductor three-dimensional and two-dimensional photonic crystals and devices, *IEEE J. Quant. Electr.* **38** (2002), 726–735.

[14] P. St. J. Russell. Photonic crystal fibers, *J. Lightwave. Technol.*, **24** (2006), 4729–4749.

[15] A. Bjarklev, J. Broeng, and A. S. Bjarklev. *Photonic Crystal Fibres* (Boston, MA: Kluwer Academic Publishers, 2003).

[16] M. Large, L. Poladin, G. Barton, and M. A. van Eijkelenborg. *Microstructured Polymer Optical Fibres* (New York: Springer, 2007).

[17] J. K. Ranka, R. S. Windeler, and A. J. Stentz. Visible continuum generation in air–silica microstructure optical fibers with anomalous dispersion at 800 nm, *Opt. Lett.* **25** (2000), 25–27.

[18] J. M. Dudley, G. Genty, and S. Coen. Supercontinuum generation in photonic crystal fiber, *Rev. Modern Phys.* **78** (2006), 1135–1184.

[19] P. Yeh, A. Yariv, and E. Marom. Theory of Bragg fiber, *J. Opt. Soc. Am.* **68** (1978), 1196–1201.

[20] S. G. Johnson, M. Ibanescu, M. Skorobogatiy, *et al.* Low-loss asymptotically single-mode propagation in large core OmniGuide fibers, *Opt. Express* **9** (2001), 748–779.

[21] B. Temelkuran, S. D. Hart, G. Benoit, J. D. Joannopoulos, and Y. Fink. Wavelength scalable optical fibers for CO_2 laser transmission, *Nature* **420** (2002), 650–653.

[22] J. C. Knight, J. Broeng, T. A. Birks, and P. St. J. Russell. Photonic band gap guidance in optical fibers, *Science* **282** (1998), 1476–1478.

[23] C. M. Smith, N. Venkataraman, M. T. Gallagher, *et al*. Low-loss hollow-core silica/air photonic bandgap fibre, *Nature* **424** (2003), 657–659.

[24] A. Hassani, A. Dupuis, and M. Skorobogatiy. Low loss porous terahertz fibers containing multiple subwave length holes, *Appl. Phys. Lett.* 92(2008), 071101.

[25] Y. Fink, J. N. Winn, S. H. Fan, *et al*. A dielectric omnidirectional reflector, *Science* **282** (1998), 1679–1682.

[26] D. Yin, H. Schmidt, J. Barber, and A. Hawkins. Integrated ARROW waveguides with hollow cores, *Opt. Express* **12** (2004), 2710–2715.

[27] S.-S. Lo, M.-S. Wang, and C.-C. Chen. Semiconductor hollow optical waveguides formed by omni-directional reflectors, *Opt. Express* **12** (2004), 6589–6593.

[28] M. Skorobogatiy. Efficient anti-guiding of TE and TM polarizations in low index core waveguides without the need of omnidirectional reflector, *Opt. Lett.* **30** (2005), 2991–2993.

[29] A. Argyros, T. A. Birks, S. G. Leon-Saval, *et al*. Photonic bandgap with an index step of one percent, *Opt. Express* **13** (2005), 309–314.

[30] M. E. Likhachev, S. L. Semjonov, M. M. Bubnov, *et al*. Development and study of Bragg fibres with a large mode field and low optical losses, *Quant. Electr.* **36** (2006), 581–586.

[31] A. Dupuis, N. Guo, B. Gauvreau, *et al*. Guiding in the visible with "colorful" solid-core Bragg fibers, *Opt. Lett*. **32** (2007), 2882–2884.

[32] J. W. Fleischer, M. Segev, N. K. Efremidis, and D. N. Christodoulides. Observation of two-dimensional discrete solitons in optically induced nonlinear photonic lattices, *Nature* **422** (2003), 147–150.

[33] I. Makasyuk, Z. Chen, and J. Yang. Bandgap guidance in optically-induced photonic lattices with a negative defect, *Phys. Rev. Lett*. **96** (2006), 223903.

[34] R. Fischer, D. Träger, D. N. Neshev, *et al*. Reduced-symmetry two-dimensional solitons in photonic lattices, *Phys. Rev. Lett*. **96** (2006), 023905.

[35] D. N. Christodoulides and E. D. Eugenieva. Blocking and routing discrete solitons in two-dimensional networks of nonlinear waveguide arrays, *Phys. Rev. Lett*. **87** (2001), 233901.

[36] A. Mekis, J. C. Chen, I. Kurland, *et al*. High transmission through sharp bends in photonic crystal waveguides, *Phys. Rev. Lett*. **77** (1996), 3787–3790.

[37] S. G. Johnson, C. Manolatou, S. Fan, *et al*. Elimination of cross talk in waveguide intersections, *Opt. Lett*. **23** (1998), 1855–1857.

[38] P. I. Borel, A. Harpøth, L. H. Frandsen, *et al*. Topology optimization and fabrication of photonic crystal structures, *Opt. Express* **12** (2004), 1996–2001.

[39] N. Moll, R. F. Mahrt, C. Bauer, *et al*. Evidence for bandedge lasing in a two-dimensional photonic bandgap polymer laser, *Appl. Phys. Lett*. **80** (2002), 734–736.

[40] Y. Akahane, T. Asano, B. S. Song, and S. Noda. High-Q photonic nanocavity in a two-dimensional photonic crystal, *Nature* **425** (2003), 944–947.

[41] K. Srinivasan, P. E. Barclay, O. Painter, *et al*. Experimental demonstration of a high quality factor photonic crystal microcavity, *Appl. Phys. Lett*. **83** (2003), 1915–1917.

[42] O. Painter, R. K. Lee, A. Scherer, *et al*. Two-dimensional photonic bandgap defect mode laser, *Science* **284** (1999), 1819–1821.

[43] H.-G. Park, S.-H. Kim, S.-H. Kwon, *et al*. Electrically driven single-cell photonic crystal laser, *Science* **305** (2004), 1444–1446.

[44] S. Noda, A. Chutinan, and M. Imada. Trapping and emission of photons by a single defect in a photonic bandgap structure, *Nature* **407** (2000), 608–610.

[45] M. Notomi, A. Shinya, S. Mitsugi, *et al*. Optical bistable switching action of Si high-Q photonic crystal nanocavities, *Opt. Express* **13** (2005), 2678–2687.

[46] S. Fan, P. R. Villeneuve, J. D. Joannopoulos, and H. A. Haus. Channel drop tunneling through localized states, *Phys. Rev. Lett.* **80** (1998), 960–963.

[47] N. Stefanou and A. Modinos. Impurity bands in photonic insulator, *Phys. Rev. B* **57** (1998), 127–133.

[48] A. Yariv, Y. Xu, R. K. Lee, and A. Scherer. Coupled-resonator optical waveguide: a proposal and analysis, *Opt. Lett.* **24** (1999), 711–713.

[49] M. Meier, A. Mekis, A. Dodabalapur, *et al*. Laser action from two-dimensional distributed feedback in photonic crystals, *Appl. Phys. Lett.* **74** (1999), 7–9.

[50] K. Sakoda, K. Ohtaka, and T. Ueta. Low threshold laser oscillation due to group velocity anomaly peculiar to two- and three dimensional photonic crystals, *Opt. Express* **4** (1999), 481–489.

[51] S.-Y. Lin, V. M. Hietala, L. Wang, and E. D. Jones. Highly dispersive photonic bandgap prism, *Opt. Lett.* **21** (1996), 1771–1773.

[52] H. Kosaka, T. Kawashima, A. Tomita, *et al*. Superprism phenomena in photonic crystals, *Phys. Rev. B* **58** (1998), R10096.

[53] H. Kosaka, T. Kawashima, A. Tomita, *et al*. Self-collimating phenomena in photonic crystals, *Appl. Phys. Lett.* **74** (1999), 1212–1214.

[54] D. N. Chigrin, S. Enoch, S. C. M. Torres, and G. Tayeb. Self-guiding in two-dimensional photonic crystals, *Opt. Express* **11** (2003), 1203–1211.

2 Hamiltonian formulation of Maxwell's equations (frequency consideration)

Electromagnetic phenomena are governed by Maxwell's equations. The topic of propagation of electromagnetic fields in general dielectric media is vast. We will only concern ourselves with periodic or otherwise symmetric dielectric structures. A description of a radiation state of a system can be found by solving Maxwell's equations and satisfying appropriate boundary conditions. It is generally true that if additional symmetries are present in a system there are conserved physical quantities that can be identified to describe the general behavior of such a system. Frequently, such conserved quantities can be used to label a particular radiation state unambiguously. We start by considering how the symmetries of a dielectric media in time and in space define the general properties of solutions. Our immediate goal is to characterize the behavior of radiation propagation in photonic crystals without the need of numerical solutions. In all the derivations we use cgs units and let the speed of light $c = 1$.

In the absence of free charges and currents, Maxwell's equations are:

$$\nabla \times \mathbf{E} + \frac{\partial \mathbf{B}}{\partial t} = 0, \tag{2.1}$$

$$\nabla \times \mathbf{H} - \frac{\partial \mathbf{D}}{\partial t} = 0, \tag{2.2}$$

$$\nabla \cdot \mathbf{B} = 0, \tag{2.3}$$

$$\nabla \cdot \mathbf{D} = 0, \tag{2.4}$$

where, respectively, $\mathbf{E}(\mathbf{r}, t)$ and $\mathbf{H}(\mathbf{r}, t)$ are the microscopic electric and magnetic fields and $\mathbf{D}(\mathbf{r}, t)$, $\mathbf{B}(\mathbf{r}, t)$ are the displacement and magnetic induction fields. Constitutive relations are taken as:

$$\mathbf{D} = \varepsilon(\mathbf{r})\mathbf{E},$$
$$\mathbf{B} = \mathbf{H}, \tag{2.5}$$

thus, assuming isotropic dielectric constant $\varepsilon(\mathbf{r})$, no nonlinearity and nonmagnetic materials, $\mu = 1$. Substituting (2.5) into (2.1–2.4) we arrive at:

$$\nabla \times \mathbf{E} + \frac{\partial \mathbf{H}}{\partial t} = 0, \tag{2.6}$$

$$\nabla \times \mathbf{H} - \varepsilon\frac{\partial \mathbf{E}}{\partial t} = 0, \tag{2.7}$$

$$\nabla \cdot \mathbf{H} = 0, \tag{2.8}$$

$$\nabla \cdot \varepsilon\mathbf{E} = 0. \tag{2.9}$$

In general, electromagnetic fields are complicated functions of time and space. However, because of the linearity of Maxwell's equations and the theorems of Fourier analysis we can represent any solution of such equations in terms of a linear combination of harmonic modes in time, which by themselves are solutions of (2.6–2.9). In the following, we will concern ourselves only with harmonic solutions of Maxwell's equations in the form:

$$\mathbf{E}(\mathbf{r}, t) = \mathbf{E}(\mathbf{r}) \exp(-i\omega t),$$
$$\mathbf{H}(\mathbf{r}, t) = \mathbf{H}(\mathbf{r}) \exp(-i\omega t); \tag{2.10}$$

also called harmonic modes. The complex-valued fields in (2.10) are introduced for mathematical convenience and one has to take real parts to obtain physical fields. Note that the introduction of a conserved quantity ω through (2.10) is only possible when the dielectric profile $\varepsilon(\mathbf{r})$ is time independent and that Maxwell's equations are invariant under the transformation $t = t' + t_0$ for any t_0. Here, for the first time, we encounter the manifestation of a general principle stating that symmetries of a system lead to conserved quantities that can be used to label solutions (modes). Substitution of (2.10) into (2.6–2.9) gives:

$$\mathbf{H} = -\frac{i}{\omega} \nabla \times \mathbf{E}, \tag{2.11}$$

$$\mathbf{E} = \frac{i}{\omega\varepsilon(\mathbf{r})} \nabla \times \mathbf{H}, \tag{2.12}$$

$$\nabla \cdot \mathbf{H} = 0, \tag{2.13}$$

$$\nabla \cdot \varepsilon(\mathbf{r})\mathbf{E} = 0. \tag{2.14}$$

To proceed with the solution of (2.11–2.14) we reduce the number of vector variables by either substituting (2.11) into (2.12) or vice versa, arriving at two alternative sets of equations:

$$\omega^2 \mathbf{E} = \frac{1}{\varepsilon(\mathbf{r})} \nabla \times (\nabla \times \mathbf{E}), \tag{2.15}$$

$$\nabla \cdot \varepsilon(\mathbf{r})\mathbf{E} = 0, \tag{2.16}$$

$$\omega^2 \mathbf{H} = \nabla \times \left(\frac{1}{\varepsilon(\mathbf{r})} \nabla \times \mathbf{H} \right), \tag{2.17}$$

$$\nabla \cdot \mathbf{H} = 0. \tag{2.18}$$

By solving for either electric (2.15, 2.16) or magnetic (2.17, 2.18) fields and finding the other field through (2.11) or (2.12), we can find the distribution of electromagnetic fields for any dielectric profile $\varepsilon(\mathbf{r})$. We will demonstrate later that, numerically, a set of equations (2.17, 2.18) (sometimes also called master equations) is easier to solve than (2.15, 2.16). Equation (2.17) presents a Hermitian eigenvalue problem with many useful orthogonality properties. Thus, in the following, we concentrate on the properties of (2.17, 2.18).

2.1   Plane-wave solution for uniform dielectrics

If a dielectric is uniform (and possibly complex) $\varepsilon(\mathbf{r}) = \varepsilon$ then (2.17), (2.18) have a simple solution. The propagating electromagnetic wave is presented by an orthogonal vector triplet $(\mathbf{E}_0, \mathbf{H}_0, \mathbf{k})$ where the solution of Maxwell's equations is:

$$\mathbf{H}(\mathbf{r}, t) = \mathbf{H}_0 \exp(i\mathbf{k} \cdot \mathbf{r} - i\omega t),$$

$$\mathbf{E}(\mathbf{r}, t) = \mathbf{E}_0 \exp(i\mathbf{k} \cdot \mathbf{r} - i\omega t),$$

$$\mathbf{E}_0 = \frac{1}{\omega \varepsilon}(\mathbf{H}_0 \times \mathbf{k}),$$

$$(\mathbf{k} \cdot \mathbf{H}_0) = 0, \tag{2.19}$$

where the transversality condition (2.18) is enforced in the form $(\mathbf{k} \cdot \mathbf{H}_0) = 0$. Frequency is connected to the propagation vector $\mathbf{k}$ by a dispersion relation:

$$\omega^2 \varepsilon - \mathbf{k}^2 = 0. \tag{2.20}$$

Thus, in the case of a uniform dielectric, a fixed real frequency ω and a set of three conserved numbers $\mathbf{k} = (k_x, k_y, k_z)$ satisfying (2.20) define unambiguously a solution of Maxwell's equations in the form of plane waves. Note that the introduction of a conserved quantity $\mathbf{k}$ in (2.19) is only possible when the dielectric profile $\varepsilon(\mathbf{r})$ is uniform and Maxwell's equations are invariant under the transformation $\mathbf{r} = \mathbf{r}' + \mathbf{r}_0$ for any vector $\mathbf{r}_0$.

It is important to understand the constraints imposed by a dispersion relation $\omega(\mathbf{k})$ like (2.20). As we will see later, many physical processes are governed by the density of radiation states $D(\omega)$ in the physical volume V at a particular frequency ω. If the number of radiation states between ω and $\omega + d\omega$ is defined as $N(\omega, \omega + d\omega)$, then:

$$N(\omega, \omega + d\omega) = D(\omega)d\omega = 2\frac{V}{(2\pi)^3} \int\limits_{\omega < \omega(\mathbf{k}) < \omega + d\omega} d^3\mathbf{k}. \tag{2.21}$$

In the case of a dispersion relation (2.20) at frequency ω, all wave vectors are confined to a 3D sphere of radius $|\mathbf{k}| = \omega\sqrt{\varepsilon}$ with $D(\omega) = \omega^2 \varepsilon^{3/2} V/\pi^2$. Another important quantity characterizing the electromagnetic state and intimately related to the dispersion relation is a group velocity, which, in photonic crystals, is also equivalent to the energy velocity. Group velocity is defined as:

$$v_g = \frac{\partial \omega(\mathbf{k})}{\partial \mathbf{k}}, \tag{2.22}$$

and in free space $v_g = 1/\sqrt{\varepsilon}$ in units of c.

Another way of visualizing (2.20) is by means of a so-called band diagram. A band diagram is a 2D plot where the frequencies of all electromagnetic states allowed by the dispersion relation $\omega(\mathbf{k})$ are plotted either along a chosen curve or a direction in a $\mathbf{k}$ space. For example, assume a lossless dielectric with purely real ε. Let us arbitrarily choose a preferred direction, as shown in Fig. 2.1(a). Denote the projection of a wave vector $\mathbf{k}$ along such a direction as β (frequently referred to as the propagation constant),

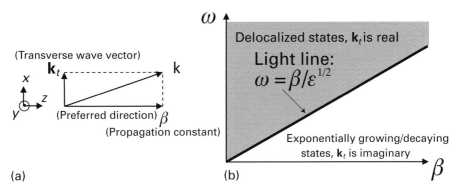

Figure 2.1 (a) Definition of a propagation constant β and a transverse wave vector $\mathbf{k}_t$. (b) Band diagram of electromagnetic states in an infinite dielectric. States above the light line are the delocalized plane waves with finite fields everywhere. States below the dielectric light line exhibit exponential divergence of their fields, and are known as evanescent states. The band diagram is symmetric with respect to the change $\beta \to -\beta$.

and in a transverse direction as $\mathbf{k}_t$. We now find all the physically allowed radiation states (β, ω). In the following, we only consider $\mathrm{Re}(\beta) \geq 0$, as the dispersion relation (2.20) is an even function of a propagation vector $\omega(\mathbf{k}) = \omega(-\mathbf{k})$. From the dispersion relation (2.20), it follows that $\mathbf{k}_t^2 = \omega^2\varepsilon - \beta^2$. For purely real $0 \leq \beta \leq \omega\sqrt{\varepsilon}$, also known as a region of w's above the light line, all the components of the transverse vectors are real and can be expressed as:

$$\mathbf{k}_t = (k_x, k_y, 0) = \sqrt{|\omega^2\varepsilon - \beta^2|}(\cos(\theta), \sin(\theta), 0), \theta \in (0, 2\pi). \qquad (2.23)$$

Such vectors form a circle in the xy plane and define plane waves delocalized in all three directions and propagating along $\mathbf{k}$ with a group velocity $1/\sqrt{\varepsilon}$. Thus, for a given ω, every point on a (β, ω) diagram (Fig. 2.1(b)), such that $0 \leq \beta \leq \omega\sqrt{\varepsilon}$ corresponds to a delocalized degenerate state (degeneracy with respect to direction of a transverse vector).

For $\beta > \omega^2\varepsilon$, however, both components of $\mathbf{k}_t$ are complex as:

$$\mathbf{k}_t = (k_x, k_y, 0) = \mathrm{i}\sqrt{|\omega^2\varepsilon - \beta^2|}(\cos(\theta), \sin(\theta), 0), \theta \in (0, 2\pi). \qquad (2.24)$$

Note, that plane wave (2.19) with complex propagation wave vector (2.24) is still a mathematical solution of Maxwell's equations. A complex-valued $\mathbf{k}_t$ defines an exponentially growing solution (2.19) in a transverse direction $\mathbf{r}_t = -|\mathbf{r}_t| \cdot (\cos(\theta), \sin(\theta), 0)$, as in this case (2.19) transforms into $\mathbf{H}(\mathbf{r}, t) = \mathbf{H}_0 \exp(\mathrm{i}\beta z - \mathrm{i}\omega t) \exp(\sqrt{|\omega^2\varepsilon - \beta^2|}|\mathbf{r}_t|)$. Typically, such a solution is considered unphysical on the grounds that it possesses infinite electromagnetic fields.

Next, choosing a purely imaginary propagation constant, $\beta = \pm i|\beta|$, all the components of transverse vectors are again real and can be expressed as:

$$\mathbf{k}_t = (k_x, k_y, 0) = \sqrt{\omega^2 \varepsilon + |\beta|^2}(\cos(\theta), \sin(\theta), 0), \theta \in (0, 2\pi). \qquad (2.25)$$

Thus, in the transverse direction the solution is still a delocalized plane wave, while in $\mathbf{z}$ it is exponentially decaying in one direction. Such waves are also known as evanescent. Note that solution (2.25) is essentially identical to (2.24) up to the interchange in vectors $\beta \cdot \mathbf{z} \leftrightarrow \mathbf{k}_t$.

Finally, a choice of a truly complex β with nonzero real and imaginary parts leads to a truly complex transverse wave vector $\mathbf{k}_t$, thus defining an exponentially divergent solution in both the transverse and $\mathbf{z}$ directions.

We conclude by noticing that solution (2.19) together with (2.20) presents a general solution of Maxwell's equations in an infinite region of a uniform dielectric with undefined boundary conditions. The classification of solutions (2.23), (2.24), (2.25) into physical and unphysical is only relevant when dealing with an infinite dielectric region. In contrast, all the solutions (2.23), (2.24), (2.25) have to be considered if used in expansion of an electromagnetic field in a finite or a semi-finite region of a uniform dielectric. For example, in the problem of electromagnetic energy propagation in the half space $z > 0$, if nontrivial boundary conditions are imposed at $z = 0$, then electromagnetic fields in the half space can contain in their expansion plane waves of the form (2.25) with $\beta = i|\beta|$, which exhibit exponentially decaying fields when $z \to +\infty$ (see Problem 2.1 of this chapter).

2.2   Methods of quantum mechanics in electromagnetism

We now revisit Maxwell's equations in the form (2.15), (2.16), (2.17), (2.18) and apply the formalism of Hamiltonian quantum mechanics to characterize their solutions. We start by recapitulating the properties of Hermitian Hamiltonians. In the quantum mechanical description of a single particle in an external potential $U(\mathbf{r})$, the two major unknowns are the particle energy E and a particle scalar $\psi(\mathbf{r})$ function. The ψ function defines the probability density $\rho(\mathbf{r})$ of finding a particle at a certain point in space as $\rho(\mathbf{r}) = |\psi(\mathbf{r})|^2$. The particle energy and its density function are related by a differential equation, known as Shrödinger equation:

$$\left[-\frac{\hbar^2}{2m}\nabla^2 + U(\mathbf{r}) \right] \psi(\mathbf{r}) = E\psi(\mathbf{r}). \qquad (2.26)$$

Equation (2.26) is said to define an eigenvalue problem, with an eigenvalue E (also called a conserved property of a state) and an eigenfunction $\psi(\mathbf{r})$ (also called a mode or a state). The operator in square brackets is said to define a Hamiltonian:

$$\hat{H} = \left[-\frac{\hbar^2}{2m}\nabla^2 + U(\mathbf{r}) \right], \qquad (2.27)$$

while the ψ function is said to define a state $|\psi_E\rangle$. In this notation (also called Dirac notation) (2.26) takes a more concise form:

$$\hat{H}|\psi_E\rangle = E|\psi_E\rangle. \tag{2.28}$$

In the following, we assume that operator $\hat{H}$ is linear (does contain $\psi(\mathbf{r})$ explicitly in its definition). We define the dot product between the two states as:

$$\langle\psi_{E'}|\psi_E\rangle = \langle\psi_E|\psi_{E'}\rangle^* = \int d\mathbf{r}\,\psi_{E'}^*(\mathbf{r})\psi_E(\mathbf{r}), \tag{2.29}$$

where the star means complex conjugation, and integration is performed over the whole volume in which the ψ function is defined (the domain of a ψ function). In the same manner as in (2.29), we define the operator average between two states as:

$$\langle\psi_{E'}|\hat{H}|\psi_E\rangle = \int d\mathbf{r}\psi_{E'}^*(\mathbf{r})\big[\hat{H}\psi_E(\mathbf{r})\big]. \tag{2.30}$$

Some Hamiltonians exhibit an additional symmetry by being Hermitian. In that case:

$$\langle\psi_{E'}|\hat{H}|\psi_E\rangle = \langle\psi_E|\hat{H}|\psi_{E'}\rangle^*. \tag{2.31}$$

For example, Hamiltonian (2.27) is Hermitian as:

$$\begin{aligned}
\langle\psi_{E'}|\hat{H}|\psi_E\rangle &= -\frac{\hbar^2}{2m}\int d\mathbf{r}\psi_{E'}^*(\mathbf{r})\nabla^2\psi_E(\mathbf{r}) + \int d\mathbf{r}\psi_{E'}^*(\mathbf{r})U(\mathbf{r})\psi_E(\mathbf{r})\\
&= \left(\frac{\hbar^2}{2m}\int d\mathbf{r}\nabla\psi_E^*(\mathbf{r})\nabla\psi_{E'}(\mathbf{r}) + \int d\mathbf{r}\psi_E^*(\mathbf{r})U(\mathbf{r})\psi_{E'}(\mathbf{r})\right)^*\\
&= \left(-\frac{\hbar^2}{2m}\int d\mathbf{r}\psi_E^*(\mathbf{r})\nabla^2\psi_{E'}(\mathbf{r}) + \int d\mathbf{r}\psi_E^*(\mathbf{r})U(\mathbf{r})\psi_{E'}(\mathbf{r})\right)^*\\
&= \langle\psi_E|\hat{H}|\psi_{E'}\rangle^* \tag{2.32}
\end{aligned}$$

where we used the identity:

$$\int_V d\mathbf{r}\psi_{E'}^*(\mathbf{r})\nabla^2\psi_E(\mathbf{r}) = \oint_S d\mathbf{s}\psi_{E'}^*(\mathbf{r})\nabla\psi_E(\mathbf{r}) - \int_V d\mathbf{r}\nabla\psi_{E'}^*(\mathbf{r})\nabla\psi_E(\mathbf{r}),$$

and a reasonable assumption that $\psi_{E'}(\mathbf{r})|_S = 0$ on a surface S surrounding the volume V (typically infinite) of the domain of ψ function, thus leading to a zero surface integral.

There are several important properties of the eigenstates of (2.28) that can be derived assuming that $\hat{H}$ is a Hermitian Hamiltonian, one of the most important being an orthogonality between different modes.

2.2.1 Orthogonality of eigenstates

Consider two eigenstates with different eigenvalues E', E such that $E' \neq E^*$. From (2.28) they satisfy:

$$\begin{aligned}
\hat{H}|\psi_E\rangle = E|\psi_E\rangle &\rightarrow \langle\psi_{E'}|\hat{H}|\psi_E\rangle = E\langle\psi_{E'}|\psi_E\rangle,\\
\hat{H}|\psi_{E'}\rangle = E'|\psi_{E'}\rangle &\rightarrow \langle\psi_E|\hat{H}|\psi_{E'}\rangle = E'\langle\psi_E|\psi_{E'}\rangle.
\end{aligned}$$

By conjugating the upper equation and using the Hermitian property of $\hat{H}$ (2.31), we get:

$$\langle \psi_E | \hat{H} | \psi_{E'} \rangle = E^* \langle \psi_E | \psi_{E'} \rangle.$$
$$\langle \psi_E | \hat{H} | \psi_{E'} \rangle = E' \langle \psi_E | \psi_{E'} \rangle$$

By subtracting one equation from the other we get:

$$(E^* - E') \langle \psi_E | \psi_{E'} \rangle = 0, \tag{2.33}$$

which can only hold if $\langle \psi_E | \psi_{E'} \rangle = 0$. Moreover, if we choose the same state $E' = E$, then from the definition of a dot product (2.29) $\langle \psi_E | \psi_E \rangle$ is purely real and larger than 0, and (2.33) can only hold if $E = E^*$, signifying a purely real eigenvalue. Thus, two different eigenstates corresponding to the same Hermitian Hamiltonian have purely real eigenvalues and they are orthogonal to each other in the sense of a dot product (2.29):

$$\langle \psi_E | \psi_{E'} \rangle = 0, \text{ when } E \neq E'. \tag{2.34}$$

Frequently, a collection of several modes can have the same eigenvalue E: in that case we call such an eigenstate degenerate. One labels different modes of the same degenerate state by introducing an additional parameter β (or a set of parameters $\beta_1, \beta_2, \dots$) such that $\psi_{E,\beta}$ label individual modes. Note that, owing to the linearity of a Hamiltonian eigen problem (2.28), any linear combination of degenerate eigenstates is also an eigenstate. Thus, assuming a countable parameter β the following will also be an eigenstate for any choice of expansion coefficients C_β:

$$\psi_E = \sum_\beta C_\beta \psi_{E,\beta}. \tag{2.35}$$

In general, one can choose expansion coefficients in (2.35) in such a way as to orthogonalize all the degenerate modes with respect to each other in accordance with (2.34). Such procedure is called a Gramm–Schmidt orthogonalization algorithm.

Finally, we note in passing that a particular form of a dot product (2.29) is crucial for establishing that all eigenvalues of (2.28), (2.31) are strictly real. In the following sections (normalizing the modes of a waveguide) we will see that it is not always possible to choose a "good" dot product, thus leading to the complex conserved numbers labeling the modes.

2.2.2   Variational principle

The variational principle theorem offers an alternative way of finding the lowest energy state (ground state) of a Hermitian Hamiltonian $\hat{H}$. Although solving the Schrödinger equation can be quite complicated, there is a simple way of understanding the qualitative features of its ground state. The variational theorem states that the ground state $|\psi_g\rangle$ psi function $\psi_g(\mathbf{r})$ with energy E_g minimizes the energy functional:

$$E_f(\psi) = \frac{\langle \psi | \hat{H} | \psi \rangle}{\langle \psi \mid \psi \rangle}. \tag{2.36}$$

Moreover, when $|\psi\rangle = |\psi_g\rangle$, the value of the functional gives the energy of the eigenstate $E_f(\psi_g) = E_g$. To prove the variational theorem, we will show that for any eigenstate $|\psi\rangle = |\psi_E\rangle$ with energy E, the Hermitian Hamiltonian $\hat{H}$ (2.36) reaches an extremum. The proof that (2.36) reaches a minimum for $|\psi\rangle = |\psi_g\rangle$ is somewhat more complicated and can be found in standard quantum mechanics textbooks. To prove that (2.36) has an extremum at $|\psi_E\rangle$ we consider the variation of a psi function $\psi(\mathbf{r}) = \psi_E(\mathbf{r}) + \delta\psi(\mathbf{r})$ around in the vicinity of an eigenstate. Then the variation of the energy functional for Hermitian Hamiltonian $\hat{H}$ is:

$$
\begin{aligned}
E_f(\psi_E + \delta\psi) - E_f(\psi_E) &= \frac{\langle \psi_E + \delta\psi | \hat{H} | \psi_E + \delta\psi \rangle}{\langle \psi_E + \delta\psi \mid \psi_E + \delta\psi \rangle} - E \\
&= \frac{\langle \psi_E | \hat{H} | \psi_E \rangle + \langle \psi_E | \hat{H} | \delta\psi \rangle + \langle \delta\psi | \hat{H} | \psi_E \rangle + O(\delta\psi^2)}{\langle \psi_E \mid \psi_E \rangle + \langle \psi_E \mid \delta\psi \rangle + \langle \delta\psi \mid \psi_E \rangle + O(\delta\psi^2)} - E \\
&= \frac{E\left(\langle \psi_E \mid \psi_E \rangle + \langle \psi_E \mid \delta\psi \rangle + \langle \delta\psi \mid \psi_E \rangle\right) + O(\delta\psi^2)}{\langle \psi_E \mid \psi_E \rangle + \langle \psi_E \mid \delta\psi \rangle + \langle \delta\psi \mid \psi_E \rangle + O(\delta\psi^2)} - E \\
&= O(\delta\psi^2). \tag{2.37}
\end{aligned}
$$

From (2.37) it follows that the variation in the energy functional (2.36) is second order with respect to the variation in the psi function of any eigenstate of a Hermitian Hamiltonian $\hat{H}$. This, in turn, means that the functional $E_f(\psi)$ reaches its extremum at any eigenstate $|\psi\rangle = |\psi_E\rangle$.

Accepting that the energy functional reaches its minimum at the ground state we now find out what general properties the ground state eigenfunction possesses. Following the same steps as in (2.32) we rewrite the operator averaging in (2.36) as:

$$
\begin{aligned}
E_f(\psi) = \frac{\langle \psi | \hat{H} | \psi \rangle}{\langle \psi \mid \psi \rangle} &= \frac{-\dfrac{\hbar^2}{2m} \displaystyle\int d\mathbf{r}\,\psi^*(\mathbf{r})\nabla^2\psi(\mathbf{r}) + \displaystyle\int d\mathbf{r}\,\psi^*(\mathbf{r})U(\mathbf{r})\psi(\mathbf{r})}{\displaystyle\int d\mathbf{r}\,|\psi(\mathbf{r})|^2} \\[2mm]
&= \frac{\dfrac{\hbar^2}{2m} \displaystyle\int d\mathbf{r}\,|\nabla\psi(\mathbf{r})|^2 + \displaystyle\int d\mathbf{r}\,U(\mathbf{r})\,|\psi(\mathbf{r})|^2}{\displaystyle\int d\mathbf{r}\,|\psi(\mathbf{r})|^2}. \tag{2.38}
\end{aligned}
$$

Assuming that the confining potential $U(\mathbf{r})$ is negative everywhere (see Fig. 2.2), then to minimize (2.38) one has to choose a function localized around the minimum of the confining potential but also smooth enough to have a small gradient. Note that using the variational theorem one can also find higher-order modes by minimizing the energy functional (2.36), while keeping a higher-order state orthogonal (2.34) to all the lower-order eigen states.

To summarize, from the variational theorem we determine that a sensible shape of a ground state (see Fig. 2.2(a)) would be a smooth function localized around the minimum of the negative confining potential. From the orthogonality relation we also find that a sensible shape for the higher-order mode would again be a sufficiently smooth function

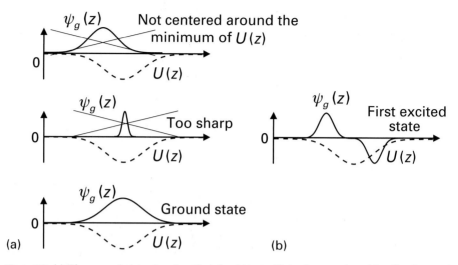

Figure 2.2 (a) The ground state eigenfunction should be sufficiently smooth and localized around the minimum of a confining potential $U(\mathbf{r})$. (b) The higher-order eigenstate should also be sufficiently smooth and localized around the minimum of a confining potential while being orthogonal to the ground state psi function.

localized around the minimum of a confining potential while changing its sign to satisfy an orthogonality relation (2.34) with a ground state.

2.2.3   Equivalence between the eigenstates of two commuting Hamiltonians

In the following, we consider two linear Hamiltonians, which are not necessarily Hermitian. Two Hamiltonians $\hat{H}_1$ and $\hat{H}_2$ are said to commute with each other if, for any function $\psi(\mathbf{r})$, the following holds:

$$(\hat{H}_1\hat{H}_2 - \hat{H}_2\hat{H}_1)\psi(\mathbf{r}) = 0. \qquad (2.39)$$

We will now demonstrate that if $|\psi_j^1\rangle$ is an eigenstate with an eigenvalue E_j^1 of a Hamiltonian $\hat{H}_1$, and if two Hamiltonians commute, $|\psi_j^1\rangle$ is then also an eigenstate of a Hamiltonian $\hat{H}_2$. Applying (2.39) to $\psi(\mathbf{r}) = \psi_j^1(\mathbf{r})$, and using (2.28) we get:

$$\hat{H}_1\left[\hat{H}_2\,|\psi_j^1\rangle\right] = \hat{H}_2\left[\hat{H}_1\,|\psi_j^1\rangle\right] \rightarrow \hat{H}_1\left[\hat{H}_2\,|\psi_j^1\rangle\right] = E_j^1\left[\hat{H}_2\,|\psi_j^1\rangle\right]. \qquad (2.40)$$

From the last equation in (2.40) it follows that $\hat{H}_2|\psi_j^1\rangle$ is an eigenfunction of a Hamiltonian $\hat{H}_1$ with an eigenvalue E_j^1. First, assume that the original $|\psi_j^1\rangle$ is nondegenerate; then $\hat{H}_2|\psi_j^1\rangle$ can differ from $|\psi_j^1\rangle$ only by a constant C, thus:

$$\hat{H}_2|\psi_j^1\rangle = C|\psi_j^1\rangle. \qquad (2.41)$$

From the definition of an eigenstate (2.28) and from (2.41) it follows that $|\psi_j^1\rangle$ is also an eigenstate of a Hamiltonian $\hat{H}_2$ with eigenvalue C.

In a more complicated case of a degenerate $|\psi_j^1\rangle$, from (2.40) and the discussion leading to (2.35) it follows that there exists a set of coefficients C_j^k such that:

$$\hat{H}_2 |\psi_j^1\rangle = \sum_{k=1}^{N} C_j^k |\psi_k^1\rangle, \tag{2.42}$$

where the index k spans all the N states $|\psi_k^1\rangle$ degenerate with $|\psi_j^1\rangle$. Coefficients C_j^k define an $(N \times N)$ matrix $\mathbf{C}$, and in a vector-matrix notation with $\overline{|\psi^1\rangle} = (|\psi_1^1\rangle, \ldots, |\psi_N^1\rangle)^T$, (2.42) can be rewritten as:

$$\hat{H}_2 \overline{|\psi^1\rangle} = \mathbf{C} \overline{|\psi^1\rangle}. \tag{2.43}$$

We now introduce a vector of expansion coefficients $\overline{A}^T = (A_1, \ldots, A_N)$ and find under what condition a linear combination:

$$|\psi\rangle = \overline{A}^T \overline{|\psi^1\rangle} = \sum_k A_k |\psi_k^1\rangle \tag{2.44}$$

will be a true eigenstate of a Hamiltonian $\hat{H}_2$ with some eigenvalue B. Particularly, multiplying (2.43) by $\overline{A}^T$ on the left, using linearity of $\hat{H}_2$, and assuming that (2.44) is an eigenstate, we arrive at:

$$\overline{A}^T \hat{H}_2 \overline{|\psi^1\rangle} = \overline{A}^T \mathbf{C} \overline{|\psi^1\rangle} \rightarrow \hat{H}_2 [\overline{A}^T \overline{|\psi^1\rangle}] = \overline{A}^T \mathbf{C} \overline{|\psi^1\rangle} \rightarrow B \overline{A}^T \overline{|\psi^1\rangle} = \overline{A}^T \mathbf{C} \overline{|\psi^1\rangle}.$$

State (2.44) will be an eigenstate of $\hat{H}_2$ if $B\overline{A}^T = \overline{A}^T \mathbf{C}$, or after transposing $B\overline{A} = \mathbf{C}^T \overline{A}$, which is a conventional eigenvalue problem with respect to the vector of expansion coefficients $\overline{A}$. Thus, choosing linear combinations of degenerate eigenstates (2.44) of a Hamiltonian $\hat{H}_1$ with expansion coefficients being the eigenvectors of a matrix $\mathbf{C}^T$ will define true eigenstates of $\hat{H}_2$ in the form (2.28).

2.2.4 Eigenstates of the operators of continuous and discrete translations and rotations

In many cases of interest in quantum mechanics and in electromagnetism, the geometry of a system under investigation exhibits some kind of translational or rotational symmetry (Fig. 2.3). Symmetry implies that a structure transforms into itself after performing a symmetry transformation. This, in turn, implies that a system Hamiltonian commutes with an operator of a corresponding symmetry operation. As established in Section 2.2.1, we can then choose the eigensolutions of a system Hamiltonian to be those of a symmetry operator, or, in the case of degenerate eigenstates, as a linear combination of these. As eigenstates of many symmetry operators are easy to find they present a natural choice for an expansion basis to find solutions of complicated Hamiltonians. Examples of systems with various translational and rotational symmetries: a waveguide of a constant cross-section having the same dielectric profile along a specific direction in space (the direction of radiation propagation); such a waveguide is said to exhibit a continuous translational symmetry in one dimension. Photonic crystals are periodic dielectric structures along certain directions, and are said to possess a discrete translational symmetry along such

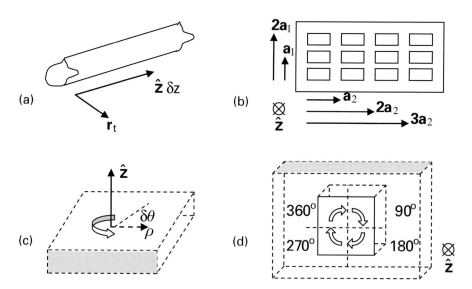

Figure 2.3 Examples of various translation and rotation symmetries. (a) Continuous translational symmetry along the $\hat{z}$ direction. (b) Discrete translational symmetry in 2D. (c) Continuous rotational symmetry in 2D slabs. (d) Discrete rotational symmetry in 2D.

directions. Certain structures, like 2D slabs or planes, possess both continuous translational symmetry along any direction in the plane and continuous rotational symmetry around an axis perpendicular to the plane. Finally, a structure like a planar square possesses only a discrete rotational symmetry around its center, with four allowed rotation angles.

Definition of the operator of translation

We define the operator of translation along vector $\delta\mathbf{r}$ acting on a function $\psi(\mathbf{r})$ as $\hat{T}_{\delta\mathbf{r}}$:

$$\hat{T}_{\delta\mathbf{r}}\psi(\mathbf{r}) = \psi(\mathbf{r} - \delta\mathbf{r}). \tag{2.45}$$

By definition, the application of a translation operator results in the translation of a function $\psi(\mathbf{r})$ along the vector $\delta\mathbf{r}$ (Fig. 2.4). The inverse of a translation operator is then $\hat{T}_{\delta\mathbf{r}}^{-1} = \hat{T}_{-\delta\mathbf{r}}$, as $\hat{T}_{\delta\mathbf{r}}\hat{T}_{-\delta\mathbf{r}}$ is an identity. We now demonstrate that for a system exhibiting translational symmetry, the translation operator commutes with a Hamiltonian (2.27). Moreover, if $\psi(\mathbf{r})$ is an eigenfunction of a Schrödinger equation (2.26), then $\hat{T}_{\delta\mathbf{r}}\psi(\mathbf{r})$ is also an eigenfunction with the same eigenvalue. Mathematically, the symmetry of a system with respect to a translation along $\delta\mathbf{r}$ is understood as the invariance of the external potential with respect to the action of a translation operator $\hat{T}_{\delta\mathbf{r}}U(\mathbf{r}) = U(\mathbf{r} - \delta\mathbf{r}) = U(\mathbf{r})$. Assuming that $\psi(\mathbf{r})$ is an eigenstate with an eigenvalue E, we are going to manipulate the Schrödinger equation:

$$\left[-\frac{\hbar^2}{2m}\nabla^2 + U(\mathbf{r}) \right]\psi(\mathbf{r}) = E\psi(\mathbf{r}),$$

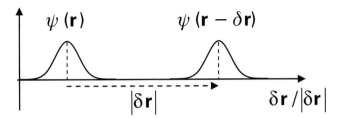

Figure 2.4 Graphical interpretation of the translation operator $\hat{T}_{\delta r}$. Application of this operator results in translation of a function $\psi(\mathbf{r})$ along the vector $\delta\mathbf{r}$.

to re-express it in terms of the function $\hat{T}_{\delta r}\psi(\mathbf{r})$. Particularly:

$$\left[-\frac{\hbar^2}{2m}\nabla^2 + U(\mathbf{r})\right]\hat{T}_{-\delta r}\hat{T}_{\delta r}\psi(\mathbf{r}) = E\hat{T}_{-\delta r}\hat{T}_{\delta r}\psi(\mathbf{r}),$$

$$\hat{T}_{\delta r}\left[-\frac{\hbar^2}{2m}\nabla^2 + U(\mathbf{r})\right]\hat{T}_{-\delta r}\left(\hat{T}_{\delta r}\psi(\mathbf{r})\right) = E(\hat{T}_{\delta r}\psi(\mathbf{r})),$$

$$\left[-\frac{\hbar^2}{2m}\hat{T}_{\delta r}\nabla^2\hat{T}_{-\delta r} + \hat{T}_{\delta r}U(\mathbf{r})\hat{T}_{-\delta r}\right]\left(\hat{T}_{\delta r}\psi(\mathbf{r})\right) = E\left(\hat{T}_{\delta r}\psi(\mathbf{r})\right). \qquad (2.46)$$

Consider now the two operator terms inside the square brackets. From the definition of the operator of translation (2.45), for any function $\phi(\mathbf{r})$ the following holds:

$$\hat{T}_{\delta r}U(\mathbf{r})\hat{T}_{-\delta r}\phi(\mathbf{r}) = \hat{T}_{\delta r}U(\mathbf{r})\phi(\mathbf{r} + \delta\mathbf{r}) = U(\mathbf{r} - \delta\mathbf{r})\phi(\mathbf{r}); \qquad (2.47)$$

moreover, from translation symmetry it follows that $U(\mathbf{r}) = U(\mathbf{r} - \delta\mathbf{r})$ and, finally,

$$\hat{T}_{\delta r}U(\mathbf{r})\hat{T}_{-\delta r}\phi(\mathbf{r}) = U(\mathbf{r})\phi(\mathbf{r}).$$

For the differential operator, a similar consideration leads to the following:

$$\hat{T}_{\delta r}\nabla^2\hat{T}_{-\delta r}\phi(\mathbf{r}) = \hat{T}_{\delta r}\nabla^2\phi(\mathbf{r} + \delta\mathbf{r})$$

$$= \hat{T}_{\delta r}\left(\frac{\partial^2}{\partial x^2} + \frac{\partial^2}{\partial y^2} + \frac{\partial^2}{\partial z^2}\right)\phi(\mathbf{r} + \delta\mathbf{r})$$

$$= \left(\frac{\partial^2}{\partial(x - \delta x)^2} + \frac{\partial^2}{\partial(y - \delta y)^2} + \frac{\partial^2}{\partial(z - \delta z)^2}\right)\phi(\mathbf{r})$$

$$= \left(\frac{\partial^2}{\partial x^2} + \frac{\partial^2}{\partial y^2} + \frac{\partial^2}{\partial z^2}\right)\phi(\mathbf{r})$$

$$= \nabla^2\phi(\mathbf{r}), \qquad (2.48)$$

where we have used the fact that $\partial^2/\partial(x - \delta x)^2 = \partial^2/\partial x^2$. Finally, by using (2.47) and (2.48) in (2.46) we arrive at the following equation:

$$\left[-\frac{\hbar^2}{2m}\nabla^2 + U(\mathbf{r})\right](\hat{T}_{\delta r}\psi(\mathbf{r})) = E(\hat{T}_{\delta r}\psi(\mathbf{r})), \qquad (2.49)$$

which implies that $\hat{T}_{\delta r}\psi(\mathbf{r})$ is also an eigenfunction of the Hamiltonian (2.27). Moreover, from (2.49) it also follows that the operator of translation and the Hamiltonian

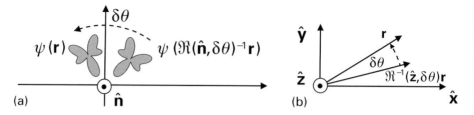

Figure 2.5 Graphical interpretation of the rotation operator $\hat{R}_{(\hat{\mathbf{n}}, \delta\theta)}$. (a) Application of this operator results in the rotation of a function $\psi(\mathbf{r})$ along the vector $\hat{\mathbf{n}}$ by angle $\delta\theta$. (b) Application of the rotation operator to a vector. The rotation is performed around $\hat{\mathbf{z}}$ axis.

of a system exhibiting translation symmetry commute with each other as for any eigenfunction $\psi(r)$:

$$\hat{H}(\hat{T}_{\delta r}\psi(\mathbf{r})) = E(\hat{T}_{\delta r}\psi(\mathbf{r})) \rightarrow \hat{H}(\hat{T}_{\delta r}\psi(\mathbf{r})) = \hat{T}_{\delta r}(E\psi(\mathbf{r})) = \hat{T}_{\delta r}(\hat{H}\psi(\mathbf{r}))$$

$$\rightarrow \hat{H}\hat{T}_{\delta r} = \hat{T}_{\delta r}\hat{H}.$$

Definition of the operator of rotation

The operator of rotation $\hat{R}_{(\hat{\mathbf{n}}, \delta\theta)}$ is characterized by the direction of a rotation axis $\hat{\mathbf{n}}$ and the angle of rotation $\delta\theta$. By definition, the application of a rotation operator results in the rotation of a function $\psi(\mathbf{r})$ along the axis $\hat{\mathbf{n}}$ through an angle $\delta\theta$ (Fig. 2.5). The rotation operator is typically defined using the rotation matrix $\Re(\hat{\mathbf{n}}, \delta\theta)$ as:

$$\hat{R}_{(\hat{\mathbf{n}}, \delta\theta)}\psi(\mathbf{r}) = \psi(\Re^{-1}(\hat{\mathbf{n}}, \delta\theta)\mathbf{r}). \tag{2.50}$$

The explicit form of $\Re(\hat{\mathbf{n}}, \delta\theta)$ depends upon the choice of coordinate system. In a rectangular coordinate system (x, y, z), with $\hat{\mathbf{n}} = \hat{\mathbf{z}}$ and counter-clockwise rotation, the original and rotated vectors are related by:

$$\begin{vmatrix} x \\ y \\ z \end{vmatrix}_{\text{rotated}} = \left[\Re(\hat{\mathbf{z}}, \delta\theta) = \begin{pmatrix} \cos(\delta\theta) & -\sin(\delta\theta) & 0 \\ \sin(\delta\theta) & \cos(\delta\theta) & 0 \\ 0 & 0 & 1 \end{pmatrix} \right] \begin{vmatrix} x \\ y \\ z \end{vmatrix}. \tag{2.51}$$

In a cylindrical coordinate system (ρ, θ, z), the rotation matrix is an identity, and the definition of a rotation operator is especially simple:

$$\hat{R}_{(\hat{\mathbf{n}}, \delta\theta)}\psi(\mathbf{r}) = \psi(\rho, \theta - \delta\theta, z). \tag{2.52}$$

Finally, as two consequent rotations around the same axis with angles $\delta\theta_1$ and $\delta\theta_2$ should be equal to a single rotation with angle $\delta\theta_1 + \delta\theta_2$ rotation matrices possess the following property:

$$\Re(\hat{\mathbf{n}}, \delta\theta_1)\Re(\hat{\mathbf{n}}, \delta\theta_2) = \Re(\hat{\mathbf{n}}, \delta\theta_1 + \delta\theta_2) = \Re(\hat{\mathbf{n}}, \delta\theta_2)\Re(\hat{\mathbf{n}}, \delta\theta_1), \tag{2.53}$$

from which it also follows that $\Re^{-1}(\hat{\mathbf{n}}, \delta\theta) = \Re(\hat{\mathbf{n}}, -\delta\theta)$.

We now demonstrate that for a system exhibiting rotational symmetry, the rotation operator commutes with Hamiltonian (2.27). Moreover, if $\psi(\mathbf{r})$ is an eigenfunction of a Schrödinger equation (2.26), then $\hat{R}_{(\hat{n},\delta\theta)}\psi(\mathbf{r})$ is also an eigenfunction with the same eigenvalue. Mathematically, the symmetry of a system with respect to a rotation $(\hat{n}, \delta\theta)$ is understood as the invariance of the external potential with respect to the action of a rotation operator

$$\hat{R}_{(\hat{n},\delta\theta)}U(\mathbf{r}) = U(\Re^{-1}_{(\hat{n},\delta\theta)}\mathbf{r}) = U(\mathbf{r}).$$

As in the case of a translation operator, we first assume that $\psi(\mathbf{r})$ is an eigenstate of (2.27) with an eigenvalue E. We then manipulate the Schrödinger equation:

$$\left[-\frac{\hbar^2}{2m}\nabla^2 + U(\mathbf{r})\right]\psi(\mathbf{r}) = E\psi(\mathbf{r}),$$

to re-express it in terms of the function $\hat{R}_{(\hat{n},\delta\theta)}\psi(\mathbf{r})$:

$$\left[-\frac{\hbar^2}{2m}\nabla^2 + U(\mathbf{r})\right]\hat{R}^{-1}_{(\hat{n},\delta\theta)}\hat{R}_{(\hat{n},\delta\theta)}\psi(\mathbf{r}) = E\hat{R}^{-1}_{(\hat{n},\delta\theta)}\hat{R}_{(\hat{n},\delta\theta)}\psi(\mathbf{r}),$$

$$\left[-\frac{\hbar^2}{2m}\hat{R}_{(\hat{n},\delta\theta)}\nabla^2\hat{R}^{-1}_{(\hat{n},\delta\theta)} + \hat{R}_{(\hat{n},\delta\theta)}U(\mathbf{r})\hat{R}^{-1}_{(\hat{n},\delta\theta)}\right]\left(\hat{R}_{(\hat{n},\delta\theta)}\psi(\mathbf{r})\right) = E(\hat{R}_{(\hat{n},\delta\theta)}\psi(\mathbf{r})). \quad (2.54)$$

Now consider the two operator terms inside the square brackets. From the definition of the operator of rotation (2.50), and using the fact that $\hat{R}^{-1}_{(\hat{n},\delta\theta)} = \hat{R}_{(\hat{n},-\delta\theta)}$, for any function $\phi(\mathbf{r})$ the following holds:

$$\hat{R}_{(\hat{n},\delta\theta)}U(\mathbf{r})\hat{R}^{-1}_{(\hat{n},\delta\theta)}\phi(\mathbf{r}) = \hat{R}_{(\hat{n},\delta\theta)}U(\mathbf{r})\phi(\Re(\hat{n}, \delta\theta)\mathbf{r}) = U(\Re^{-1}(\hat{n}, \delta\theta)\mathbf{r})\phi(\mathbf{r}), \quad (2.55)$$

moreover, from rotational symmetry it follows that $U(\mathbf{r}) = U(\Re^{-1}(\hat{n}, \delta\theta)\mathbf{r})$, and finally:

$$\hat{R}_{(\hat{n},\delta\theta)}U(\mathbf{r})\hat{R}^{-1}_{(\hat{n},\delta\theta)}\phi(\mathbf{r}) = U(\mathbf{r})\phi(\mathbf{r}).$$

For the differential operator, it is advantageous to work in a cylindrical coordinate system, where the axis of rotation $\hat{n}$ is directed along $\hat{z}$. Using the form of the rotation operator as in (2.52) we arrive at the following:

$$\hat{R}_{(\hat{n},\delta\theta)}\nabla^2\hat{R}^{-1}_{(\hat{n},\delta\theta)}\phi(\mathbf{r}) = \hat{R}_{(\hat{n},\delta\theta)}\nabla^2\phi(\rho, \theta + \delta\theta, z)$$

$$= \hat{R}_{(\hat{n},\delta\theta)}\left(\frac{1}{\rho}\frac{\partial}{\partial\rho}\left(\rho\frac{\partial}{\partial\rho}\right) + \frac{1}{\rho^2}\frac{\partial^2}{\partial\theta^2} + \frac{\partial^2}{\partial z^2}\right)\phi(\rho, \theta + \delta\theta, z)$$

$$= \left(\frac{1}{\rho}\frac{\partial}{\partial\rho}\left(\rho\frac{\partial}{\partial\rho}\right) + \frac{1}{\rho^2}\frac{\partial^2}{\partial(\theta - \delta\theta)^2} + \frac{\partial^2}{\partial z^2}\right)\phi(\rho, \theta, z)$$

$$= \nabla^2\phi(\mathbf{r}), \quad (2.56)$$

where we have used the fact that $\partial^2/\partial(\theta - \delta\theta)^2 = \partial^2/\partial\theta^2$. Finally, by using (2.55) and (2.56) in (2.54) we arrive at the following equation:

$$\left[-\frac{\hbar^2}{2m}\nabla^2 + U(\mathbf{r})\right](\hat{R}_{(\hat{n},\delta\theta)}\psi(\mathbf{r})) = E(\hat{R}_{(\hat{n},\delta\theta)}\psi(\mathbf{r})), \quad (2.57)$$

which implies that $\hat{R}_{(\hat{n},\delta\theta)}\psi(\mathbf{r})$ is also an eigenfunction of Hamiltonian (2.27). More-over, from (2.49) it also follows that the operator of rotation and the Hamiltonian of a system exhibiting translation symmetry commute with each other as for any eigenfunction $\psi(r)$

$$\hat{H}(\hat{R}_{(\hat{n},\delta\theta)}\psi(\mathbf{r})) = E(\hat{R}_{(\hat{n},\delta\theta)}\psi(\mathbf{r})) \rightarrow \hat{H}(\hat{R}_{(\hat{n},\delta\theta)}\psi(\mathbf{r})) = \hat{R}_{(\hat{n},\delta\theta)}(E\psi(\mathbf{r})) = \hat{R}_{(\hat{n},\delta\theta)}(\hat{H}\psi(\mathbf{r}))$$
$$\rightarrow \hat{H}\hat{R}_{(\hat{n},\delta\theta)} = \hat{R}_{(\hat{n},\delta\theta)}\hat{H}.$$

One-dimensional continuous translational symmetry

Assume that the system Hamiltonian transforms into itself for any translation $\delta\mathbf{r} = \hat{\mathbf{z}}dz$ along the unitary vector $\hat{\mathbf{z}}$, see Fig. 2.3(a). An example of such a system can be a waveguide of a constant cross-section directed along $\hat{\mathbf{z}}$. The eigenstates of a translation operator (2.45) then satisfy:

$$\hat{T}_{\delta\mathbf{r}}\psi(\mathbf{r}) = \psi(\mathbf{r} - \hat{\mathbf{z}}\delta z) = C(\delta z)\psi(\mathbf{r}). \tag{2.58}$$

One can verify by substitution that eigenfunctions satisfying (2.58) are

$$\psi_\beta(\mathbf{r}) = \exp(i\beta z)U(\mathbf{r}_t) \tag{2.59}$$

for any choice of a complex scalar β, and $(\mathbf{r} = (\mathbf{r}_t + \hat{\mathbf{z}}z), \mathbf{r}_t \perp \hat{\mathbf{z}})$

Two-dimensional discrete translational symmetry and 1D continuous translational symmetry

Assume that a system Hamiltonian transforms into itself for any translation along vector $\delta\mathbf{r}$ of the form $\delta\mathbf{r} = \delta\mathbf{r}_t + \hat{\mathbf{z}}\delta z = \bar{a}_1 N_1 + \bar{a}_2 N_2 + \hat{\mathbf{z}}\delta z$, where vectors $(\bar{a}_1, \bar{a}_2) \perp \hat{\mathbf{z}}$ are noncollinear and N_1, N_2 are any integers, see Fig. 2.3(b). The eigenstates of a translation operator (2.45) then satisfy:

$$\hat{T}_{\delta\mathbf{r}}\psi(\mathbf{r}) = \psi(\mathbf{r} - \bar{a}_1 N_1 - \bar{a}_2 N_2 - \hat{\mathbf{z}}\delta z) = C(N_1, N_2, \delta z)\psi(\mathbf{r}). \tag{2.60}$$

One can verify by substitution that eigenfunctions satisfying (2.60) are $\psi(\mathbf{r}) = \exp(i\beta z + i\mathbf{k}_t\mathbf{r}_t)$, with corresponding eigenvalues:

$$C(N_1, N_2, \delta z) = \exp(-i\beta\delta z - i\mathbf{k}_t\delta\mathbf{r}_t). \tag{2.61}$$

As in (2.59), β is a conserved number characterizing the mode. Note, however, that eigenvalue (2.61) is now degenerate. Particularly, for any vector of the form $\mathbf{k}_t + \mathbf{G}$ where $\mathbf{G} = \bar{b}_1 P_1 + \bar{b}_2 P_2$, P_1, P_2 are integers, and:

$$\bar{b}_1 = 2\pi \frac{\bar{a}_2 \times \hat{\mathbf{z}}}{\bar{a}_1 \cdot (\bar{a}_2 \times \hat{\mathbf{z}})}; \bar{b}_2 = 2\pi \frac{\hat{\mathbf{z}} \times \bar{a}_1}{\bar{a}_1 \cdot (\bar{a}_2 \times \hat{\mathbf{z}})},$$

the corresponding eigenvalue is:

$$\begin{aligned}
C(\mathbf{k}_t + \mathbf{G}) &= \exp(-i\beta\delta z - i(\mathbf{k}_t + \mathbf{G})\delta\mathbf{r}_t) \\
&= \exp(-i\beta\delta z - i\mathbf{k}_t\delta\mathbf{r}_t)\exp(-i\mathbf{G}(\bar{a}_1 N_1 + \bar{a}_2 N_2)) \\
&= \exp(-i\beta\delta z - i\mathbf{k}_t\delta\mathbf{r}_t)\exp(-i2\pi(N_1 + N_2)) \\
&= \exp(-i\beta\delta z - i\mathbf{k}_t\delta\mathbf{r}_t) \\
&= C(\mathbf{k}_t).
\end{aligned}$$

Thus, the eigenstate of a translation operator (2.60) is degenerate for a set of transverse wave vectors of the form $\mathbf{k}_t + \mathbf{G}$. From the discussion of Section 2.2.3 it follows that the eigenstate of a Hamiltonian possessing a discrete translational symmetry can be labeled with a vector of conserved parameters $(\mathbf{k}_t, \beta)$, and can be expressed using a linear combination of all the degenerate states having the same eigenvalue (2.61):

$$\psi_{\beta,\mathbf{k}_t}(\mathbf{r}) = \exp(i\beta z + i\mathbf{k}_t\mathbf{r}_t) \sum_{\mathbf{G}} A_\beta(\mathbf{G}) \exp(i\mathbf{G}\mathbf{r}_t) = \exp(i\beta z + i\mathbf{k}_t\mathbf{r}_t) U_\beta(\mathbf{r}_t)$$

$$U_\beta(\mathbf{r}_t + \bar{a}_1 N_1 + \bar{a}_2 N_2) = U_\beta(\mathbf{r}_t), \tag{2.62}$$

where the expansion coefficients $A_\beta(\mathbf{G})$ define a periodic function $U_\beta(\mathbf{r}_t)$ of transverse coordinates. The form (2.62) of an eigensolution for Hamiltonians with periodic potentials is a well-known result and is known as the Bloch theorem.

Continuous rotational symmetry

Assume that a system Hamiltonian transforms into itself for any rotation angle $\delta\theta$ around vector $\hat{z}$, see Fig. 2.3 (c). The eigenstates of a rotation operator (2.50) written in a cylindrical coordinate system satisfy:

$$\hat{R}_{\delta\theta}\psi(\rho, \theta, z) = \psi(\rho, \theta - \delta\theta, z) = C(\delta\theta)\psi(\rho, \theta, z). \tag{2.63}$$

One can verify by substitution that the eigenfunctions satisfying (2.63) are:

$$\psi(\rho, \theta, z) = \exp(im\theta)U_m(\rho, z), \tag{2.64}$$

where $U_m(\rho, z)$ is some function of coordinates (ρ, z), and the eigenvalues are:

$$C(\delta\theta) = \exp(-im\delta\theta). \tag{2.65}$$

Note that rotation by 2π should not change the form of a solution (2.64), thus necessitating a conserved parameter m to be integer.

Discrete rotational symmetry C_N

Assume that a system Hamiltonian transforms into itself for a set of discrete rotations $\delta\theta = 2\pi k/N$, $k = [0, N - 1]$ around vector $\hat{z}$, see Fig. 2.3 (d). The eigenstates of a rotation operator (2.50) written in a cylindrical coordinate system should then satisfy:

$$\hat{R}_{\delta\theta}\psi(\rho, \theta, z) = \psi\left(\rho, \theta - \frac{2\pi}{N}k, z\right) = C(k)\psi(\rho, \theta, z). \tag{2.66}$$

One can verify by substitution that the eigenfunctions satisfying (2.66) are again:

$$\psi(\rho, \theta, z) = \exp(im\theta)U_m(\rho, z), \tag{2.67}$$

while the eigenvalues are:

$$C(k) = \exp\left(-im\frac{2\pi}{N}k\right). \tag{2.68}$$

As in (2.64), m is a conserved integer characterizing the mode, however the eigenvalue (2.68) is the same for any $m = m + Np$, p an integer, $m = [0, \ldots, N - 1]$. Thus, the eigenstate (2.67) characterized by an integer m is a degenerate one. From the discussion in Section 2.2.3 it follows that the eigenstates of a Hamiltonian possessing a discrete rotational symmetry can be labeled with a conserved integer $m = [0, \ldots, N - 1]$ and can be expressed by a linear combination of all the degenerate states having the same eigenvalue (2.68):

$$\psi_m(\rho, \theta, z) = \exp(im\theta) \sum_p A(p) \exp(ipN\theta) U_{m+pN}(\rho, z) = \exp(im\theta) U_m(\rho, \theta, z)$$

$$U_m\left(\rho, \theta + \tfrac{2\pi}{N}k, z\right) = U_m(\rho, \theta, z),$$

$$k - \text{integer}, \, m = [0, \ldots, N - 1], \qquad (2.69)$$

where the expansion coefficients $A(p)$ define a periodic in θ function $U_m(\rho, \theta, z)$ with a period $2\pi/N$.

An important conclusion suggested by the form (2.69) is that all the solutions of a system exhibiting discrete rotational symmetry C_N fall into one of the N classes labeled by an integer $m = [0, \ldots, N - 1]$.

2.3   Properties of the harmonic modes of Maxwell's equations

At the beginning of Section 2 we have demonstrated that Maxwell's equations can be rewritten in terms of either only electric or only magnetic fields, thus reducing the total number of unknowns from six to three. We first consider a system of equations (2.17), (2.18). As discussed in Section 2.1, equation $\nabla \cdot \mathbf{H} = 0$ enforces transversality between the direction of plane wave propagation and the direction of a field vector. Particularly, inside any region with uniform dielectric the fields can be presented as a sum of plane waves $\mathbf{H}(\mathbf{r}, t) = \mathbf{H}_0 \exp(i\mathbf{k} \cdot \mathbf{r} - i\omega t)$ satisfying the transversality condition $(\mathbf{k} \cdot \mathbf{H}_0) = 0$. As long as we remember to impose such transversality on a solution we can concentrate on equation (2.17). By analogy with the Hamiltonian formulation of quantum mechanics we are going to introduce a Hamiltonian formulation of Maxwell's equations. Thus, equation (2.17) $\omega^2 \mathbf{H}_\omega = \nabla \times (\varepsilon^{-1}(\mathbf{r}) \nabla \times \mathbf{H}_\omega)$, with respect to a 3D vector of magnetic field $\mathbf{H}_\omega(\mathbf{r})$, defines an eigenvalue problem:

$$\hat{H} \, |\mathbf{H}_\omega\rangle = \omega^2 \, |\mathbf{H}_\omega\rangle, \qquad (2.70)$$

with a Hamiltonian:

$$\hat{H} = \nabla \times \left(\frac{1}{\varepsilon(\mathbf{r})} \nabla \times\right), \qquad (2.71)$$

and an eigenvalue ω^2. We define a dot product similarly to (2.28):

$$\langle \mathbf{H}_\phi | \mathbf{H}_\varphi \rangle = \int_V d\mathbf{r} \mathbf{H}_\phi^*(\mathbf{r}) \cdot \mathbf{H}_\varphi(\mathbf{r}), \qquad (2.72)$$

where the states $|\mathbf{H}_\phi\rangle$, $|\mathbf{H}_\varphi\rangle$ correspond to the arbitrary vector functions $\mathbf{H}_\phi(\mathbf{r})$, $\mathbf{H}_\varphi(\mathbf{r})$ and integration is performed over the volume of the domain of $\mathbf{H}_\phi(\mathbf{r})$ and $\mathbf{H}_\varphi(\mathbf{r})$. In what follows, we demonstrate that the Maxwell Hamiltonian (2.71) is Hermitian when the dielectric profile $\varepsilon(\mathbf{r})$ is purely real. Particularly, we demonstrate that:

$$\langle \mathbf{H}_\phi | \hat{H} | \mathbf{H}_\varphi \rangle = \langle \mathbf{H}_\varphi | \hat{H} | \mathbf{H}_\phi \rangle^*, \tag{2.73}$$

for any choice of states $|\mathbf{H}_\phi\rangle$, $|\mathbf{H}_\varphi\rangle$ (not necessarily eigenstates), where the operator average between two states is defined similarly to (2.30) as:

$$\langle \mathbf{H}_\phi | \hat{H} | \mathbf{H}_\varphi \rangle = \int_V d\mathbf{r} \mathbf{H}_\phi^*(\mathbf{r}) \cdot [\hat{H} \mathbf{H}_\varphi(\mathbf{r})]. \tag{2.74}$$

To prove that the Maxwell Hamiltonian is Hermitian. WE write (2.74) as:

$$\langle \mathbf{H}_\phi | \hat{H} | \mathbf{H}_\varphi \rangle = \int_V d\mathbf{r} \mathbf{H}_\phi^* \cdot [\hat{H} \mathbf{H}_\varphi] = \int_V d\mathbf{r} \mathbf{H}_\phi^* \cdot \left[\nabla \times \left(\frac{1}{\varepsilon} \nabla \times \mathbf{H}_\varphi \right) \right], \tag{2.75}$$

where integration is over the volume V of a generally three-dimensional space. For any vector functions $\mathbf{a}$ and $\mathbf{b}$ the following vector identity holds:

$$\mathbf{b} \cdot (\nabla \times \mathbf{a}) = \mathbf{a} \cdot (\nabla \times \mathbf{b}) + \nabla \cdot (\mathbf{a} \times \mathbf{b}).$$

Now, using $\mathbf{b} = \mathbf{H}_\phi^*$ and $\mathbf{a} = \varepsilon^{-1} \nabla \times \mathbf{H}_\varphi$, one can write (2.75) as:

$$\langle \mathbf{H}_\phi | \hat{H} | \mathbf{H}_\varphi \rangle = \int_V d\mathbf{r} \left(\frac{1}{\varepsilon} \nabla \times \mathbf{H}_\varphi \right) \cdot (\nabla \times \mathbf{H}_\phi^*) + \int_V d\mathbf{r} \nabla \cdot \left[\left(\frac{1}{\varepsilon} \nabla \times \mathbf{H}_\varphi \right) \times \mathbf{H}_\phi^* \right]. \tag{2.76}$$

Using the divergence theorem, one can rewrite the second integral in (2.76) as a surface integral:

$$\langle \mathbf{H}_\phi | \hat{H} | \mathbf{H}_\varphi \rangle = \int_V d\mathbf{r} \left(\frac{1}{\varepsilon} \nabla \times \mathbf{H}_\varphi \right) \cdot (\nabla \times \mathbf{H}_\phi^*) + \int_S d\mathbf{S} \left[\left(\frac{1}{\varepsilon} \nabla \times \mathbf{H}_\varphi \right) \times \mathbf{H}_\phi^* \right]. \tag{2.77}$$

Assuming an infinitely large volume V, the magnitude of electromagnetic fields on the surface ∂V will tend to zero for a localized field. Assuming that $(\varepsilon^{-1} \nabla \times \mathbf{H}_\varphi) \times \mathbf{H}_\phi^*$ tends to zero more rapidly than the area of integration $\oint_S d\mathbf{S}$, the surface integral in (2.77) can thus be taken to zero, leading to:

$$\langle \mathbf{H}_\phi | \hat{H} | \mathbf{H}_\varphi \rangle = \int_V d\mathbf{r} \left(\frac{1}{\varepsilon} \nabla \times \mathbf{H}_\varphi \right) \cdot (\nabla \times \mathbf{H}_\phi^*) = \int_V d\mathbf{r} \left(\frac{1}{\varepsilon} \nabla \times \mathbf{H}_\phi^* \right) \cdot (\nabla \times \mathbf{H}_\varphi). \tag{2.78}$$

Using vector identity one more time, now with $\mathbf{b} = (1/\varepsilon) \nabla \times \mathbf{H}_1^*$ and $\mathbf{a} = \mathbf{H}_2$ the integral (2.78) can be rewritten as:

$$\langle \mathbf{H}_\phi | \hat{H} | \mathbf{H}_\varphi \rangle = \int_V d\mathbf{r} \mathbf{H}_2 \cdot \left[\nabla \times \left(\frac{1}{\varepsilon} \nabla \times \mathbf{H}_\phi^* \right) \right] + \int_V d\mathbf{r} \nabla \cdot \left[\mathbf{H}_\varphi \times \left(\frac{1}{\varepsilon} \nabla \times \mathbf{H}_\phi^* \right) \right]. \tag{2.79}$$

Again using the divergence theorem, the second integral in (2.79) can be written as a surface integral and then taken to zero for the reasons discussed before, thus leading to:

$$\langle \mathbf{H}_\phi | \hat{H} | \mathbf{H}_\varphi \rangle = \int_V d\mathbf{r} \mathbf{H}_\varphi \cdot \left[\nabla \times \left(\frac{1}{\varepsilon} \nabla \times \mathbf{H}_\phi^* \right) \right]. \tag{2.80}$$

Finally, taking complex conjugate of (2.80) and assuming that the dielectric constant is real we get:

$$\langle \mathbf{H}_\phi | \hat{H} | \mathbf{H}_\varphi \rangle = \left\{ \int_V d\mathbf{r}^3 \mathbf{H}_\varphi^* \cdot \left[\nabla \times \left(\frac{1}{\varepsilon} \nabla \times \mathbf{H}_\phi \right) \right] \right\}^* = \langle \mathbf{H}_\varphi | \hat{H} | \mathbf{H}_\phi \rangle^*, \quad (2.81)$$

which concludes the proof.

Owing to the similarity of the eigenvalue problems in quantum mechanics and the Hamiltonian formulation of Maxwell's equations, as well as similarity of the dot product operators (2.28), (2.72) and the operator averages (2.30), (2.73), many of the conclusions drawn in Section 2.2 will be directly applicable to Maxwell's equations. Finally, one can also demonstrate that the Maxwell Hamiltonian defined by (2.15) is not Hermitian even for real dielectric profiles, which is the main reason why we chose to concentrate on the Hermitian Maxwell Hamiltonian (2.71).

2.3.1   Orthogonality of electromagnetic modes

The orthogonality condition (2.34) derived for the eigenmodes of a Hermitian operator in Section 2.2.1 holds without changes in a case of the eigenmodes of a Hermitian Maxwell Hamiltonian (2.71). In particular, the integrable electromagnetic eigenstates $|\mathbf{H}_\omega\rangle$, $|\mathbf{H}_{\omega'}\rangle$ with eigenfrequencies ω' and ω are orthogonal in the sense of:

$$\langle \mathbf{H}_{\omega'} | \mathbf{H}_\omega \rangle = 0, \, \omega' \neq \omega. \quad (2.82)$$

2.3.2   Eigenvalues and the variational principle

We will now demonstrate that the eigenvalue ω^2 of a Hermitian operator (2.71) is strictly real and positive. To demonstrate this, we consider an eigenstate $|\mathbf{H}_\omega\rangle$ that satisfies $\hat{H} | \mathbf{H}_\omega \rangle = \omega^2 | \mathbf{H}_\omega \rangle$. Multiplying the left-hand and right-hand sides by $\langle \mathbf{H}_\omega |$, one gets:

$$\langle \mathbf{H}_\omega | \hat{H} | \mathbf{H}_\omega \rangle = \omega^2 \langle \mathbf{H}_\omega | \mathbf{H}_\omega \rangle . \quad (2.83)$$

As operator (2.71) is Hermitian, after conjugation of (2.83) and using (2.73) we arrive at:

$$\langle \mathbf{H}_\omega | \hat{H} | \mathbf{H}_\omega \rangle = (\omega^2)^* \langle \mathbf{H}_\omega | \mathbf{H}_\omega \rangle^* . \quad (2.84)$$

From the definition of a dot product it follows that

$$\langle \mathbf{H}_\omega | \mathbf{H}_\omega \rangle = \langle \mathbf{H}_\omega | \mathbf{H}_\omega \rangle^* = \int_V d\mathbf{r} \, |\mathbf{H}_\omega(\mathbf{r})|^2 > 0. \quad (2.85)$$

Comparing (2.83) and (2.84) we get $\omega^2 = (\omega^2)^*$, and, as a consequence, ω^2 is strictly real. Taking into account (2.83) and (2.85), to demonstrate that $\omega^2 > 0$, we are only left to show that an expectation value $\langle \mathbf{H} | \hat{H} | \mathbf{H} \rangle$ of a Maxwell Hamiltonian for any state $|\mathbf{H}\rangle$

(not necessarily an eigenstate) is strictly positive. From (2.74) we get:

$$\langle H|\hat{H}|H\rangle = \int_V d\mathbf{r} H^*(\mathbf{r}) \cdot \nabla \times \left(\frac{1}{\varepsilon(\mathbf{r})} \nabla \times H(\mathbf{r})\right) = \int_V d\mathbf{r} \frac{1}{\varepsilon(\mathbf{r})} |\nabla \times H(\mathbf{r})|^2 > 0, \quad (2.86)$$

where similarly to (2.76), (2.77) we used vector identity and a divergence theorem to extract a surface integral and equate it to zero. Note also that, from (2.86) and (2.12), an expectation value of the Maxwell Hamiltonian for an eigenstate $|H_\omega\rangle$ equals:

$$\langle H_\omega|\hat{H}|H_\omega\rangle = \int_V d\mathbf{r} \frac{1}{\varepsilon(\mathbf{r})} |\nabla \times H_\omega(\mathbf{r})|^2 \underset{(2.12)}{=} \int_V d\mathbf{r} \frac{\omega^2}{\varepsilon(\mathbf{r})} |D_\omega(\mathbf{r})|^2, \quad (2.87)$$

where $D_\omega(\mathbf{r})$ is an eigenstate displacement field. The expectation value (2.87) is proportional to the electric energy $E_D = 1/(8\pi) \int_V d\mathbf{r} \varepsilon^{-1}(\mathbf{r})|D_\omega(\mathbf{r})|^2$ of the modal field, which, for a harmonic mode, is also equal to the field magnetic energy $E_H = 1/(8\pi) \int_V d\mathbf{r}|H_\omega(\mathbf{r})|^2$.

In the same manner as in quantum mechanics, the variational theorem holds in the case of a Hermitian Maxwell Hamiltonian stating that the lowest frequency harmonic mode (ground state) $|H_g\rangle$ with eigenvalue ω_g^2 minimizes the electromagnetic energy functional:

$$E_f(H) = \frac{\langle H|\hat{H}|H\rangle}{\langle H | H\rangle}. \quad (2.88)$$

From (2.87) it follows that for an eigenstate $|H_\omega\rangle$, the energy functional (2.88) is proportional to the eigenstate electric energy:

$$E_f(H_\omega) = \frac{\langle H_\omega|\hat{H}|H_\omega\rangle}{\langle H_\omega | H_\omega\rangle} = \frac{\int_V d\mathbf{r} \frac{\omega^2}{\varepsilon(\mathbf{r})} |D_\omega(\mathbf{r})|^2}{\int_V d\mathbf{r} |H_\omega(\mathbf{r})|^2}. \quad (2.89)$$

Minimization of (2.89) in search of a ground state of a Hermitian Maxwell Hamiltonian suggests a field pattern with the state displacement field concentrated in the regions of high dielectric constant. The same conclusion holds for higher frequency eigenstates while requiring them to be orthogonal to the modes of lower frequencies. Thus, the variational principle alone allows us to picture the distribution of the electromagnetic fields in many systems. For example, consider an electromagnetic wave propagating in a slab waveguide along the $\hat{z}$ direction with a vector of the electric field parallel to the slabs along the $\hat{x}$ direction (see Fig. 2.6(a)). According to the variational theorem, the lowest eigenfrequency mode will have a displacement vector concentrated mostly in the core region, which has a larger dielectric constant than the cladding region (see Fig. 2.6(b)). The second-lowest frequency mode will then have to be orthogonal to the lowest frequency mode, thus forcing it to change the sign of its displacement field in the core region (see Fig. 2.6(c)). Field discontinuities in displacement fields on the slab boundaries (see Figs. 2.6(b),(c)) occur because the modal electric fields parallel to the

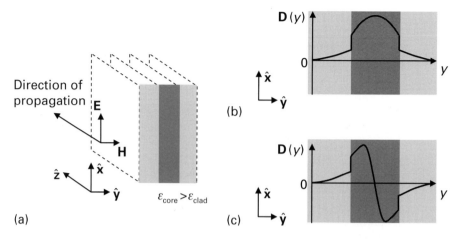

Figure 2.6 Transverse electric modes in a slab waveguide (the vector of the electric field is parallel to the slabs). (a) Geometry of a slab waveguide. (b) Sketch of the displacement field distribution in the lowest frequency guided mode. (c) Sketch of the displacement field distribution in the second lowest frequency guided mode.

slab interfaces are continuous across the boundaries of the slabs while the dielectric constant is discontinuous.

2.3.3   Absence of the fundamental length scale in Maxwell's equations

An important property of Maxwell's equations in dielectric media is the absence of a fundamental length scale. In particular, eigenmodes calculated for one structure can be trivially mapped onto solutions in a scaled structure. In more detail, consider Maxwell's equations in the form (2.17) and an eigenstate with frequency ω:

$$\omega^2 \mathbf{H}_\omega(\mathbf{r}) = \nabla_r \times \left(\frac{1}{\varepsilon(\mathbf{r})} \nabla_r \times \mathbf{H}_\omega(\mathbf{r}) \right). \tag{2.90}$$

Assume that we uniformly scale down the structure by a factor s (s larger than 1 makes the structure smaller) such that a new dielectric profile can be expressed using the unscaled one as $\tilde{\varepsilon}(\mathbf{r}) = \varepsilon(\mathbf{r}s)$ or, inversely, $\varepsilon(\mathbf{r}) = \tilde{\varepsilon}(\mathbf{r}/s)$. Substituting the scaled profile into (2.90) we get:

$$\omega^2 \mathbf{H}_\omega(\mathbf{r}) = \nabla_r \times \left(\frac{1}{\tilde{\varepsilon}(\mathbf{r}/s)} \nabla_r \times \mathbf{H}_\omega(\mathbf{r}) \right). \tag{2.91}$$

Making a coordinate transformation $\mathbf{r}' = \mathbf{r}/s$ in (2.91) we get:

$$\omega^2 \mathbf{H}_\omega(\mathbf{r}'s) = \nabla_{r's} \times \left(\frac{1}{\tilde{\varepsilon}(\mathbf{r}')} \nabla_{r's} \times \mathbf{H}_\omega(\mathbf{r}'s) \right). \tag{2.92}$$

Using the linear properties of a gradient $\nabla_{r's} = s^{-1}\nabla_{r'}$ we finally derive:

$$(\omega s)^2 \mathbf{H}_\omega(\mathbf{r}'s) = \nabla_{r'} \times \left(\frac{1}{\tilde{\varepsilon}(\mathbf{r}')} \nabla_{r'} \times \mathbf{H}_\omega(\mathbf{r}'s) \right). \tag{2.93}$$

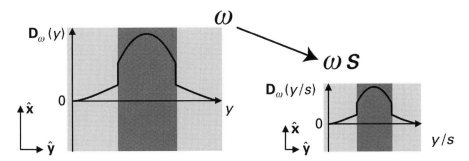

Figure 2.7 Eigenmodes of a scaled dielectric profile for the lowest-frequency guided state of a slab waveguide, shown in Fig. 2.6.

Equation (2.93) is identical in form to (2.90) but unlike (2.90) it defines eigenmodes of a scaled profile $\tilde{\varepsilon}$, which, from (2.93), are just a scaled version of the original eigenmodes having a scaled frequency. In particular, by reducing the scale by a factor s the eigenfrequency increases by the same factor, while the field extent in space shrinks by that factor (see Fig. 2.7).

2.4   Symmetries of electromagnetic eigenmodes

In the same way as in quantum mechanics we will demonstrate that when an electromagnetic Hamiltonian possesses a certain symmetry, such symmetry will be reflected in the form of the solution. We will start with time-reversal invariance, then we will consider continuous symmetries that are characteristic of various waveguides and finish with discrete translational and rotational symmetries, which are the symmetries of photonic crystals. In our treatment of the modal symmetries we always find the general form of the magnetic vector field $\mathbf{H}(\mathbf{r})$. One can then apply the equation $\mathbf{E} = \mathrm{i}/(\omega\varepsilon(\mathbf{r})) \cdot \nabla \times \mathbf{H}$ to find the related form of the electric field. It is straightforward to demonstrate that in all the cases presented below, the general form of the solution for the electric field will be the same as that for the magnetic field.

2.4.1   Time-reversal symmetry

For real dielectric profiles (no losses), the dispersion relation possesses additional so-called time-reversal symmetry. By complex conjugation of (2.70) and using the fact that ω is real, as established in Section 2.3.2, it follows that if $\mathbf{H}_\omega(\mathbf{r})$ is an eigenstate of a Hamiltonian (2.71), then $\mathbf{H}_\omega^*(\mathbf{r})$ is also an eigenstate of the same Hamiltonian with the same ω. As we will see in the next four sections, field solutions in systems exhibiting translational and rotational symmetries can be frequently labeled by conserved parameters. When an eigensolution has a form proportional to a complex exponential with respect to one of the conserved parameters, say k_2:

$$\mathbf{H}_{\omega(k_2)}(x_1, x_2, x_3) \sim \exp(\mathrm{i}k_2x_2)\mathbf{U}_{\omega(k_2)}(x_1, x_2, x_3), \tag{2.94}$$

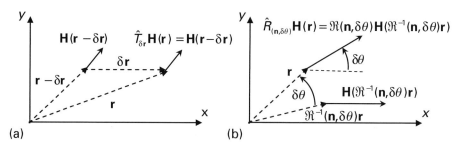

Figure 2.8 Action of operators of translation and rotation on a vector field. (a) Operator of translation shifts the vector field without changing the vector direction. (b) Operator of rotation rotates both the vector field distribution and the direction of a vector.

then due to time-reversal symmetry, after conjugation of (2.94) we again get the eigenstate with the same frequency:

$$\mathbf{H}^*_{\omega(k_2)}(x_1, x_2, x_3) \sim \exp(-ik_2x_2)\mathbf{U}^*_{\omega(k_2)}(x_1, x_2, x_3). \tag{2.95}$$

The form of (2.95) suggests that $\mathbf{H}^*_{\omega}(\mathbf{r})$ is an eigenstate of (2.71) but with a negated value of a conserved parameter $-k_2$. From this, it follows that for systems with real dielectric profiles:

$$\omega(k_2) = \omega(-k_2). \tag{2.96}$$

2.4.2 Definition of the operators of translation and rotation

Derivations of the functional forms of the electromagnetic fields reflecting various continuous and discrete translational symmetries are essentially identical to the derivations in quantum mechanics, as the form of the translation operator is the same. For the case of rotational symmetries, Maxwell's equations transform in a somewhat more complicated fashion than Schrödinger equations and we will re-derive the appropriate formulas. In the following, we assume that underlying space is 3D, meaning that solutions are described by a 3D spatial distribution of a 3D vector of a magnetic field.

As with the case of quantum mechanics, we will first introduce operators of translation and rotation, and then demonstrate that the electromagnetic Hamiltonian commutes with such operators when the dielectric is invariant with respect to the action of such operators.

Translation operator

In analogy to quantum mechanics, we define an operator of translation by $\delta\mathbf{r}$ acting on a vector function $\mathbf{H}(\mathbf{r})$ as $\hat{T}_{\delta\mathbf{r}}$:

$$\hat{T}_{\delta\mathbf{r}}\mathbf{H}(\mathbf{r}) = \mathbf{H}(\mathbf{r} - \delta\mathbf{r}). \tag{2.97}$$

By definition, the application of a translation operator results in the translation of a vector function $\mathbf{H}(\mathbf{r})$ along the vector $\delta\mathbf{r}$ without changing the vector direction (Fig. 2.8(a)). Assuming that the dielectric function is invariant with respect to the action of a translation operator $\hat{T}_{\delta\mathbf{r}}\varepsilon(\mathbf{r}) = \varepsilon(\mathbf{r} - \delta\mathbf{r}) = \varepsilon(\mathbf{r})$, one can easily show that the electromagnetic Hamiltonian $\hat{H}$ (2.71) commutes with the operator of translation $\hat{T}_{\delta\mathbf{r}}$.

Rotation operator

The operator of rotation acting on a vector function $\mathbf{H}(\mathbf{r})$ is denoted as $\hat{\mathbf{R}}_{(\hat{\mathbf{n}},\delta\theta)}$, where the unitary vector $\hat{\mathbf{n}}$ defines the rotation axis and $\delta\theta$ defines the rotation angle. The definition of $\hat{\mathbf{R}}_{(\hat{\mathbf{n}},\delta\theta)}$ is somewhat different from the case of quantum mechanics (2.50), as both the direction of a vector function as well as its position in space is changed:

$$\hat{\mathbf{R}}_{(\hat{\mathbf{n}},\delta\theta)}\mathbf{H}(\mathbf{r}) = \Re(\hat{\mathbf{n}}, \delta\theta)\mathbf{H}(\Re(\hat{\mathbf{n}}, \delta\theta)^{-1}\mathbf{r}). \tag{2.98}$$

The rotation matrix $\Re(\hat{\mathbf{n}}, \delta\theta)$ is defined in the same way as in quantum mechanics (2.51, 2.53). By definition, the application of a rotation operator results in the rotation of both a vector function distribution and a vector direction (Fig. 2.8(b)). We will now demonstrate that if the scalar dielectric function is invariant with respect to the action of a rotation operator $\hat{R}_{(\hat{\mathbf{n}},\delta\theta)}\varepsilon(\mathbf{r}) = \varepsilon(\Re^{-1}_{(\hat{\mathbf{n}},\delta\theta)}\mathbf{r}) = \varepsilon(\mathbf{r})$ then the electromagnetic Hamiltonian $\hat{H}$ (2.71) commutes with the operator of rotations $\hat{\mathbf{R}}_{(\hat{\mathbf{n}},\delta\theta)}$. To differentiate rotation operators of scalar and vector functions, we will use the $\hat{R}$ and $\hat{\mathbf{R}}$ symbols correspondingly.

In what follows, we will use the cylindrical coordinate system with its axis $\hat{\mathbf{z}}$ directed along the axis of rotation $\hat{\mathbf{n}}$. In this case, the rotation matrix $\Re(\hat{\mathbf{n}}, \delta\theta)$ is an identity and

$$\hat{\mathbf{R}}_{(\hat{\mathbf{n}},\delta\theta)}\mathbf{H}(\rho, \theta, z) = \mathbf{H}(\rho, \theta - \delta\theta, z), \tag{2.99}$$

where $\mathbf{H} = \hat{\rho}H_\rho + \hat{\theta}H_\theta + \hat{\mathbf{z}}H_z$. Moreover, the invariance of the dielectric function with respect to rotation $(\hat{\mathbf{n}}, \delta\theta)$ can now be simply expressed as $\hat{R}_{(\hat{\mathbf{n}},\delta\theta)}\varepsilon(\rho, \theta, z) = \varepsilon(\rho, \theta - \delta\theta, z) = \varepsilon(\rho, \theta, z)$.

Following the same path as in the case of quantum mechanics, we first assume that $\mathbf{H}(\mathbf{r})$ is an eigenstate of (2.71) with an eigenvalue ω^2. We then manipulate the master equation (2.70):

$$\left[\nabla \times \left(\frac{1}{\varepsilon(\mathbf{r})}\nabla\times\right)\right]\mathbf{H}(\mathbf{r}) = \omega^2\mathbf{H}(\mathbf{r}),$$

to re-express it in terms of the function $\hat{\mathbf{R}}_{(\hat{\mathbf{n}},\delta\theta)}\mathbf{H}(\mathbf{r})$:

$$\left[\nabla \times \left(\frac{1}{\varepsilon(\mathbf{r})}\nabla\times\right)\right]\hat{\mathbf{R}}^{-1}_{(\hat{\mathbf{n}},\delta\theta)}\hat{\mathbf{R}}_{(\hat{\mathbf{n}},\delta\theta)}\mathbf{H}(\mathbf{r}) = \omega^2\hat{\mathbf{R}}^{-1}_{(\hat{\mathbf{n}},\delta\theta)}\hat{\mathbf{R}}_{(\hat{\mathbf{n}},\delta\theta)}\mathbf{H}(\mathbf{r}),$$

$$\left[\hat{\mathbf{R}}_{(\hat{\mathbf{n}},\delta\theta)}\nabla \times \left(\frac{1}{\varepsilon(\mathbf{r})}\nabla\times\right)\hat{\mathbf{R}}^{-1}_{(\hat{\mathbf{n}},\delta\theta)}\right](\hat{\mathbf{R}}_{(\hat{\mathbf{n}},\delta\theta)}\mathbf{H}(\mathbf{r})) = \omega^2(\hat{\mathbf{R}}_{(\hat{\mathbf{n}},\delta\theta)}\mathbf{H}(\mathbf{r})). \tag{2.100}$$

We will now write a differential operator in square brackets in a cylindrical coordinate system with its axis $\hat{\mathbf{z}}$ directed along the axis of rotation $\hat{\mathbf{n}}$. The well-known form of a

$\nabla \times$ operator in cylindrical coordinates acting on a vector function $\boldsymbol{\varphi}(\mathbf{r})$ is:

$$\nabla_{(\rho,\theta,z)} \times \boldsymbol{\varphi}(\rho, \theta, z) = \begin{vmatrix} \hat{\boldsymbol{\rho}} & \hat{\boldsymbol{\theta}} & \hat{\mathbf{z}} \\ \rho & & \rho \\ \dfrac{\partial}{\partial \rho} & \dfrac{\partial}{\partial \theta} & \dfrac{\partial}{\partial z} \\ \varphi_\rho & \rho\varphi_\theta & \varphi_z \end{vmatrix}, \tag{2.101}$$

where $||$ signifies the matrix determinant. From (2.101), and from the fact that $\partial/\partial\,(\theta - \delta\theta) = \partial/\partial\theta$ for any constant $\delta\theta$, the following also holds:

$$\nabla_{(\rho,\theta,z)} \times \boldsymbol{\varphi}(\rho, \theta, z) = \nabla_{(\rho,\theta-\delta\theta,z)} \times \boldsymbol{\varphi}(\rho, \theta, z). \tag{2.102}$$

Finally, from the definition of the operator of rotation (2.98) it follows that $\hat{\mathbf{R}}_{(\hat{n},\delta\theta)}^{-1} = \hat{\mathbf{R}}_{(\hat{n},-\delta\theta)}$. The operator on the left-hand side of (2.100) can then be simplified as:

$$\hat{\mathbf{R}}_{(\hat{n},\delta\theta)} \nabla \times \left(\frac{1}{\varepsilon(\mathbf{r})} \nabla \times \right) \hat{\mathbf{R}}_{(\hat{n},\delta\theta)}^{-1} \boldsymbol{\varphi}(\mathbf{r}) =$$

$$= \hat{\mathbf{R}}_{(\hat{n},\delta\theta)} \nabla_{(\rho,\theta,z)} \left(\frac{1}{\varepsilon(\rho,\theta,z)} \left(\nabla_{(\rho,\theta,z)} \times \boldsymbol{\varphi}(\rho, \theta + \delta\theta, z) \right) \right)$$

$$= \nabla_{(\rho,\theta-\delta\theta,z)} \times \left(\frac{1}{\varepsilon(\rho,\theta-\delta\theta,z)} \left(\nabla_{(\rho,\theta-\delta\theta,z)} \times \boldsymbol{\varphi}(\rho, \theta, z) \right) \right)$$

$$= \nabla_{(\rho,\theta,z)} \times \left(\frac{1}{\varepsilon(\rho,\theta,z)} \left(\nabla_{(\rho,\theta,z)} \times \boldsymbol{\varphi}(\rho, \theta, z) \right) \right)$$

$$= \nabla \times \left(\frac{1}{\varepsilon(\mathbf{r})} \nabla \times \right) \boldsymbol{\varphi}(\mathbf{r}), \tag{2.103}$$

where we used the rotational symmetry of the dielectric function $\varepsilon(\rho, \theta - \delta\theta, z) = \varepsilon(\rho, \theta, z)$, as well as (2.102). Finally, by using (2.103) in (2.100) we arrive at the following equation:

$$\left[\nabla \times \left(\frac{1}{\varepsilon(\mathbf{r})} \nabla \times \right) \right] (\hat{\mathbf{R}}_{(\hat{n},\delta\theta)} \mathbf{H}(\mathbf{r})) = \omega^2 (\hat{\mathbf{R}}_{(\hat{n},\delta\theta)} \mathbf{H}(\mathbf{r})), \tag{2.104}$$

which implies that $\hat{\mathbf{R}}_{(\hat{n},\delta\theta)} \mathbf{H}(\mathbf{r})$ is also an eigenfunction of the Hamiltonian (2.71). Moreover, from (2.104) it also follows that the operator of rotation and the Hamiltonian of a system exhibiting rotational symmetry commute with each other as for any eigenfield $\mathbf{H}(\mathbf{r})$:

$$\hat{H}(\hat{\mathbf{R}}_{(\hat{n},\delta\theta)} \mathbf{H}(\mathbf{r})) = \omega^2 (\hat{\mathbf{R}}_{(\hat{n},\delta\theta)} \mathbf{H}(\mathbf{r})) \rightarrow \hat{H}(\hat{\mathbf{R}}_{(\hat{n},\delta\theta)} \mathbf{H}(\mathbf{r})) = \hat{\mathbf{R}}_{(\hat{n},\delta\theta)}(\omega^2 \mathbf{H}(\mathbf{r})) = \hat{\mathbf{R}}_{(\hat{n},\delta\theta)}(\hat{H} \mathbf{H}(\mathbf{r}))$$

$$\rightarrow \hat{H} \hat{\mathbf{R}}_{(\hat{n},\delta\theta)} = \hat{\mathbf{R}}_{(\hat{n},\delta\theta)} \hat{H}.$$

2.4.3   Continuous translational and rotational symmetries

The dielectric profiles presented in Fig. 2.9(a)–(d) exhibit various continuous symmetries.

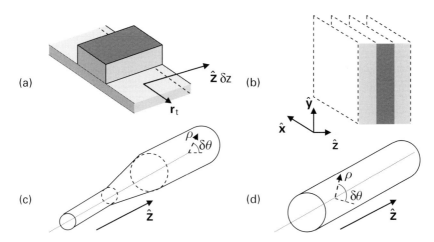

Figure 2.9 Examples of various continuous symmetries. (a) 1D continuous translation symmetry along $\hat{\mathbf{z}}$. (b) 2D continuous translation symmetry along $\hat{\mathbf{x}}$ and $\hat{\mathbf{y}}$. (c) Uniaxial rotational symmetry around $\hat{\mathbf{z}}$. (d) Uniaxial rotational symmetry around $\hat{\mathbf{z}}$ + 1D continuous translation symmetry along $\hat{\mathbf{z}}$.

One-dimensional continuous translational symmetry

Such symmetry is typically presented in integrated waveguides, optical fibers, and long scatterers of constant cross-section (Fig. 2.9(a)). The general form of a solution consistent with 1D translation symmetry is:

$$\mathbf{H}_{k_z}(\mathbf{r}) = \exp(ik_z z)\mathbf{U}_{k_z}(\mathbf{r}_t), \tag{2.105}$$

where $\mathbf{U}_{k_z}(\mathbf{r}_t)$ is a vector function of the transverse coordinates only. For historical reasons, to label solution (2.105) instead of a propagation vector $\hat{\mathbf{z}}k_z$ one frequently uses the notion of a propagation constant β defined as $\beta = k_z$.

Two-dimensional continuous translational symmetry

Such symmetry is typical for planar waveguides, mirrors, and multilayer structures (Fig. 2.9(b)). The general form of a solution consistent with 2D translation symmetry is:

$$\mathbf{H}_{k_t}(\mathbf{r}) = \exp(i\mathbf{k}_t\mathbf{r})\mathbf{U}_{k_t}(z), \tag{2.106}$$

where $\mathbf{U}_{k_t}(z)$ is a vector function of the z coordinate only, and $\mathbf{k}_t$ is a vector in the plane of a structure.

Three-dimensional continuous translational symmetry

This is symmetry of a uniform space. The general form of a solution consistent with 3D translation symmetry (compare with (2.19)) is:

$$\mathbf{H}_{\mathbf{k}}(\mathbf{r}) = \exp(i\mathbf{k}\mathbf{r})\mathbf{U}_{\mathbf{k}}, \tag{2.107}$$

where $\mathbf{U}_{\mathbf{k}}$ is a vector constant, and $\mathbf{k}$ is any 3D vector. Note that the only information missing from (2.107) is a transversality condition $(\mathbf{k} \cdot \mathbf{U}_{\mathbf{k}}) = 0$.

Uniaxial rotational symmetry

This is symmetry of tapered fibers, fiber lenses, and general lenses with rotational symmetry around a single axis (Fig. 2.9(c)). To find the general form of a solution consistent with uniaxial rotational symmetry, we first find the eigenmodes of an operator of continuous rotations. From (2.98), the eigenmodes of a rotational operator satisfy:

$$\Re(\hat{\mathbf{n}}, \delta\theta)\mathbf{H}(\Re(\hat{\mathbf{n}}, \delta\theta)^{-1}\mathbf{r}) = C(\hat{\mathbf{n}}, \delta\theta)\mathbf{H}(\mathbf{r}). \tag{2.108}$$

In a cylindrical coordinate system where $\hat{\mathbf{z}} = \hat{\mathbf{n}}$, $\Re(\hat{\mathbf{n}}, \delta\theta)$ is an identity matrix, the eigenequation (2.108) can be written as:

$$\Re(\hat{\mathbf{n}}, \delta\theta)\mathbf{H}(\Re(\hat{\mathbf{n}}, \delta\theta)^{-1}\mathbf{r}) = \mathbf{H}(\rho, \theta - \delta\theta, z) = C(\delta\theta)\mathbf{H}(\rho, \theta, z). \tag{2.109}$$

One can verify by substitution that in a cylindrical coordinate system, the general form of a nondegenerate vector eigenmode satisfying (2.109) is:

$$\mathbf{H}_m(\rho, \theta, z) = \exp(im\theta)\mathbf{U}_m(\rho, z) \tag{2.110}$$

where $\mathbf{U}_m(\rho, z)$ is a vector function of the radial and longitudinal coordinates, $C(\delta\theta) = \exp(-im\delta\theta)$, and m is an integer to guarantee that (2.110) is single-valued.

Uniaxial rotational symmetry and 1D translational symmetry (fibers)

This is the symmetry of conventional index guided optical fibers, multilayer Bragg fibers, and any other fiber drawn from a cylindrical preform (Fig. 2.9(d)). In the case when two distinct symmetry operations, $\hat{S}_1$ and $\hat{S}_2$, commute with each other, the eigenstates of one symmetry operator can be constructed from the eigenstates of another symmetry operator. Thus, to find the eigenstates satisfying both symmetries, one starts by finding the eigenstates of one of the two operators, and then one finds what other additional restrictions should be imposed on a form of a solution to satisfy the eigenvalue equation for the other symmetry operator. Finally, if degeneracy is present, the eigenstate of a system Hamiltonian will be a linear combination of the degenerate eigenstates common to both symmetry operators.

As found in the case of uniaxial rotational symmetry, in the cylindrical coordinate system with $\hat{\mathbf{z}} = \hat{\mathbf{n}}$, nondegenerate eigenstates reflecting such a symmetry have the form $\mathbf{H}_m(\rho, \theta, z) = \exp(im\theta)\mathbf{U}_m(\rho, z)$. Now, for such states also to exhibit translational symmetry along the $\hat{\mathbf{z}}$ direction they should additionally satisfy:

$$\hat{T}_{\hat{z}\delta z}\mathbf{H}_m(\mathbf{r}) = \mathbf{H}_m(\rho, \theta, z - \delta z) = C(\delta z)\mathbf{H}_m(\mathbf{r})$$
$$\exp(im\theta)\mathbf{U}_m(\rho, z - \delta z) = C(\delta z)\exp(im\theta)\mathbf{U}_m(\rho, z), \tag{2.111}$$
$$\mathbf{U}_m(\rho, z - \delta z) = C(\delta z)\mathbf{U}_m(\rho, z)$$

where from the last equation it follows that $\mathbf{U}_m(\rho, z) = \exp(ik_z z)\mathbf{U}_{m,k_z}(\rho)$ and $\mathbf{U}_{m,k_z}(\rho)$ is a vector function of a radial coordinate only. Finally, a general form of an eigenstate exhibiting a uniaxial rotational symmetry plus 1D translational symmetry is:

$$\mathbf{H}_{m,k_z}(\rho, \theta, z) = \exp(im\theta)\exp(ik_z z)\mathbf{U}_{m,k_z}(\rho). \tag{2.112}$$

Two-dimensional continuous translational symmetry from the point of view of continuous rotational symmetry

It is interesting to note that 2D continuous translational symmetry, characteristic to slab waveguides, mirrors, and multilayer structures, can also be considered from the point of view of continuous rotational symmetry. This leads to an alternative form of a general solution of (2.70). As described earlier, the general form of a solution in a system exhibiting 2D translational symmetry is

$$\mathbf{H}_{\mathbf{k}_t}(\mathbf{r}) = \exp(i\mathbf{k}_t \mathbf{r})\mathbf{U}_{\mathbf{k}_t}(z). \tag{2.113}$$

It is reasonable to assume that in a system exhibiting continuous rotational symmetry, the eigenvalue $\omega^2(\mathbf{k}_t)$ of a system Hamiltonian (2.71) depends only on the absolute value of a transverse wavevector, and not on its direction (this will be proved in Section 2.4.6), that is, $\omega^2(\mathbf{k}_t) = \omega^2(|\mathbf{k}_t|)$. Therefore, solution (2.113) is degenerate with respect to the direction of a transverse wave vector. As established in Section 2.4.2 (discussion of rotation operator), if (2.113) is an eigensolution of a Hamiltonian (2.71), then

$$\hat{\mathbf{R}}_{(\hat{z},\tilde{\theta})}\mathbf{H}_{\mathbf{k}_t}(\mathbf{r}) = \exp\left(i\mathbf{k}_t\mathfrak{R}^{-1}_{(\hat{z},\tilde{\theta})}\mathbf{r}\right)\mathfrak{R}_{(\hat{z},\tilde{\theta})}\mathbf{U}_{\mathbf{k}_t}(z), \tag{2.114}$$

is also a solution with the same eigenvalue for any $\tilde{\theta}$. Therefore, a general solution can be written in terms of a linear combination of all the degenerate solutions:

$$\mathbf{H}(\mathbf{r})_{\mathbf{k}_t}\Big|_{\text{Cartesian}} = \int_0^{2\pi} d\tilde{\theta}\, A(\tilde{\theta})\hat{\mathbf{R}}_{(\hat{z},\tilde{\theta})}\mathbf{H}_{\mathbf{k}_t}(\mathbf{r}) = \int_0^{2\pi} d\tilde{\theta}\, A(\tilde{\theta})\exp\left(i\mathbf{k}_t\mathfrak{R}^{-1}_{(\hat{z},\tilde{\theta})}\mathbf{r}\right)\mathfrak{R}_{(\hat{z},\tilde{\theta})}\mathbf{U}_{\mathbf{k}_t}(z). \tag{2.115}$$

Note that (2.115) is written in Cartesian coordinates. To rewrite it in cylindrical coordinates we use the fact that $\mathbf{H}(\mathbf{r})|_{\text{Cartesian}} = \mathfrak{R}_{(\hat{z},\theta)}\mathbf{H}(\rho,\theta,z)|_{\text{cylindrical}}$. This allows us to rewrite (2.115) in cylindrical coordinates as:

$$\mathbf{H}(\rho,\theta,z)_{k_\rho}\Big|_{\text{cylindrical}} = \int_0^{2\pi} d\tilde{\theta}\, A(\tilde{\theta})\exp(i\mathbf{k}_t\mathfrak{R}^{-1}_{(\hat{z},\tilde{\theta})}\mathbf{r})\mathfrak{R}^{-1}_{(\hat{z},\theta)}\mathfrak{R}_{(\hat{z},\tilde{\theta})}\mathbf{U}_{\mathbf{k}_t}(z)$$

$$= \int_0^{2\pi} d\tilde{\theta}\, A(\tilde{\theta})\exp(ik_\rho\rho\cos(\theta-\tilde{\theta}))\mathfrak{R}_{(\hat{z},\tilde{\theta}-\theta)}\mathbf{U}_{k_\rho}(z) \tag{2.116}$$

where without loss of generality we suppose that in the Cartesian coordinate system $\mathbf{k}_t = (k_\rho, 0, 0)$, and $\mathbf{U}_{\mathbf{k}_t}(z)$ is relabeled as $\mathbf{U}_{k_\rho}(z)$.

Alternatively, in a system with continuous rotational symmetry, a general solution can be chosen in the form (2.110),

$$\mathbf{H}_m(\rho,\theta,z) = \exp(im\theta)\mathbf{U}_m(\rho,z).$$

To cast (2.116) into this form, we choose expansion coefficients as

$$A(\tilde{\theta}) = A_0/(2\pi)\exp(im\tilde{\theta}),$$

$$A_0 = -\exp(im \cdot \pi/2).$$

In this case (2.116) transforms as:

$$\mathbf{H}(\rho, \theta, z)_{m,k_\rho}|_{\text{cylindrical}} =$$

$$= \frac{A_0}{2\pi} \int_0^{2\pi} d\tilde{\theta}\, \exp(\mathrm{i}k_\rho\rho\cos(\theta - \tilde{\theta}) + \mathrm{i}m\tilde{\theta})\Re_{(\hat{z},\tilde{\theta}-\theta)}\mathbf{U}_{k_\rho}(z)$$

$$= \exp(\mathrm{i}m\theta)\frac{A_0}{2\pi} \int_0^{2\pi} d\tilde{\theta}\, \exp(\mathrm{i}k_\rho\rho\cos(\tilde{\theta} - \theta) + \mathrm{i}m(\tilde{\theta} - \theta))\Re_{(\hat{z},\tilde{\theta}-\theta)}\mathbf{U}_{k_\rho}(z)$$

$$= \exp(\mathrm{i}m\theta)\left[\frac{A_0}{2\pi} \int_0^{2\pi} d\tilde{\tilde{\theta}}\, \exp(\mathrm{i}k_\rho\rho\cos(\tilde{\tilde{\theta}}) + \mathrm{i}m\tilde{\tilde{\theta}})\Re_{(\hat{z},\tilde{\tilde{\theta}})}\right]\mathbf{U}_{k_\rho}(z). \tag{2.117}$$

By using an explicit form of the rotation matrix in Cartesian coordinate system, and the integral representation of the Bessel function:

$$J_m(x) = \frac{-\exp\left(-\mathrm{i}m\frac{\pi}{2}\right)}{2\pi} \int_0^{2\pi} d\theta\, \exp(\mathrm{i}x\cos(\theta) + \mathrm{i}m\theta)$$

$$\frac{J_{m-1}(x) + J_{m+1}(x)}{2} = \frac{-\exp\left(-\mathrm{i}m\frac{\pi}{2}\right)}{2\pi} \int_0^{2\pi} d\theta\, \sin(\theta)\exp(\mathrm{i}x\cos(\theta) + \mathrm{i}m\theta),$$

$$\frac{J_{m-1}(x) - J_{m+1}(x)}{2\mathrm{i}} = \frac{-\exp\left(-\mathrm{i}m\frac{\pi}{2}\right)}{2\pi} \int_0^{2\pi} d\theta\, \cos(\theta)\exp(\mathrm{i}x\cos(\theta) + \mathrm{i}m\theta),$$

we can simplify (2.117):

$$\mathbf{H}(\rho, \theta, z)_{m,k_\rho}|_{\text{cylindrical}} =$$

$$= \exp(\mathrm{i}m\theta)\left[\frac{A_0}{2\pi} \int_0^{2\pi} d\tilde{\tilde{\theta}}\, \exp(\mathrm{i}k_\rho\rho\cos(\tilde{\tilde{\theta}}) + \mathrm{i}m\tilde{\tilde{\theta}})\Re_{(\hat{z},\tilde{\tilde{\theta}})}\right]\mathbf{U}_{k_\rho}(z)$$

$$= \exp(\mathrm{i}m\theta)\left[\frac{A_0}{2\pi} \int_0^{2\pi} d\tilde{\tilde{\theta}}\, \exp(\mathrm{i}k_\rho\rho\cos(\tilde{\tilde{\theta}}) + \mathrm{i}m\tilde{\tilde{\theta}})\begin{pmatrix} \cos(\tilde{\tilde{\theta}}) & -\sin(\tilde{\tilde{\theta}}) & 0 \\ \sin(\tilde{\tilde{\theta}}) & \cos(\tilde{\tilde{\theta}}) & 0 \\ 0 & 0 & 1 \end{pmatrix}\right]\mathbf{U}_{k_\rho}(z)$$

$$= \exp(\mathrm{i}m\theta)\begin{pmatrix} \dfrac{J_{m-1}(k_\rho\rho) - J_{m+1}(k_\rho\rho)}{2} & -\dfrac{J_{m-1}(k_\rho\rho) + J_{m+1}(k_\rho\rho)}{2} & 0 \\ \dfrac{J_{m-1}(k_\rho\rho) + J_{m+1}(k_\rho\rho)}{2} & \dfrac{J_{m-1}(k_\rho\rho) - J_{m+1}(k_\rho\rho)}{2} & 0 \\ 0 & 0 & J_m(k_\rho\rho) \end{pmatrix}\mathbf{U}_{k_\rho}(z)$$

$$\tag{2.118}$$

We therefore conclude that an alternative form of a general solution for the case of a system with 2D translational symmetry is:

$$\mathbf{H}_{m,k_\rho}(\rho, \theta, z) = \exp(\mathrm{i}m\theta)\mathbf{M}_m(k_\rho\rho)\mathbf{U}_{k_\rho}(z), \tag{2.119}$$

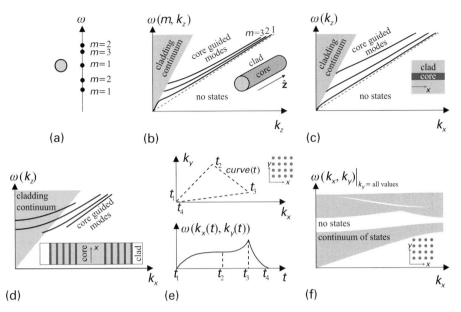

Figure 2.10 Various types of band diagrams, which are used to present dispersion relations of the eigenstates of electromagnetic systems.

where two conservative numbers (m, k_ρ) characterize a solution, $\mathbf{M}_m(k_\rho\rho)$ is a 3×3 matrix of a specific form (2.118), and $\mathbf{U}_{k_\rho}(z)$ is a vectorial function of the z coordinate only.

2.4.4   Band diagrams

Depending on the number and nature of the conserved parameters, different types of band diagrams are possible. As introduced in Section 2.1, band diagrams serve as a tool to visualize the phase space of the allowed electromagnetic modes of a system. Particularly, one most frequently uses 2D plots depicting the dependence of frequency of the allowed states as a function of a single continuous parameter.

If the symmetry of a system does not allow any conserved parameters, or if it only allows discrete conserved parameters (such as integer m in (2.110)), then a corresponding band diagram ω becomes a discrete collection of points along the frequency axis. As an example, Fig. 2.10(a) shows a band diagram of the eigenstates of a circular resonator labeled by the corresponding values of the angular momenta. Distinct states can share the same value of angular momentum m.

If there are several conserved numbers among which one is discrete while another one is continuous (such as (m, k_z) in (2.112) or (m, k_ρ) in (2.119)), then one typically fixes the value of a discrete parameter and then plots the dispersion relation as a function of a single continuous parameter. In this case, the band diagram appears as a collection of continuous "bands" labeled by different values of a discrete parameter. As an example, in Fig. 2.10(b), a band diagram of modes of a circular fiber is presented with different bands labeled by the different values of angular momentum m. Using the considerations of Section 2.1 we also conclude that there are no guided states below the light line of a dielectric with the highest value of a dielectric constant (the light line of a fiber core in this example).

Note that even if all the conserved quantities describing the symmetry of a system are continuous, one can still have discrete bands in the band diagram. Compare, for example, two different descriptions of solutions in systems exhibiting 2D translational symmetry. Thus, the modes of a planar multilayer system (Fig. 2.9(b)) can be labeled either by two continuous parameters (k_x, k_y) as in (2.106), or by one discrete and one continuous parameter (m, k_ρ) as in (2.119) (hence exhibiting discrete bands). Generally, whenever there is confinement even in a single spatial direction (for example, confinement along the $\hat{z}$ direction in Fig. 2.9(b)) solutions typically form discrete bands. Another example of this is presented in Fig. 2.10(c), where eigenfrequencies of the modes of a slab waveguide are plotted against k_x, assuming $k_y = 0$.

The dielectric profile of a practical electromagnetic system always features an infinite dielectric region called cladding, which supports a continuum of delocalized radiation states. Far into the cladding region, solutions of Maxwell's equations can be described in terms of plane waves characterized by the corresponding diagram of Fig. 2.1(b). In the presence of structural or material imperfections, such as waveguide bending, surface roughness, regions of material crystallinity, etc., localized eigenstates of a system can couple to the radiation continuum of the cladding, thus resulting in scattering losses. When plotting band diagrams of complex systems it is always useful to add a band diagram of a cladding continuum, as modes that are truly localized by the structure will lie below such a continuum. As an example, consider a planar photonic crystal waveguide with a core surrounded by the periodic reflector and, finally, a low-refractive-index cladding (see Fig. 2.10(d)). As we will see later in the book, for the frequencies falling into the bandgap of a periodic reflector, the waveguide core can support guided modes located inside the continuum of cladding states. When excited, such states could be guided in the waveguide core, however, they will be prone to "leaking" into the cladding due to imperfections in the confining reflector.

In the case when there are several continuous parameters labeling solutions (like (k_x, k_y) in (2.106) or (2.127)), two types of band diagrams are typically employed. The first type of band diagram presents the dispersion relation along the given one-dimensional curve in the multidimensional continuous parameter space (Fig. 2.10(e)). The second type of band diagram presents all the dispersion relations plotted on the same graph as a function of a single continuous parameter for all possible values of the other continuous parameters. This is a so-called "projected" band diagram (Fig. 2.10(f)). Both band diagram types are considered in detail in the following sections.

2.4.5 Discrete translational and rotational symmetries

In what follows, we consider the general structure of the electromagnetic solutions in systems exhibiting various discrete symmetries. We particularly consider modes of the dielectric profiles corresponding to a fiber Bragg grating (Fig. 2.11(ai)), planar photonic crystal waveguide (Fig. 2.11(aii)), a diffraction grating (Fig. 2.11(b)), an ideal

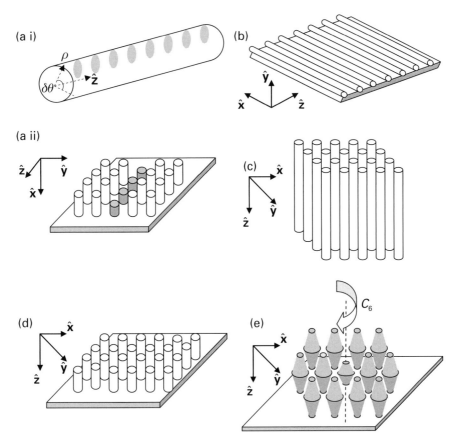

Figure 2.11 Examples of various discrete translational symmetries. (a) Discrete translations in 1D (fiber Bragg gratings, photonic planar crystal waveguides). (b) Discrete translations in 1D + continuous translations in 1D (planar Bragg gratings). (c) Discrete translations in 2D + continuous translations in 1D (2D photonic crystals). (d) Discrete translations in 2D (realistic 2D photonic crystals in slab geometry). (e) Discrete rotational symmetry (point defects in a periodic 2D lattice).

two-dimensional photonic crystal (Fig. 2.11(c)), a photonic crystal slab (Fig. 2.11(d)), and a resonator embedded into a photonic crystal lattice (Fig. 2.11(e)).

One-dimensional discrete translational symmetry

Such symmetry is presented in a periodic sequence of identical finite size scatterers. Examples of structures exhibiting such a symmetry are fiber Bragg gratings (see Fig. 2.11(a)), and slab photonic crystal waveguides (see Fig. 2.11(a)). In fiber Bragg gratings, in particular, the periodic modulation of the dielectric profile is written into a fiber core along the direction of light propagation. For most frequencies, a fiber Bragg grating guides the light in the same way as a normal fiber does, while in some frequency ranges, called stop bands, a fiber Bragg grating exhibits strong changes in its transmission properties as a function of wavelength. For example, inside a stop band, a fiber

Bragg grating can completely reflect the light back towards the source, thus enabling such applications as frequency rejection filters. A periodic sequence of weakly coupled high-quality resonators has also been proposed to generate slow light (modes with ultra-low values of group velocity) to be used as delay lines.

The general form of a nondegenerate solution consistent with 1D discrete symmetry can be derived in exactly the same fashion as in the case of quantum mechanics. Nevertheless, we will present the derivation again to introduce the notions of the Bravais lattice and the Brillouin zone. Thus, assume that a system Hamiltonian transforms into itself $\varepsilon(\mathbf{r} - \delta\mathbf{r}) = \varepsilon(\mathbf{r})$ for any discrete translations along vector $\delta\mathbf{r}$ of the form $\delta\mathbf{r} = \bar{a}_1 N_1$ where vector $\bar{a}_1 \parallel \hat{\mathbf{z}}$ and N_1 is an integer, see Fig. 2.11(a). All the points in space described by displacements $\delta\mathbf{r} = \bar{a}_1 N_1$ form a periodic array, also called a Bravais lattice. The eigenstates of a translation operator (2.97) then satisfy:

$$\hat{T}_{\delta\mathbf{r}}\mathbf{H}(\mathbf{r}) = \mathbf{H}(\mathbf{r} - \bar{a}_1 N_1) = C(N_1)\mathbf{H}(\mathbf{r}). \tag{2.120}$$

One can verify by substitution that eigenfunctions satisfying (2.120) are $\mathbf{H}(\mathbf{r}) = \exp(ik_z z)\mathbf{U}_{k_z}(\mathbf{r}_t)$ where $(\mathbf{r} = (\mathbf{r}_t + z\hat{\mathbf{z}}), \mathbf{r}_t \perp \hat{\mathbf{z}})$ and $\mathbf{U}_{k_z}(\mathbf{r}_t)$ is a vector function of transverse coordinates only. The corresponding eigenvalues are:

$$C(N_1) = \exp(-ik_z (\bar{a}_1 \hat{\mathbf{z}}) N_1). \tag{2.121}$$

The eigenvalue (2.121) is, however, the same for any $k_z\hat{\mathbf{z}} + \mathbf{G}$ where $\mathbf{G} = \bar{b}_1 P_1$, P_1 is an integer, and $\bar{b}_1 = \hat{\mathbf{z}} \cdot 2\pi / |\bar{a}_1|$, which are called reciprocal lattice vectors. Thus, the eigenstate of a translation operator is degenerate for a set of wave vectors of the form $k_z\hat{\mathbf{z}} + \mathbf{G}$. That means that a Bloch state (eigenstate for periodic systems) of a Hamiltonian possessing a 1D discrete translational symmetry can be labeled by a continuous conserved parameter k_z, and can be expressed as a linear combination of all the degenerate states having the same eigenvalue (2.121):

$$\mathbf{H}_{k_z}(\mathbf{r}) = \exp(ik_z z) \sum_{\mathbf{G}} A(k_z\hat{\mathbf{z}} + \mathbf{G})\exp(i(\mathbf{G}\hat{\mathbf{z}})z)\mathbf{U}_{k_z\hat{\mathbf{z}}+\mathbf{G}}(\mathbf{r}_t) = \exp(ik_z z)\mathbf{U}_{k_z}(\mathbf{r})$$

$$\mathbf{U}_{k_z}(\mathbf{r} + \bar{a}_1 N_1) = \mathbf{U}_{k_z}(\mathbf{r}), \tag{2.122}$$

where the expansion coefficients $A(k_z\hat{\mathbf{z}} + \mathbf{G})$ define a periodic function $\mathbf{U}_{k_z}(\mathbf{r})$ along a symmetry direction. Note that Bloch states with $k_z\hat{\mathbf{z}}$ and $k_z\hat{\mathbf{z}} + \mathbf{G}'$ are identical for any $\mathbf{G}'$. Indeed:

$$\mathbf{H}_{k_z\hat{\mathbf{z}}+\mathbf{G}'}(\mathbf{r}) = \exp(i(k_z + (\mathbf{G}'\hat{\mathbf{z}}))z) \sum_{\mathbf{G}} A(k_z\hat{\mathbf{z}} + \mathbf{G}' + \mathbf{G})\exp(i(\mathbf{G}\hat{\mathbf{z}})z)\mathbf{U}_{k_z\hat{\mathbf{z}}+\mathbf{G}'+\mathbf{G}}(\mathbf{r}_t)$$

$$= \exp(ik_z z) \sum_{\mathbf{G}} A(k_z\hat{\mathbf{z}} + \mathbf{G}' + \mathbf{G})\exp(i((\mathbf{G}' + \mathbf{G})\hat{\mathbf{z}})z)\mathbf{U}_{k_z\hat{\mathbf{z}}+\mathbf{G}'+\mathbf{G}}(\mathbf{r}_t)$$

$$\underset{(\mathbf{G}'+\mathbf{G})\to\mathbf{G}}{=} \mathbf{H}_{k_z\hat{\mathbf{z}}}(\mathbf{r}). \tag{2.123}$$

Thus, we only need a part of the reciprocal lattice space, namely $k_z \in (-|\bar{b}_1|/2, |\bar{b}_1|/2]$ to label all the nondegenerate Bloch states. The smallest region of reciprocal space needed to label all the Bloch states is called a first Brillouin zone.

Note also that owing to equivalency of all the $k_z\hat{\mathbf{z}} + \mathbf{G}'$ eigenstates, the dispersion relation satisfies $\omega(k_z) = \omega(k_z + (\mathbf{G}'\hat{\mathbf{z}}))$ for any $\mathbf{G}'$. Finally, from time-reversal symmetry,

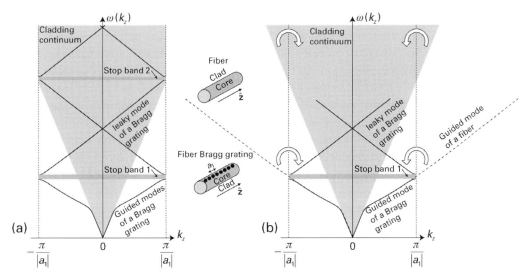

Figure 2.12 (a) Schematic band diagram of a fiber Bragg grating. (b) Construction of a band diagram for a 1D periodic system.

$\omega(k_z) = \omega(-k_z)$ and one can only consider $k_z \in [0, |\bar{b}_1|/2]$, also known as an irreducible Brillouin zone.

As periodic structures present a focus of our work we will explain in detail the properties of their band diagrams in the corresponding sections. At this point our goal is rather to highlight the differences between systems with continuous and discrete symmetries.

A typical band diagram for a fiber Bragg grating is presented in Fig. 2.12(a). At the edges of the first Brillouin zone the curvature of the bands is zero, which is a consequence of the time-reversal symmetry (see Problem 2.2 of this section). The eigenstates of a fiber Bragg grating can be classified as either the true guides states if they are located below the cladding continuum, or as the "leaky" radiative states if they are located in the cladding continuum. In the case of a weak Bragg grating, the periodic perturbation of the dielectric profile of a fiber core is small, thus, far from the Brillouin zone edges dispersion relations of the fiber Bragg grating modes should be similar to those of a uniform fiber. This suggests that to get the first approximation to the band diagram of a weak fiber Bragg grating one can first plot the dispersion relations of the fiber modes, and then simply reflect them into the first Brillouin zone (Fig. 2.12(b)). At the points of intersection with the edges of the first Brillouin zone small frequency gaps (stop bands) will appear. Finally, at higher frequencies the true guided modes of a fiber are reflected into the continuum of the cladding states, thus becoming the "leaky" radiative modes of a fiber Bragg grating.

One-dimensional discrete translational symmetry and 1D continuous symmetry

Such symmetry is characteristic to structures made of a periodic sequence of scatterers extended in one dimension. A representative of such structures is a diffraction grating (see Fig. 2.11(b)) where the periodic modulation of dielectric profile is written as lines into a solid substrate. The transmission (or reflection) properties of diffractive gratings are strongly dependent on the angle of radiation incidence, as well as the wavelength of the incident light. This property of diffraction gratings enables, for example, the spatial separation of distinct wavelength components traveling in the same nonmonochromatic beam.

The symmetry of diffraction gratings implies that its system Hamiltonian transforms into itself $\varepsilon(\mathbf{r} - \delta\mathbf{r}) = \varepsilon(\mathbf{r})$ for any translation along the vector $\delta\mathbf{r}$ of the form $\delta\mathbf{r} = \bar{a}_1 N_1 + \hat{\mathbf{x}}\delta x$, where vector $\bar{a}_1 \parallel \hat{\mathbf{z}}$ and N_1 is an integer. The general form of a field solution can be readily derived as:

$$\mathbf{H}_{k_x,k_z}(\mathbf{r}) = \exp(ik_z z + ik_x x)\mathbf{U}_{k_x,k_z}(y,z)$$
$$\mathbf{U}_{k_x,k_z}(y, z + (\bar{a}_1\hat{\mathbf{z}})\,N_1) = \mathbf{U}_{k_x,k_z}(y, z)$$

(2.124)

Two-dimensional discrete translational symmetry and 1D continuous symmetry

Systems possessing this symmetry are called 2D photonic crystals. The optical response of 2D photonic crystals can be readily computed without resorting to demanding numerical simulations. Practical implementations of 2D photonic crystals include photoinduced lattices and photonic crystal fibers (see Fig. 2.11(c)); there the direction of light propagation coincides with the direction of continuous translational symmetry $\hat{\mathbf{z}}$. For fibers, in particular, fiber core is implemented as a continuous defect in the $\hat{\mathbf{z}}$ direction of a 2D periodic lattice; light in the core is then confined by the bandgap of a periodic cladding. Band diagrams of 2D photonic crystals also present a departure point for understanding the band diagrams of more practical slab photonic crystals; there light propagation is confined strictly to the xy plane of a photonic crystal (see Fig. 2.11(d)).

The symmetry considered implies that a system Hamiltonian transforms into itself $\varepsilon(\mathbf{r} - \delta\mathbf{r}) = \varepsilon(\mathbf{r})$ for any translation along the vector $\delta\mathbf{r}$ of the form $\delta\mathbf{r} = \mathbf{r}_t + \hat{\mathbf{z}}\delta z = \bar{a}_1 N_1 + \bar{a}_2 N_2 + \hat{\mathbf{z}}\delta z$, where vectors $(\bar{a}_1, \bar{a}_2) \perp \hat{\mathbf{z}}$ are noncollinear and N_1, N_2 are any integers. Then, the general form of a field solution can be easily derived to be:

$$\mathbf{H}_{\mathbf{k}_t,k_z}(\mathbf{r}) = \exp(ik_z z + i\mathbf{k}_t\mathbf{r}_t)\mathbf{U}_{\mathbf{k}_t,k_z}(\mathbf{r}_t)$$
$$\mathbf{U}_{\mathbf{k}_t,k_z}(\mathbf{r}_t + \bar{a}_1 N_1 + \bar{a}_2 N_2) = \mathbf{U}_{\mathbf{k}_t,k_z}(\mathbf{r}_t)$$

(2.125)

where N_1, N_2 are any integers. We define basis vectors of reciprocal space as:

$$\bar{b}_1 = 2\pi\frac{\bar{a}_2 \times \hat{\mathbf{z}}}{\bar{a}_1 \cdot (\bar{a}_2 \times \hat{\mathbf{z}})}; \bar{b}_2 = 2\pi\frac{\hat{\mathbf{z}} \times \bar{a}_1}{\bar{a}_1 \cdot (\bar{a}_2 \times \hat{\mathbf{z}})}.$$

(2.126)

As in the case of 1D discrete translation symmetry, the spatial points described by the displacements $\delta\mathbf{r} = \bar{a}_1 N_1 + \bar{a}_2 N_2$ form a Bravais lattice. The Bloch states with $\mathbf{k}_t$ and $\mathbf{k}_t + \mathbf{G}$ are identical for any $\mathbf{G} = \bar{b}_1 P_1 + \bar{b}_2 P_2$, thus, only a finite volume of the reciprocal phase space can be used to label the modes (the first Brillouin zone).

Band diagrams of the 2D photonic crystals will be considered in detail later. In passing, the band diagram most frequently used to describe bandgaps of 2D photonic crystals is $\omega(k_z = 0, \mathbf{k}_t =$ the Brillouin zone edge) (see Fig. 2.13(a)), where eigenfrequencies are presented along the edge of an irreducible Brillouin zone (see definition at the end of Section 2.4.6). The reasoning for choosing the Brillouin zone edge to visualize

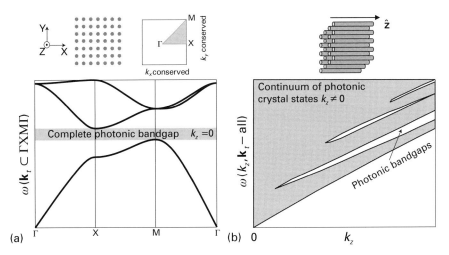

Figure 2.13 Band diagrams for 2D photonic crystals. (a) Band diagram to study the modes propagating strictly in the plane of a photonic crystal $k_z = 0$ (planar photonic crystal circuitry). (b) Projected band diagram to study states guided along the direction of continuous translational symmetry $k_z \neq 0$ (photonic crystal fibers).

the bandgaps is that all the frequencies corresponding to the interior points of a first Brillouin zone typically fall in between the lowest and the highest frequencies of the states at the irreducible zone edge. As the goal, frequently, is to establish the presence of a complete bandgap (a bandgap for all the transverse propagation directions) then one can do that just by checking the points on the Brillouin zone edge. Another type of band diagram is typically used to characterize guided modes of photonic crystal fibers. It presents a band diagram projected onto a k_z direction of a propagation wavevector, namely $\omega(k_z = \text{const}, \mathbf{k}_t = \text{all allowed})$ (see Fig. 2.13(b)). In the resultant band diagram a continuum of cladding states is interrupted with bandgap regions. In the presence of a continuous defect (waveguide core), states guided by the defect appear as discrete bands inside the bandgap regions.

Two-dimensional discrete translational symmetry

This is symmetry of a practical implementation of a 2D photonic crystal in the form of a planar dielectric slab with a 2D periodic pattern imprinted in it (see Fig. 2.11(d)). Such structures are typically realized by means of electronic beam lithography. In the vertical direction, optical confinement is achieved by assuring that the effective refractive index of the slab is higher than that of the cladding. The symmetry considered implies that a system Hamiltonian transforms into itself $\varepsilon(\mathbf{r} + \delta\mathbf{r}) = \varepsilon(\mathbf{r})$ for any translation along vector $\delta\mathbf{r}$ of the form $\delta\mathbf{r} = \bar{a}_1 N_1 + \bar{a}_2 N_2$, where vectors $(\bar{a}_1, \bar{a}_2) \perp \hat{\mathbf{z}}$ are noncollinear and N_1, N_2 are any integers. Then, the general form of a field solution can be now easily derived as:

$$\mathbf{H}_{\mathbf{k}_t}(\mathbf{r}) = \exp(i\mathbf{k}_t \mathbf{r}_t) \mathbf{U}_{\mathbf{k}_t}(\mathbf{r}_t, z)$$
$$\mathbf{U}_{\mathbf{k}_t}(\mathbf{r}_t + \bar{a}_1 N_1 + \bar{a}_2 N_2, z) = \mathbf{U}_{\mathbf{k}_t}(\mathbf{r}_t, z)$$

$$(2.127)$$

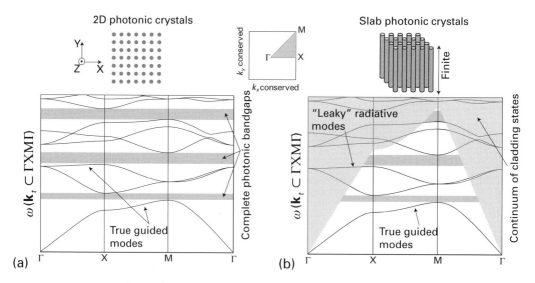

Figure 2.14 Comparison between the band diagrams of (a) 2D photonic crystals and (b) realistic slab photonic crystals.

The band diagram for the case of a 2D photonic crystal slab (Fig. 2.14(b)) is similar to that of a 2D photonic crystal with continuous translation symmetry in the transverse direction (see Fig. 2.14(a)). One important difference between the two diagrams is in the presence of a continuum of cladding states for a slab photonic crystal. Photonic crystal modes situated in the cladding continuum are inherently "leaky" and can be efficiently irradiated into the cladding by the structural imperfections of a photonic crystal.

Three-dimensional discrete translational symmetry

Substantial experimental progress has been made in creating 3D photonic crystal structures using, among other techniques, self-assembly, lithography-assisted stacking, holography, and multiphoton polymer polymerization techniques. The major remaining challenges include difficulty in maintaining long-range order of a photonic crystal lattice, as well as a limited ability to introduce carefully designed defects into the lattice structure.

The symmetry considered implies that a system Hamiltonian transforms into itself $\varepsilon(\mathbf{r} - \delta\mathbf{r}) = \varepsilon(\mathbf{r})$ for any translation along vector $\delta\mathbf{r}$ of the form $\delta\mathbf{r} = \bar{a}_1 N_1 + \bar{a}_2 N_2 + \bar{a}_3 N_3$, where vectors $(\bar{a}_1, \bar{a}_2, \bar{a}_3)$ are noncollinear and N_1, N_2, N_3 are any integers. Then, the general form of a field solution can be easily derived as:

$$\mathbf{H_k(r)} = \exp(i\mathbf{kr})\mathbf{U_k(r)}$$
$$\mathbf{U_k}(\mathbf{r} + \bar{a}_1 N_1 + \bar{a}_2 N_2 + \bar{a}_3 N_3) = \mathbf{U_k(r)} \quad (2.128)$$

where N_1, N_2, N_3 are any integers, and the basis vectors of reciprocal space are:

$$\bar{b}_1 = 2\pi \frac{\bar{a}_2 \times \bar{a}_3}{\bar{a}_1 \cdot (\bar{a}_2 \times \bar{a}_3)}; \bar{b}_2 = 2\pi \frac{\bar{a}_3 \times \bar{a}_1}{\bar{a}_1 \cdot (\bar{a}_2 \times \bar{a}_3)}; \bar{b}_3 = 2\pi \frac{\bar{a}_1 \times \bar{a}_2}{\bar{a}_1 \cdot (\bar{a}_2 \times \bar{a}_3)}. \quad (2.129)$$

All the points in space described by displacements $\delta\mathbf{r} = \bar{a}_1 N_1 + \bar{a}_2 N_2 + \bar{a}_3 N_3$ form a Bravais lattice. Bloch states with $\mathbf{k}$ and $\mathbf{k} + \mathbf{G}$ are identical for any $\mathbf{G} = \bar{b}_1 P_1 + \bar{b}_2 P_2 + \bar{b}_3 P_3$, thus only a first Brillouin zone need be used to label the modes.

Normalization of Bloch modes

In the case of discrete translational symmetries, the orthogonality relation between the Bloch modes becomes more restrictive than (2.82). In particular, consider two Bloch modes characterized by wave vectors $\mathbf{k}$ and $\mathbf{k}'$ (confined to the first Brillouin zone), and belonging to the bands m and m'. Owing to the Hermitian nature of the Maxwell Hamiltonian, the following general othrogonality relation (2.82) holds:

$$\int_{\infty} d\mathbf{r}\, \mathbf{H}^*_{\omega_m(\mathbf{k})}(\mathbf{r}) \mathbf{H}_{\omega_{m'}(\mathbf{k}')}(\mathbf{r}) \sim \delta_{\omega_m(\mathbf{k}),\omega_{m'}(\mathbf{k}')}, \tag{2.130}$$

where δ is a Kronecker delta, and integration is performed over the whole space. Owing to the particular form of a Bloch solution, an additional orthogonality condition with respect to the Bloch wave vectors also holds. Thus, using the general form of a Bloch solution:

$$\mathbf{H}_{\omega_m(\mathbf{k})} = \exp(i\mathbf{k}\mathbf{r})\mathbf{U}_{\omega_m(\mathbf{k})}(\mathbf{r})$$
$$\mathbf{U}_{\omega_m(\mathbf{k})}(\mathbf{r} + \mathbf{R}) = \mathbf{U}_{\omega_m(\mathbf{k})}(\mathbf{r}); \quad \mathbf{R} - \text{any lattice vector} \tag{2.131}$$

and after substitution into (2.130) we get:

$$\int_{\infty} d\mathbf{r}\, \mathbf{H}^*_{\omega_m(\mathbf{k})}(\mathbf{r}) \mathbf{H}_{\omega_{m'}(\mathbf{k}')}(\mathbf{r}) = \int_{\infty} d\mathbf{r}\, \exp\left(i(\mathbf{k}' - \mathbf{k})\mathbf{r}\right) \mathbf{U}^*_{\omega_m(\mathbf{k})}(\mathbf{r})\mathbf{U}_{\omega_{m'}(\mathbf{k}')}(\mathbf{r})$$

$$= \sum_{\mathbf{R}} \exp\left(i(\mathbf{k}' - \mathbf{k})\mathbf{R}\right) \int_{\text{unit cell}} d\mathbf{r}\, \mathbf{U}^*_{\omega_m(\mathbf{k})}(\mathbf{r})\mathbf{U}_{\omega_{m'}(\mathbf{k}')}(\mathbf{r})$$

$$= \frac{(2\pi)^d}{V_{\text{unit cell}}} \delta(\mathbf{k}' - \mathbf{k}) \int_{\text{unit cell}} d\mathbf{r}\, \mathbf{U}^*_{\omega_m(\mathbf{k})}(\mathbf{r})\mathbf{U}_{\omega_{m'}(\mathbf{k}')}(\mathbf{r})$$

$$\text{from (12.130)} = C\delta_{\omega_m(\mathbf{k}),\omega_{m'}(\mathbf{k}')}\delta(\mathbf{k}' - \mathbf{k}), \tag{2.132}$$

where $V_{\text{unit cell}}$ is the volume of a unit lattice cell, d is the dimensionality of a problem, and C is a normalization constant. Thus, the overlap integral between the two Bloch modes is zero if either of the frequencies of their wave vectors are different.

Discrete rotational symmetry C_N

Assume that a system Hamiltonian transforms into itself for a set of discrete rotations $\delta\theta = 2\pi k/N, k = [0, N-1]$ around vector $\hat{\mathbf{z}}$ (example of a point defect in a 2D periodic lattice, Fig. 2.11(e)). The eigenstates of a rotation operator (2.98) written in a cylindrical coordinate system should then satisfy:

$$\hat{\mathbf{R}}_{(\hat{z},\delta\theta)}\mathbf{H}(\rho, \theta, z) = \mathbf{H}\left(\rho, \theta - \frac{2\pi}{N}k, z\right) = C(k)\mathbf{H}(\rho, \theta, z). \tag{2.133}$$

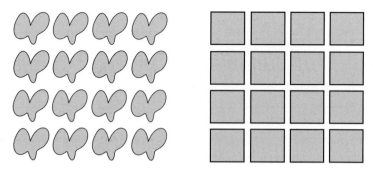

Figure 2.15 Structure exhibiting 2D discrete translational symmetry (left). Structure exhibiting 2D discrete translational symmetry + C_4 rotational symmetry (right).

One can verify by substitution that eigenfunctions satisfying (2.133) are:

$$\mathbf{H}(\rho, \theta, z) = \exp(im\theta)\mathbf{U}_m(\rho, z), \tag{2.134}$$

with corresponding eigenvalues:

$$C(k) = \exp\left(-im\frac{2\pi}{N}k\right). \tag{2.135}$$

The eigenvalue (2.135) is, however, the same for any $m = m + Np$, where p is as integer and $m = [0, \ldots, N-1]$. Thus, the eigenstate (2.134) characterized by an integer m is a degenerate one. Finally, the eigenstates of a Hamiltonian possessing a discrete rotational symmetry can be labeled with a conserved integer $m = [0, \ldots, N-1]$ and can be expressed as a linear combination of all the degenerate states having the same eigenvalue (2.135):

$$\mathbf{H}_m(\rho, \theta, z) = \exp(im\theta)\sum_p A(p)\exp(ipN\theta)\mathbf{U}_{m+pN}(\rho, z) = \exp(im\theta)\mathbf{U}_m(\rho, \theta, z)$$

$$\mathbf{U}_m\left(\rho, \theta + \frac{2\pi}{N}k, z\right) = \mathbf{U}_m(\rho, \theta, z),$$

$$k - \text{integer}, m = [0, \ldots, N-1], \tag{2.136}$$

where $\mathbf{U}_m(\rho, \theta, z)$ is a periodic in θ function with a period $2\pi/N$.

2.4.6   Discrete translational symmetry and discrete rotational symmetry

Suppose that, in addition to a discrete translational symmetry, a structure also possesses a discrete rotational symmetry described by the rotational matrix $\Re(\hat{\mathbf{n}}, \delta\theta)$. An example of such a system can be a square lattice of circular or square rods, as shown in Fig. 2.11(c) or Fig. 2.15. Then, the Maxwell Hamiltonian commutes with such a rotational operator $\hat{\mathbf{R}}_{(\hat{\mathbf{n}},\delta\theta)}\hat{H} = \hat{H}\hat{\mathbf{R}}_{(\hat{\mathbf{n}},\delta\theta)}$. Consider a particular solution labeled with a wave vector $\mathbf{k}$ and satisfying Maxwell's equations:

$$\hat{H}\,|\mathbf{H}_\mathbf{k}\rangle = \omega^2(\mathbf{k})\,|\mathbf{H}_\mathbf{k}\rangle$$
$$\hat{\mathbf{R}}_{(\hat{\mathbf{n}},\delta\theta)}\hat{H}\,|\mathbf{H}_\mathbf{k}\rangle = \hat{H}[\hat{\mathbf{R}}_{(\hat{\mathbf{n}},\delta\theta)}\,|\mathbf{H}_\mathbf{k}\rangle] = \omega^2(\mathbf{k})[\hat{\mathbf{R}}_{(\hat{\mathbf{n}},\delta\theta)}\,|\mathbf{H}_\mathbf{k}\rangle]' \tag{2.137}$$

from which it follows that the rotated eigenstate $\hat{\mathbf{R}}_{(\hat{\mathbf{n}}, \delta\theta)}|\mathbf{H}_{\mathbf{k}}\rangle$ is also an eigenstate with the same eigenvalue $\omega^2(\mathbf{k})$. Now we demonstrate that eigenstate $\hat{\mathbf{R}}_{(\hat{\mathbf{n}}, \delta\theta)}|\mathbf{H}_{\mathbf{k}}\rangle$ is nothing else but an eigenstate with a properly rotated wave vector $\hat{\mathbf{R}}_{(\hat{\mathbf{n}}, \delta\theta)}|\mathbf{H}_{\mathbf{k}}\rangle = |\mathbf{H}_{\Re(\hat{\mathbf{n}}, \delta\theta)\mathbf{k}}\rangle$ (up to a multiplicative constant). Namely, from the definition of a rotation operator and the general form of a Bloch state it follows that:

$$\hat{\mathbf{R}}_{(\hat{\mathbf{n}}, \delta\theta)} |\mathbf{H}_{\mathbf{k}}\rangle = \exp(i\mathbf{k}(\Re^{-1}(\hat{\mathbf{n}}, \delta\theta)\mathbf{r}))\Re(\hat{\mathbf{n}}, \delta\theta)\mathbf{U}_{\mathbf{k}}(\Re^{-1}(\hat{\mathbf{n}}, \delta\theta)\mathbf{r}). \quad (2.138)$$

From the basic properties of a vector dot product we conclude that $\mathbf{k}(\Re^{-1}(\hat{\mathbf{n}}, \delta\theta)\mathbf{r}) = (\Re(\hat{\mathbf{n}}, \delta\theta)\mathbf{k})\mathbf{r}$. As the rotated system maps the dielectric profile of a crystal onto itself, then for any lattice vector $\mathbf{R}$, $\Re^{-1}(\hat{\mathbf{n}}, \delta\theta)\mathbf{R}$ can be different from $\mathbf{R}$ only by a vector of lattice translations. As $\mathbf{U}_{\mathbf{k}}(\mathbf{r})$ is periodic with respect to any lattice translation, it follows that $(\Re(\hat{\mathbf{n}}, \delta\theta))\mathbf{U}_{\mathbf{k}}(\Re^{-1}(\hat{\mathbf{n}}, \delta\theta)\mathbf{r})$ is also periodic. Therefore, $\hat{\mathbf{R}}_{(\hat{\mathbf{n}}, \delta\theta)}|\mathbf{H}_{\mathbf{k}}\rangle = \exp(i(\Re(\hat{\mathbf{n}}, \delta\theta)\mathbf{k})\mathbf{r})$, $\Re(\hat{\mathbf{n}}, \delta\theta)\mathbf{U}_{\mathbf{k}}(\Re^{-1}\hat{\mathbf{n}}, \delta\theta)\mathbf{r}$ has the general form of a Bloch state, however, with a rotated wave vector. Thus, $\hat{\mathbf{R}}_{(\hat{\mathbf{n}}, \delta\theta)}|\mathbf{H}_{\mathbf{k}}\rangle = |\mathbf{H}_{\Re(\hat{\mathbf{n}}, \delta\theta)\mathbf{k}}\rangle$ and $\omega(\mathbf{k}) = \omega(\Re(\hat{\mathbf{n}}, \delta\theta)\mathbf{k})$. We conclude that in the case of a periodic system exhibiting discrete rotational symmetry, the dispersion relation $\omega(\mathbf{k})$ of the system eigenstates possesses the same discrete rotational symmetry as system Hamiltonian:

$$\hat{\mathbf{R}}_{(\hat{\mathbf{n}}, \delta\theta)}\hat{H} = \hat{H}\hat{\mathbf{R}}_{(\hat{\mathbf{n}}, \delta\theta)} \rightarrow \omega(\mathbf{k}) = \omega(\Re(\hat{\mathbf{n}}, \delta\theta)\mathbf{k}). \quad (2.139)$$

Rotational symmetry substantially reduces the complexity of finding independent solutions within the first Brillouin zone. In particular, with discrete rotational symmetry present, one only has to find solutions in a section of a first Brillouin zone, called the irreducible Brillouin zone, which is unrelated to the rest of the Brillouin zone by any of the discrete rotations. The dispersion relation in the rest of a Brillouin zone is then given by (2.139).

Finally, we note that the derivation of this result can be repeated for the case of systems exhibiting 2D continuous translational symmetry, such as slab and multilayer waveguides (see Section 2.4.3, discussion of 2D continuous translational symmetry). Indeed, such systems exhibit continuous rotational symmetry around the axis $\hat{\mathbf{n}}$ perpendicular to the multilayer plane. Then, from (2.139), it follows that $\omega(\mathbf{k}_t) = \omega(\Re(\hat{\mathbf{n}}, \delta\theta)\mathbf{k}_t)$, for any angle of rotation $\delta\theta$. This, in turn, implies that $\omega(\mathbf{k}_t)$ is a function of the transverse wave vector amplitude only $\omega(\mathbf{k}_t) = \omega(|\mathbf{k}_t|)$.

2.4.7   Inversion symmetry, mirror symmetry, and other symmetries

Symmetries beyond translational and rotational symmetries further restrict the general form of a solution. Group theory tells us that for 2D periodic structures there exist 17 different space symmetry groups consisting of operations of discrete rotations, reflections, inversions, etc., while the symmetry of any 3D crystalline structure falls into one of 230 space symmetry groups. Moreover, group theory concludes that each symmetry group describing a system defines a number of distinct (in frequency) states (irreducible representations), whose forms are compliant with all the symmetry operations of a symmetry group. Some states can be degenerate with the number of independent degenerate solutions equal to the dimension of irreducible representations. Thus, at points of high symmetry (typically the edges, and especially the corners of a first Brillouin zone) solutions tend to exhibit degeneracy while becoming nondegenerate away from the points of high symmetry. We will conclude our description of various symmetries by considering inversion and mirror symmetries. These symmetries are important as they are frequently present and they allow a simple characterization of modes.

Inversion symmetry

Inversion is present when a dielectric profile transforms into itself under spatial inversion with respect to a coordinate center: $\varepsilon(\mathbf{r}) = \varepsilon(-\mathbf{r})$. If the Maxwell Hamiltonian commutes with the operator of inversion then its eigenmodes $\mathbf{H}(\mathbf{r})$ can be chosen as the eigenmodes of an operator of inversion $O_I \mathbf{H}(\mathbf{r}) = \alpha \mathbf{H}(\mathbf{r})$. Applying the operation of inversion twice to the same state, one arrives at the original state, thus $\mathbf{H}(\mathbf{r}) = O_I[O_I \mathbf{H}(\mathbf{r})] = \alpha^2 \mathbf{H}(\mathbf{r})$, from which it follows that $\alpha = \pm 1$. Note that transformations of the corresponding electric field can be found using (2.12):

$$
\begin{aligned}
O_I \mathbf{E}(\mathbf{r}) = O_I \left(\frac{i}{\omega \varepsilon(\mathbf{r})} \Delta \times \mathbf{H}(\mathbf{r}) \right) &= -\frac{i}{\omega \varepsilon(-\mathbf{r})} \Delta \times \mathbf{H}(-\mathbf{r}) \\
&= -\frac{i}{\omega \varepsilon(\mathbf{r})} \Delta \times (O_I \mathbf{H}(\mathbf{r})) = -\alpha \left(\frac{i}{\omega \varepsilon(\mathbf{r})} \Delta \times \mathbf{H}(\mathbf{r}) \right) \\
&= -\alpha \mathbf{E}(\mathbf{r})
\end{aligned}
\tag{2.140}
$$

from which it follows that all the modes of a system possessing inversion symmetry can be classified as odd or even according to their symmetries:

$$
\begin{aligned}
\mathbf{H}(\mathbf{r}) &= \mathbf{H}(-\mathbf{r}); \quad \mathbf{E}(\mathbf{r}) = -\mathbf{E}(-\mathbf{r}) \text{ for even modes} \\
\mathbf{H}(\mathbf{r}) &= -\mathbf{H}(-\mathbf{r}); \quad \mathbf{E}(\mathbf{r}) = \mathbf{E}(-\mathbf{r}) \text{ for odd modes}
\end{aligned}
\tag{2.141}
$$

Mirror symmetry

Mirror symmetry is present when a dielectric profile transforms into itself under spatial mirror reflection for any plane perpendicular to a certain direction in space: $\varepsilon(x, y, z) = \varepsilon(x, y, -z)$, for any z. For example, the 2D photonic crystal in Fig. 2.11(c) has a mirror symmetry plane perpendicular to $\hat{z}$. Directly from Maxwell's equations, it can be verified that if $\mathbf{H}(\mathbf{r})$, $\mathbf{E}(\mathbf{r})$ are the eigenfields, then using reflection symmetry of a dielectric constant the following fields are also eigenfields:

$$
-\mathbf{H_t}(\mathbf{r_t}, -z), H_z(\mathbf{r_t}, -z); \quad \mathbf{E_t}(\mathbf{r_t}, -z), -E_z(\mathbf{r_t}, -z).
\tag{2.142}
$$

Using (2.142) we can thus define the form of a reflection operator O_{σ_z} commuting with the Maxwell Hamiltonian as:

$$
\begin{aligned}
O_{\sigma_z} \mathbf{H}(\mathbf{r}) &= (-\mathbf{H_t}(\mathbf{r_t}, -z), H_z(\mathbf{r_t}, -z)) \\
O_{\sigma_z} \mathbf{E}(\mathbf{r}) &= (\mathbf{E_t}(\mathbf{r_t}, -z), -E_z(\mathbf{r_t}, -z))
\end{aligned}
\tag{2.143}
$$

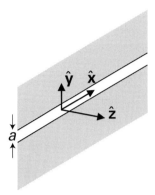

Figure P2.1.1 Infinite slit of width a.

One can verify that eigenvalues of the operator (2.143) are ± 1 and the following symmetries hold for the two possible types of solution:

$$\text{even}: O_{\sigma_z} \mathbf{H}(\mathbf{r}) = \mathbf{H}(\mathbf{r}) \rightarrow (-\mathbf{H}_t(\mathbf{r}_t, -z), H_z(\mathbf{r}_t, -z)) = (\mathbf{H}_t(\mathbf{r}_t, z), H_z(\mathbf{r}_t, z))$$

$$\text{odd}: O_{\sigma_z} \mathbf{H}(\mathbf{r}) = -\mathbf{H}(\mathbf{r}) \rightarrow (-\mathbf{H}_t(\mathbf{r}_t, -z), H_z(\mathbf{r}_t, -z)) = (-\mathbf{H}_t(\mathbf{r}_t, z), -H_z(\mathbf{r}_t, z))$$
$$(2.144)$$

In the special case of a 2D system, symmetries (2.144) become very restrictive with respect to the possible polarization of propagating light. As we have demonstrated earlier, 2D photonic crystals are characterized by a discrete symmetry in 2D plus a continuous symmetry in 1D. The general form of a solution for such a symmetry (2.125) (the same for the electric field vector) indicates that if one considers electromagnetic states propagating strictly in the plane of a photonic crystal with $k_z = 0$, then for such states:

$$\mathbf{H}_{k_t,0}(\mathbf{r}_t, z) = \mathbf{H}_{k_t,0}(\mathbf{r}_t, -z)$$
$$\mathbf{E}_{k_t,0}(\mathbf{r}_t, z) = \mathbf{E}_{k_t,0}(\mathbf{r}_t, -z)$$
$$(2.145)$$

Substituting (2.145) into (2.144) leads us to conclude that some of the components of the fields should be identically zero for the modes propagating strictly in the plane of a 2D photonic crystal. Thus, all such modes can be classified according to their polarizations as TE and TM modes:

$$\text{TM}: (0, H_z(\mathbf{r}_t, z)); (\mathbf{E}_t(\mathbf{r}_t, z), 0)$$
$$\text{TE}: (\mathbf{H}_t(\mathbf{r}_t, z), 0); (0, E_z(\mathbf{r}_t, z))$$
$$(2.146)$$

2.5 Problems

2.1 Excitation of evanescent waves by a subwavelength slit

Consider an infinitely long slit of width a confined to the plane $(\hat{\mathbf{x}}, \hat{\mathbf{y}})$ and directed along the axis $\hat{\mathbf{x}}$ (see Fig. P2.1.1).

The boundary conditions in the plane of a slit $z = 0$ are as follows. Inside the slit, the magnetic field $\mathbf{H}(y, z = 0)$ is parallel to the slit and uniform $H_0 \hat{\mathbf{x}} \exp(-i\omega t)$, while outside the slit, the magnetic field is zero.

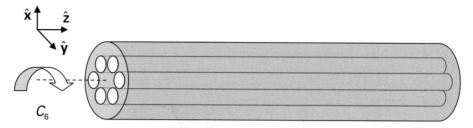

Figure P2.3.1 Example of a photonic crystal fiber exhibiting 1D continuous translational symmetry $+ C_6$ discrete rotational symmetry.

(a) Expanding the solution of Maxwell's equations in the half space $z \geq 0$ in terms of the outgoing and evanescent plane waves, find a complete solution of the problem satisfying the above-mentioned boundary conditions.

(b) Assuming that the slit is subwavelength, $a \ll (\lambda = 2\pi c/\omega)$, find the dependence of the evanescent field $\mathbf{H}_{\text{evanescent}}(y = 0, z)$ for $z \sim \lambda$.

2.2   Zero derivative of the dispersion relation at the edge of a Brillouin zone

Consider a system exhibiting discrete translational symmetry along the $\hat{z}$ axis with spatial period a. In this case, as demonstrated in Section 2.4.5 (see the subsection on 1D discrete translational symmetry) the general form of a solution is given by:

$$\mathbf{H}_{\omega(k_z),k_z}(\mathbf{r}) = \exp(ik_z z)\mathbf{U}_{k_z}(\mathbf{r})$$

$$\mathbf{U}_{k_z}(\mathbf{r} + a\hat{z}N) = \mathbf{U}_{k_z}(\mathbf{r}), \; \omega\left(k_z + \frac{2\pi}{a}p\right) = \omega(k_z). \tag{P2.2.1}$$

Moreover, time-reversal symmetry, $\omega(k_z) = \omega(-k_z)$, limits the choice of the distinct wave vectors to the first Brillouin zone $k_z = [0, \pi/a]$. Demonstrate that at the edge of the Brillouin zone, the derivative of the dispersion relation $d\omega(k_z)/dk_z|_{\pi/a}$ is zero. Hint:

$$\left.\frac{d\omega(k_z)}{dk_z}\right|_{\frac{\pi}{a}} = \lim_{\delta\to+0}\frac{\omega\left(\frac{\pi}{a}\right) - \omega\left(\frac{\pi}{a} - \delta\right)}{\delta} = \lim_{\delta\to+0}\frac{\omega\left(\frac{\pi}{a} + \delta\right) - \omega\left(\frac{\pi}{a}\right)}{\delta}. \tag{P2.2.2}$$

2.3   One-dimensional continuous translational symmetry and C_N discrete rotational symmetry

Find the general form of a solution of Maxwell's equations for the system exhibiting 1D continuous translational symmetry plus C_N discrete rotational symmetry. An example of a system having such symmetry is a photonic crystal fiber, shown in Fig. P2.3.1.

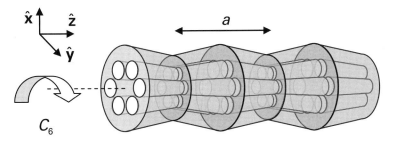

Figure P2.4.1 Example of a system exhibiting 1D discrete translational symmetry plus C_6 discrete rotational symmetry: the case of a Bragg grating written into a photonic crystal fiber.

2.4 One-dimensional discrete translational symmetry and C_N discrete rotational symmetry

Find the general form of a solution of Maxwell's equations for the system exhibiting 1D discrete translational symmetry plus C_N discrete rotational symmetry. This is a symmetry of Bragg gratings written into photonic crystal fibers, shown schematically in Fig. P2.4.1.

2.5 Polarization of modes of circularly symmetric fibers

In what follows we consider modes of a circularly symmetric fiber, shown schematically in Fig. 2.9(d). Such a system exhibits continuous rotational and translational symmetries. The dielectric function of a fiber cross-section is assumed to be invariant under the rotation around the fiber axis.

(a) Directly from Maxwell's equations (2.11), (2.12) verify that if $\mathbf{H}(\rho, \theta, z)$, $\mathbf{E}(\rho, \theta, z)$ are the eigenfields of a fiber, the following fields are also eigenfields of the same frequency:

$$
\begin{aligned}
& H_\rho(\rho, -\theta, z), \; -H_\theta(\rho, -\theta, z), \; H_z(\rho, -\theta, z) \\
& -E_\rho(\rho, -\theta, z), \; E_\theta(\rho, -\theta, z), \; -E_z(\rho, -\theta, z)
\end{aligned}
. \tag{P2.5.1}
$$

(b) For the fiber mode fields in the form

$$
\begin{aligned}
\mathbf{H}_m(\rho, \theta, z) &= \exp(im\theta)\exp(ik_z z)\mathbf{h}_m(\rho) \\
\mathbf{E}_m(\rho, \theta, z) &= \exp(im\theta)\exp(ik_z z)\mathbf{e}_m(\rho)
\end{aligned}
, \tag{P2.5.2}
$$

demonstrate that from (a) it follows that modes of the opposite angular momenta m, $-m$ are degenerate. Using (P2.5.1) find expressions for the fields $\mathbf{H}_{-m}(\rho, \theta, z)$ and $\mathbf{E}_{-m}(\rho, \theta, z)$ in terms of $\mathbf{h}_m(\rho)$ and $\mathbf{e}_m(\rho)$.

(c) Define the action of an angle reflection operator $\hat{\sigma}_\theta$ on the vector fields $\mathbf{H}(\rho, \theta, z)$, $\mathbf{E}(\rho, \theta, z)$ as:

$$
\begin{aligned}
\hat{\sigma}_\theta \mathbf{H}(\mathbf{r}) &= (H_\rho(\rho, -\theta, z), \; -H_\theta(\rho, -\theta, z), \; H_z(\rho, -\theta, z)) \\
\hat{\sigma}_\theta \mathbf{E}(\mathbf{r}) &= (-E_\rho(\rho, -\theta, z), \; E_\theta(\rho, -\theta, z), \; -E_z(\rho, -\theta, z))
\end{aligned}
. \tag{P2.5.3}
$$

From (P2.5.1) it follows that such an operator commutes with the Maxwell Hamiltonian. From (P2.5.3) one can then verify that eigenvalues of the $\hat{\sigma}_\theta$ operator are $\sigma = \pm 1$, thus defining two possible symmetries of the fiber mode fields. Expanding the eigenmodes of an angle reflection operator $\hat{\sigma}_\theta$ into the linear combination of two degenerate eigenmodes in the form (P2.5.2) with m, $-m$:

$$\mathbf{H}(\rho, \theta, z) = A\mathbf{H}_m(\rho, \theta, z) + B\mathbf{H}_{-m}(\rho, \theta, z), \qquad (\text{P2.5.4})$$

find explicit dependence of the eigenfield components on the angle θ for each of the two polarizations.

(d) For the $m = 0$ value of angular momentum, for each of the two polarizations, which of the field components become zero? (This is a case of so-called TE- and TM-polarized modes.)

3 One-dimensional photonic crystals – multilayer stacks

In this chapter, we will consider reflective properties of planar multilayers, and guidance by multilayer waveguides. We will first introduce a transfer-matrix method to find electromagnetic solutions for a system with an arbitrary number of planar dielectric layers. We will then investigate the reflection properties of a single dielectric interface. Next, we will solve the problems of reflection from a multilayer stack, guidance inside a dielectric stack (planar waveguides), and finally, propagation perpendicular to an infinitely periodic multilayer stack. We will then describe omnidirectional reflectors that reflect radiation completely for all angles of incidence and all states of polarization. Next, we will discuss bulk and surface defect states of a multilayer. We will conclude by describing guidance in the low-refractive-index core waveguides.

Figure 3.1 presents a schematic of a planar multilayer. Each stack $j = [1 \ldots N]$ is characterized by its thickness d_j and an index of refraction n_j. The indices of the first and last half spaces (claddings) are denoted n_0 and n_{N+1}. The positions of the interfaces (except for $j = 0$) along the $\hat{\mathbf{z}}$ axis are labeled $z_j, j = [1 \ldots N + 1]$, whereas z_0 can be chosen arbitrarily inside of a first half space. In the following, we assume that the incoming plane wave has a propagation vector $\mathbf{k}$ confined to the xz plane. The planar multilayer possesses mirror symmetry with respect to the mirror plane xz. From the discussion in Section 2.4.6 it follows that electromagnetic solutions of a planar multilayer can be classified as having TE or TM polarizations with the vector of electric field either directed perpendicular to the plane xz, for TE polarization, or parallel to it, for TM polarization.

3.1 Transfer matrix technique

The general form of a solution for a multilayer stack (from (2.106), where $\mathbf{k}_t = (k_x, 0)$), is $\mathbf{H}_{k_x}(\mathbf{r}) = \exp(ik_x x)\mathbf{U}_{k_x}^H(z); \mathbf{E}_{k_x}(\mathbf{r}) = \exp(ik_x x)\mathbf{U}_{k_x}^E(z)$. In each of the layers, the dielectric profile is uniform, thus the fields can be represented as a sum of two counter-propagating plane waves with the same projection of a propagation vector k_x.

3.1.1 Multilayer stack, TE polarization

Defining A_j and B_j to be the expansion coefficients of the electric field component $E_y^j(x, y, z)$ in terms of the forward- and backward- (along the $\hat{\mathbf{z}}$ axis) propagating waves

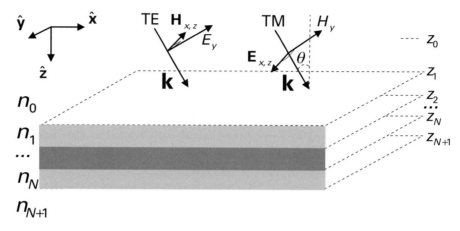

Figure 3.1 Propagation in a planar multilayer, showing the directions of electromagnetic fields for the TE and TM polarizations.

inside a layer j, we write:

$$E_y^j(x, y, z) = \exp(ik_x x)\left(A_j \exp\left(ik_z^j(z - z_j)\right) + B_j \exp\left(-ik_z^j(z - z_j)\right)\right). \qquad (3.1)$$

The corresponding magnetic field components from (2.19) are:

$$H_x^j(x, y, z) = -\exp(ik_x x)\frac{k_z^j}{\omega}\left(A_j \exp\left(ik_z^j(z - z_j)\right) - B_j \exp\left(-ik_z^j(z - z_j)\right)\right)$$

$$H_z^j(x, y, z) = \exp(ik_x x)\frac{k_x}{\omega}\left(A_j \exp\left(ik_z^j(z - z_j)\right) + B_j \exp\left(-ik_z^j(z - z_j)\right)\right), \qquad (3.2)$$

where $k_x^2 + (k_z^j)^2 = \omega^2 \varepsilon_j$. The solutions in each of the adjacent slabs $j - 1$ and j have to be related to each other by the boundary condition of continuity of the field components parallel to the interface j. At the interface j positioned at z_j, the condition of field continuity results in the following equations:

$$H_x^{j-1}(x, y, z_j) = H_x^j(x, y, z_j)$$

$$E_y^{j-1}(x, y, z_j) = E_y^j(x, y, z_j)$$

$$\begin{pmatrix} -\dfrac{k_z^{j-1}}{\omega}\exp\left(ik_z^{j-1}(z_j - z_{j-1})\right) & \dfrac{k_z^{j-1}}{\omega}\exp\left(-ik_z^{j-1}(z_j - z_{j-1})\right) \\ \exp\left(ik_z^{j-1}(z_j - z_{j-1})\right) & \exp\left(-ik_z^{j-1}(z_j - z_{j-1})\right) \end{pmatrix} \begin{pmatrix} A_{j-1} \\ B_{j-1} \end{pmatrix} =$$

$$\begin{pmatrix} -\dfrac{k_z^j}{\omega} & \dfrac{k_z^j}{\omega} \\ 1 & 1 \end{pmatrix} \begin{pmatrix} A_j \\ B_j \end{pmatrix}, \qquad (3.3)$$

which can be rewritten in terms of a so-called transfer matrix $M_{j-1,j}$ relating the expansion coefficients in the adjacent layers:

$$M_{j-1,j}\begin{pmatrix} A_{j-1} \\ B_{j-1} \end{pmatrix} = \begin{pmatrix} A_j \\ B_j \end{pmatrix}$$

$$M_{j-1,j} = \frac{1}{2}\begin{pmatrix} \left(1 + \dfrac{k_z^{j-1}}{k_z^j}\right)\exp\left(ik_z^{j-1}d_{j-1}\right) & \left(1 - \dfrac{k_z^{j-1}}{k_z^j}\right)\exp\left(-ik_z^{j-1}d_{j-1}\right) \\ \left(1 - \dfrac{k_z^{j-1}}{k_z^j}\right)\exp\left(ik_z^{j-1}d_{j-1}\right) & \left(1 + \dfrac{k_z^{j-1}}{k_z^j}\right)\exp\left(-ik_z^{j-1}d_{j-1}\right) \end{pmatrix}.$$

$$(3.4)$$

Given the expansion coefficients A_0 and B_0 in the first half space, the expansion coefficients A_j and B_j in any layer j can be found by multiplication of all the transfer matrices in the intermediate layers:

$$\begin{pmatrix} A_j \\ B_j \end{pmatrix} = M_{j-1,j}\ldots M_{1,2}M_{0,1}\begin{pmatrix} A_0 \\ B_0 \end{pmatrix}. \qquad (3.5)$$

3.1.2   Multilayer stack, TM polarization

Defining A_i and B_i to be the expansion coefficients of the magnetic field component $H_y^j(x, y, z)$ in terms of the forward- and backward- (along the $\hat{z}$ axis) propagating waves inside a layer j, we write:

$$H_y^j(x, y, z) = \exp(ik_x x)\left(A_j \exp\left(ik_z^j(z - z_j)\right) + B_j \exp\left(-ik_z^j(z - z_j)\right)\right). \qquad (3.6)$$

The corresponding electric field components from (2.19) are:

$$E_x^j(x, y, z) = \exp(ik_x x)\frac{k_z^j}{\omega \varepsilon_j}\left(A_j \exp\left(ik_z^j(z - z_j)\right) - B_j \exp\left(-ik_z^j(z - z_j)\right)\right)$$

$$E_z^j(x, y, z) = -\exp(ik_x x)\frac{k_x}{\omega \varepsilon_j}\left(A_j \exp\left(ik_z^j(z - z_j)\right) + B_j \exp\left(-ik_z^j(z - z_j)\right)\right), \qquad (3.7)$$

where $k_x^2 + (k_z^j)^2 = \omega^2 \varepsilon_j$. The solutions in each of the adjacent slabs $j - 1$ and j have to be related to each other by the boundary condition of continuity of the field components parallel to the interface j. At the interface j positioned at z_j, the condition of field continuity results in the following equations:

$$E_x^{j-1}(x, y, z_j) = E_x^j(x, y, z_j)$$

$$H_y^{j-1}(x, y, z_j) = H_y^j(x, y, z_j)$$

$$\begin{pmatrix} \dfrac{k_z^{j-1}}{\omega \varepsilon_{j-1}}\exp\left(ik_z^{j-1}(z_j - z_{j-1})\right) & -\dfrac{k_z^{j-1}}{\omega \varepsilon_{j-1}}\exp\left(-ik_z^{j-1}(z_j - z_{j-1})\right) \\ \exp\left(ik_z^{j-1}(z_j - z_{j-1})\right) & \exp\left(-ik_z^{j-1}(z_j - z_{j-1})\right) \end{pmatrix}\begin{pmatrix} A_{j-1} \\ B_{j-1} \end{pmatrix} =$$

$$\begin{pmatrix} \dfrac{k_z^j}{\omega \varepsilon_j} & -\dfrac{k_z^j}{\omega \varepsilon_j} \\ 1 & 1 \end{pmatrix}\begin{pmatrix} A_j \\ B_j \end{pmatrix}, \qquad (3.8)$$

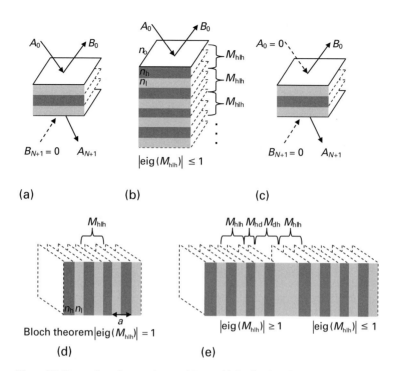

Figure 3.2 Examples of scattering problems. (a) Reflection from a finite multilayer.
(b) Reflection from a semi-infinite periodic multilayer. (c) Finite multilayer waveguide.
(d) Infinite periodic multilayer. (e) Infinite periodic multilayer with a defect.

which can be rewritten in terms of a so-called transfer matrix $M_{j-1,j}$ relating the expansion coefficients in the adjacent layers:

$$M_{j-1,j} \begin{pmatrix} A_{j-1} \\ B_{j-1} \end{pmatrix} = \begin{pmatrix} A_j \\ B_j \end{pmatrix}$$

$$M_{j-1,j} = \frac{1}{2} \begin{pmatrix} \left(1 + \dfrac{k_z^{j-1}\varepsilon_j}{k_z^j \varepsilon_{j-1}}\right) \exp\left(ik_z^{j-1}d_{j-1}\right) & \left(1 - \dfrac{k_z^{j-1}\varepsilon_j}{k_z^j \varepsilon_{j-1}}\right) \exp\left(-ik_z^{j-1}d_{j-1}\right) \\ \left(1 - \dfrac{k_z^{j-1}\varepsilon_j}{k_z^j \varepsilon_{j-1}}\right) \exp\left(ik_z^{j-1}d_{j-1}\right) & \left(1 + \dfrac{k_z^{j-1}\varepsilon_j}{k_z^j \varepsilon_{j-1}}\right) \exp\left(-ik_z^{j-1}d_{j-1}\right) \end{pmatrix}.$$

$$(3.9)$$

Given the expansion coefficients A_0 and B_0 in the first half space, the expansion coefficients A_j and B_j in any layer j can again be found by multiplication of all the transfer matrices in the intermediate layers as in (3.5).

3.1.3   Boundary conditions

Depending upon the nature of the scattering problem and the geometry of the multilayer, the boundary conditions vary (see Fig. 3.2). We will consider five cases of interest:

reflection from a finite multilayer (dielectric mirrors), reflection from a semi-infinite periodic multilayer (dielectric photonic crystal mirror, omnidirectional reflector), guiding within a finite multilayer (slab waveguides), guiding in the interior of an infinitely periodic multilayer (planar Bragg gratings), and guiding in the defect of an infinitely periodic multilayer (planar photonic crystal waveguides, hollow waveguides).

3.2   Reflection from a finite multilayer (dielectric mirror)

In the problem of scattering from N dielectric layers surrounded by two cladding regions (see Fig. 3.2 (a)), A_0 is a known coefficient of the incoming plane wave and is usually assumed to be unity, B_0 is a coefficient of the reflected wave, and A_{N+1} is a coefficient of the transmitted wave. The coefficient $B_{N+1} = 0$, as there is no incoming wave from the other side of the multilayer. From (3.5), it follows that incoming, reflection, and transmission coefficients are related by the following relation:

$$\begin{pmatrix} A_{N+1} \\ 0 \end{pmatrix} = M_{N,N+1} \dots M_{1,2} M_{0,1} \begin{pmatrix} 1 \\ B_0 \end{pmatrix}, \tag{3.10}$$

which can easily be solved by matrix rearrangement followed by inversion. In particular,

$$\begin{pmatrix} A_{N+1} \\ 0 \end{pmatrix} = M_{N,N+1} \dots M_{1,2} M_{0,1} \begin{pmatrix} 1 \\ B_0 \end{pmatrix} = \begin{pmatrix} a_{1,1} & a_{1,2} \\ a_{2,1} & a_{2,2} \end{pmatrix} \begin{pmatrix} 1 \\ B_0 \end{pmatrix}$$

$$\rightarrow B_0 = -\frac{a_{2,1}}{a_{2,2}}; A_{N+1} = \frac{a_{1,1}a_{2,2} - a_{1,2}a_{2,1}}{a_{2,2}}. \tag{3.11}$$

As an example, consider scattering from a single dielectric interface when coming from a region of low index of refraction n_l into a high index of refraction n_h. For a single interface, product (3.10) contains a single transfer matrix. Using explicit forms of the transfer matrices for TE (3.4) and TM (3.9) polarizations we get for the corresponding reflection coefficients:

$$B_0^{TE} = \frac{k_z^{n_l} - k_z^{n_h}}{k_z^{n_l} + k_z^{n_h}}; B_0^{TM} = \frac{\varepsilon_h k_z^{n_l} - \varepsilon_l k_z^{n_h}}{\varepsilon_h k_z^{n_l} + \varepsilon_l k_z^{n_h}}$$

$$k_z^{n_h} = \sqrt{\omega^2 \varepsilon_h - k_x^2}; k_z^{n_l} = \sqrt{\omega^2 \varepsilon_l - k_x^2}. \tag{3.12}$$

The TE power reflection coefficient $|B_0^{TE}|^2$ is never zero for any angle of incidence, and it increases monotonically, $|B_0^{TE}| \geq (n_h - n_l)/(n_h + n_l)$, as the angle of incidence becomes larger. The TM power reflection coefficient $|B_0^{TM}|^2$, however, becomes zero when the angle of incidence of an incoming plane wave reaches a so-called Brewster's angle, $\theta_B = \tan^{-1}(k_x/k_z^{n_l})_B = \tan^{-1}(n_h/n_l)$, at which there is no back-reflection of TM polarization.

In Fig. 3.3, we present a sketch of the reflection coefficients of a dielectric interface with $n_h > n_l$. Note that the reflection of TE-polarized light improves for a larger index contrast. For TM polarization, however, at the Brewster's angle, reflection becomes zero. For grazing angles of incidence ($\theta \approx 90°$), the reflection for both polarizations becomes close to unity.

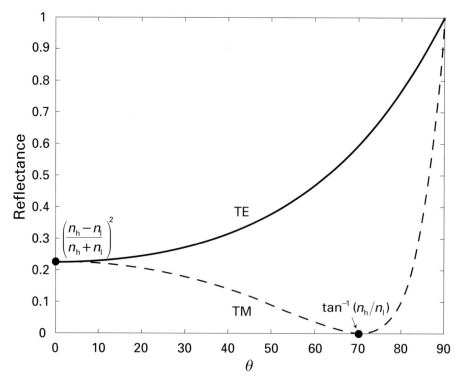

Figure 3.3 Single interface dielectric mirrors exhibit strong angular and polarization dependence ($n_l = 1; n_h = 2.8$). The reflectance of a mirror becomes more uniform for all polarizations and a wider range of angles as the index contrast increases; however, owing to the presence of a Brewster's angle for TM polarization, there is always an angle where the reflection of TM polarized light is zero.

To conclude, the reflection property of a single dielectric interface is a sensitive function of the angle of incidence of light as well as the light polarization. While high index contrast improves the reflection of TE-polarized light for all angles of incidence, efficient reflection of TM-polarized light for all angles of incidence is problematic because of the existence of a Brewster's angle.

3.3   Reflection from a semi-infinite multilayer (dielectric photonic crystal mirror)

In the next example we consider a semi-infinite periodic multilayer made of a repeated bilayer (see Fig. 3.2 (b)). Each bilayer is made of layers of low refractive index, n_l, and high refractive index, n_h, materials with corresponding thicknesses d_l and d_h. Given the coefficients of the incident and reflected waves A_0, B_0 the expansion coefficients A_j and B_j in any layer j can be found using (3.5). In the case of a finite multilayer reflector, to relate coefficients A_0, B_0 we have used a condition of no incoming plane wave from the opposite side of the multilayer (3.10). In the case of a semi-infinite periodic multilayer this condition should be modified.

To find the appropriate boundary condition we rewrite (3.5) in such a way as to account explicitly for the periodicity of a structure. In the following we assume that the high index layer borders with a semi-infinite cladding material. There are only three types of transfer matrix involved. The first one M_{oh} relates the fields in the uniform half space and a high-index material. The second transfer matrix M_{hl} relates the fields in the high- and low-index materials. Finally, the third transfer matrix M_{lh} relates the fields in the low- and high-index materials. Thus, for the layer $j = 2N + 1$, where N is the number of bilayers in between the uniform half space and a layer of interest, we can rewrite (3.5) as:

$$\begin{pmatrix} A_j \\ B_j \end{pmatrix} = \underbrace{M_{\mathrm{lh}} M_{\mathrm{hl}} \dots M_{\mathrm{lh}} M_{\mathrm{hl}}}_{2N} M_{\mathrm{oh}} \begin{pmatrix} A_0 \\ B_0 \end{pmatrix} = (M_{\mathrm{lh}} M_{\mathrm{hl}})^N M_{\mathrm{oh}} \begin{pmatrix} A_0 \\ B_0 \end{pmatrix}. \quad (3.13)$$

For a physical solution, the field coefficients in (3.13) have to be finite for any layer $j \to +\infty$ to avoid unphysical infinite energy flux. Consider in more detail the properties of a bilayer matrix $M_{\mathrm{hlh}} = M_{\mathrm{lh}} M_{\mathrm{hl}}$. Defining V_{hlh} to be a nondegenerate matrix of M_{hlh} eigenvectors, and

$$\Lambda_{\mathrm{hlh}} = \begin{pmatrix} \lambda_1 & 0 \\ 0 & \lambda_2 \end{pmatrix}$$

to be a diagonal matrix of M_{hlh} eigenvalues we can write $M_{\mathrm{hlh}} = V_{\mathrm{hlh}} \Lambda_{\mathrm{hlh}} V_{\mathrm{hlh}}^{-1}$. Substitution of this form into (3.13) gives:

$$\begin{pmatrix} A_j \\ B_j \end{pmatrix} = V_{\mathrm{hlh}} \Lambda_{\mathrm{hlh}}^N V_{\mathrm{hlh}}^{-1} M_{\mathrm{oh}} \begin{pmatrix} A_0 \\ B_0 \end{pmatrix} = V_{\mathrm{hlh}} \begin{pmatrix} \lambda_1^N & 0 \\ 0 & \lambda_2^N \end{pmatrix} V_{\mathrm{hlh}}^{-1} M_{\mathrm{oh}} \begin{pmatrix} A_0 \\ B_0 \end{pmatrix}. \quad (3.14)$$

To guarantee that, for any $N \to +\infty$, the expansion coefficients A_{2N+1} and B_{2N+1} are finite one has to choose incidence and reflection coefficients A_0, B_0 in such a way as to excite only the eigenvalues with magnitudes less than or equal to 1. As we will demonstrate shortly, the eigenvalues of a transfer matrix are related to each other as $\lambda_1 \lambda_2 = 1$. Therefore, only two cases are possible. In the first case, one of the eigenvalues has a magnitude smaller than 1, while the other eigenvalue has a magnitude larger than 1. As will be seen in what follows, this corresponds to the case of a true reflector. In the second case, both eigenvalues have magnitudes equal to 1. This case will be treated in more detail in the following section.

In what follows we concentrate on the case of a true reflector, assuming $|\lambda_1| < 1$, $|\lambda_2| > 1$. The incidence and reflection coefficients exciting λ_1 can then be chosen as:

$$\begin{pmatrix} A_0 \\ B_0 \end{pmatrix} = M_{\mathrm{oh}}^{-1} V_{\mathrm{hlh}} \begin{pmatrix} \alpha \\ 0 \end{pmatrix},$$

which can be verified by its direct substitution into (3.14). The field expansion coefficients in the high-refractive-index layer $j = 2N + 1$ will then be:

$$\begin{pmatrix} A_j \\ B_j \end{pmatrix} = \alpha \lambda_1^N \overline{V}_{\mathrm{hlh}}^{-1}, \quad (3.15)$$

where the value of the coefficient α is typically chosen to normalize the amplitude of an incident field $|A_0| = 1$. Note the exponentially fast decay of the field coefficients (3.15) inside of the reflector when $N \to +\infty$, signifying that electromagnetic fields are decaying exponentially inside a periodic reflector. In turn, this means that deep inside the reflector ($N \to +\infty$) there is no energy flux perpendicular to the reflector, thus, in the absence of material losses, reflection is complete and $|B_0|^2 = |A_0|^2$. The number of bilayers inside the reflector after which the field is strongly attenuated can be estimated as $N_{\text{attenuation}} \sim 1/\log(|\lambda_1^{-1}|)$, where $\log(|\lambda_1^{-1}|)$ can be defined as a field decay rate.

We now consider optimization of the multilayer reflective properties. Despite the fact that, in practice, multilayer reflectors are finite, they can still be used for efficient reflection if the number of bilayers in them exceeds $N_{\text{attenuation}}$. The optimization of the reflector will then be understood in terms of the reduction of $N_{\text{attenuation}}$, or, alternatively, in terms of the reduction of the absolute value of λ_1. In particular, for a given angle of incidence (for a given k_x) we will now find what layer thicknesses maximize the field decay rate (minimize $|\lambda_1|$) inside a multilayer. The following consideration is the same for TE and TM modes. We define the following ratios and phases:

$$r_{\text{TE}} = \frac{k_z^l}{k_z^h}; \ r_{\text{TM}} = \frac{\varepsilon_h k_z^l}{\varepsilon_l k_z^h}$$

(3.16)

$$\phi_h = \left(k_z^h d_h\right); \ \phi_l = \left(k_z^l d_l\right).$$

In this notation, the bilayer transfer matrices can be written as:

$$M_{\text{hlh}}^{\text{TE,TM}} = \frac{1}{4} \begin{pmatrix} \left(1 + r^{\text{TE,TM}}\right) \exp\left(i\phi_l\right) & \left(1 - r^{\text{TE,TM}}\right) \exp\left(-i\phi_l\right) \\ \left(1 - r^{\text{TE,TM}}\right) \exp\left(i\phi_l\right) & \left(1 + r^{\text{TE,TM}}\right) \exp\left(-i\phi_l\right) \end{pmatrix} \times$$

$$\begin{pmatrix} \left(1 + r_{\text{TE,TM}}^{-1}\right) \exp\left(i\phi_h\right) & \left(1 - r_{\text{TE,TM}}^{-1}\right) \exp\left(-i\phi_h\right) \\ \left(1 - r_{\text{TE,TM}}^{-1}\right) \exp\left(i\phi_h\right) & \left(1 + r_{\text{TE,TM}}^{-1}\right) \exp\left(-i\phi_h\right) \end{pmatrix}.$$

(3.17)

The quadratic equation defining eigenvalues of the bilayer transfer matrix (3.17) is then:

$$\lambda^2 - \lambda(2\cos(\phi_h)\cos(\phi_l) - \left(r_{\text{TE,TM}} + r_{\text{TE,TM}}^{-1}\right)\sin(\phi_h)\sin(\phi_l)) + 1 = 0.$$

(3.18)

Now we find the largest and the smallest possible eigenvalues by finding a point of extremum of an eigenvalue in (3.18) with respect to the phases ϕ_h and ϕ_l. In particular, we find the values of the phases such that $\partial\lambda/\partial\phi_l = \partial\lambda/\partial\phi_h = 0$. By differentiating (3.18) with respect to each of the phases we arrive at the following system of equations:

$$\frac{\partial\lambda}{\partial\phi_h} = -\frac{\lambda(2\sin(\phi_h)\cos(\phi_l) + \left(r_{\text{TE,TM}} + r_{\text{TE,TM}}^{-1}\right)\cos(\phi_h)\sin(\phi_l))}{(2\lambda - (2\cos(\phi_h)\cos(\phi_l) - \left(r_{\text{TE,TM}} + r_{\text{TE,TM}}^{-1}\right)\sin(\phi_h)\sin(\phi_l)))} = 0$$

$$\frac{\partial\lambda}{\partial\phi_l} = -\frac{\lambda(2\sin(\phi_l)\cos(\phi_h) + \left(r_{\text{TE,TM}} + r_{\text{TE,TM}}^{-1}\right)\cos(\phi_l)\sin(\phi_h))}{(2\lambda - (2\cos(\phi_h)\cos(\phi_l) - \left(r_{\text{TE,TM}} + r_{\text{TE,TM}}^{-1}\right)\sin(\phi_h)\sin(\phi_l)))} = 0.$$

(3.19)

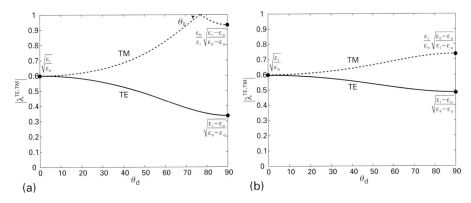

Figure 3.4 TE, TM field decay rates per bilayer for different quarter-wave stacks designed to operate at an angle of incidence θ_d. (a) $\varepsilon_o \leq \varepsilon_h \varepsilon_l/(\varepsilon_h + \varepsilon_l)$ $n_1 = 1.5; n_h = 2.5; n_0 = 1.0$ (air) . (b) $\varepsilon_o > \varepsilon_h \varepsilon_l/(\varepsilon_h + \varepsilon_l)$ $(n_1 = 1.5; n_h = 2.5; n_0 = 1.32$ (water)).

For any designated incidence angle, the solution of (3.19) is simply:

$$\phi_h = \frac{\pi}{2}(2p_h + 1); \; \phi_l = \frac{\pi}{2}(2p_l + 1)$$

$$\rightarrow k_z^h d_h = \frac{\pi}{2}(2p_h + 1); \; k_z^l d_l = \frac{\pi}{2}(2p_l + 1), \tag{3.20}$$

for any integer p_h, p_l. The condition (3.20) presents a quarter-wave stack condition generalized for a designated angle of incidence θ_d. For the quarter-wave stack multilayers, magnitudes of the eigenvalues are then either $|\lambda_1^{TE,TM}| = r_{TE,TM}$ or $|\lambda_2^{TE,TM}| = r_{TE,TM}^{-1}$.

For TE-polarized light, the smallest eigenvalue is:

$$|\lambda^{TE}| = r_{TE} = \frac{k_z^l}{k_z^h} = \sqrt{\frac{\omega^2 \varepsilon_l - k_x^2}{\omega^2 \varepsilon_h - k_x^2}} = \sqrt{\frac{\varepsilon_l - \varepsilon_o \sin^2(\theta_d)}{\varepsilon_h - \varepsilon_o \sin^2(\theta_d)}} < 1, \forall \theta_d \subset \left[0, \frac{\pi}{2}\right]. \tag{3.21}$$

Thus, for any choice of $\varepsilon_o < \varepsilon_l < \varepsilon_h$ and for any designated angle of incidence θ_d the choice of a quarter-wave multilayer stack (3.20) guarantees a complete reflection of a TE-polarized wave. In Fig. 3.4 (a) we present $|\lambda_1|$ of a quarter-wave stack reflector as a function of a designated angle of incidence $\theta_d \subset [0, \pi/2)$. Note that for TE-polarization, the efficiency of the quarter-wave reflector increases for grazing design angles $(\theta_d \sim 90°)$.

For the TM-polarized light, the consideration is somewhat more complicated. Analysis of eigenvalues for TM polarization shows that depending on the value of a cladding dielectric constant, ε_o, the smallest eigenvalue can be either r_{TM} or r_{TM}^{-1}. Thus, when $\varepsilon_o < \varepsilon_l \varepsilon_h/(\varepsilon_l + \varepsilon_h)$,

$$|\lambda^{TM}| = \frac{1}{r_{TM}} = \frac{\varepsilon_l k_z^h}{\varepsilon_h k_z^l} = \frac{\varepsilon_l}{\varepsilon_h} \sqrt{\frac{\varepsilon_h - \varepsilon_o \sin^2(\theta_d)}{\varepsilon_l - \varepsilon_o \sin^2(\theta_d)}} < 1, \forall \theta_d \subset \left[0, \frac{\pi}{2}\right], \tag{3.22}$$

otherwise, when $\varepsilon_1\varepsilon_h/(\varepsilon_1 + \varepsilon_h) \leq \varepsilon_o < \varepsilon_1$:

$$
\begin{cases}
|\lambda^{\mathrm{TM}}| = \dfrac{1}{r_{\mathrm{TM}}} = \dfrac{\varepsilon_1 k_z^h}{\varepsilon_h k_z^l} = \dfrac{\varepsilon_1}{\varepsilon_h}\sqrt{\dfrac{\varepsilon_h - \varepsilon_o\sin^2(\theta_d)}{\varepsilon_1 - \varepsilon_o\sin^2(\theta_d)}} < 1,\ \forall \theta_d \subset [0, \theta_o] \\[4mm]
|\lambda^{\mathrm{TM}}| = r_{\mathrm{TM}} = \dfrac{\varepsilon_h k_z^l}{\varepsilon_1 k_z^h} = \dfrac{\varepsilon_h}{\varepsilon_1}\sqrt{\dfrac{\varepsilon_1 - \varepsilon_o\sin^2(\theta_d)}{\varepsilon_h - \varepsilon_o\sin^2(\theta_d)}} < 1,\ \forall \theta_d \subset \left[\theta_o, \dfrac{\pi}{2}\right]
\end{cases}
\tag{3.23}
$$

$$
\theta_o = \sin^{-1}\sqrt{\dfrac{\varepsilon_1\varepsilon_h}{\varepsilon_o(\varepsilon_1 + \varepsilon_h)}}.
$$

In Fig. 3.4 (a) we demonstrate $|\lambda_1|$ for TM polarization as a function of a designated angle of incidence when $\varepsilon_o \leq \varepsilon_h\varepsilon_1/(\varepsilon_h + \varepsilon_1)$.

As seen from Fig. 3.4(a), in this regime, for any value of a designated angle of incidence $|\lambda^{\mathrm{TE}}| \leq |\lambda^{\mathrm{TM}}| < 1$, thus it is possible to design a semi-infinite reflector that reflects both polarizations simultaneously. Note also that the penetration depth of TM polarized light into a multilayer is always larger than that of TE polarization for any designated angle $|\lambda^{\mathrm{TE}}| \leq |\lambda^{\mathrm{TM}}|$.

In Fig. 3.4 (b) we demonstrate $|\lambda_1|$ for TM polarization as a function of a designated angle of incidence when $\varepsilon_o > \varepsilon_h\varepsilon_1/(\varepsilon_h + \varepsilon_1)$. As seen from Fig. 3.4 (b), for the designated angle θ_o, $|\lambda_{\mathrm{TM}}| = 1$ and there is no TM field decay into the multilayer. Moreover, for any designated angle in the vicinity of θ_o, the TM field will extend greatly into the multilayer, therefore no efficient multilayer exists in the vicinity of θ_o. As, in practice, one always deals with a finite number of bilayers, this also signifies that there will be a range of angles of incidence for which it will be impossible to design an efficient TM reflector.

To summarize, for a TE-polarized wave and any design incidence angle, a corresponding semi-infinite quarter-wave periodic multilayer will reflect light completely. If the cladding refractive index is low enough, $\varepsilon_o \leq \varepsilon_h\varepsilon_1/(\varepsilon_h + \varepsilon_1)$, the same multilayer will also completely reflect TM-polarized light. In the case when the cladding index is not low enough $\varepsilon_o > \varepsilon_h\varepsilon_1/(\varepsilon_h + \varepsilon_1)$, there will be an angle of incidence for TM-polarized light $\theta_o = \sin^{-1}\sqrt{\varepsilon_1\varepsilon_h/\varepsilon_o(\varepsilon_1 + \varepsilon_h)}$ for which it will be impossible to design an efficient periodic reflector. For both TE and TM polarizations and a given angle of incidence, the most efficient reflector (with the least penetration of the fields inside a multilayer) is a quarter-wave stack, defined by (3.20). The field penetration into the multilayer is always higher for TM polarization than for TE polarization, and decreases for both polarizations as the index contrast $\varepsilon_h/\varepsilon_1$ increases. Note also from (3.15) that for a finite number N of bilayers the radiation energy flux passing through a finite size reflector will be proportional to $|\lambda_1|^N$.

3.3.1   Omnidirectional reflectors I

In the previous section we have demonstrated that for a given frequency $\omega_0 = 2\pi/\lambda_0$ and a designated angle of incidence θ_d onto a cladding-reflector interface, the most efficient periodic reflector is a quarter-wave stack with thicknesses defined by (3.20). In particular, for the thinnest stack the individual layer thicknesses are:

$$d_{1,h} = \frac{\lambda_0}{4\sqrt{n_{1,h}^2 - n_0^2 \sin^2(\theta_d)}}, \tag{3.24}$$

where the refractive indices of the reflector layers are n_1, n_h, and the cladding refractive index is n_0.

Although designed for a particular frequency and incidence angle, a quarter-wave reflector (3.24) remains efficient even when used at other frequencies and incidence angles. In fact, the refractive index contrast plays a key role in determining whether the periodic reflector remains efficient when used outside of its design regime. Consider, for example, a quarter-wave reflector designed for a normal angle of incidence $\theta_d = 0$. In what follows we consider $n_1 = 1.5$, while n_h can take any value in the interval $[1.5, 3.2]$. For a given value of n_h we then investigate efficiency of a quarter-wave stack (3.24) for incidence angles different from a design one $\theta \neq \theta_d = 0$. In particular, we scan the value of an incidence angle $\theta \subset [0, 90]$ and record whether an infinite reflector is efficient or not. To judge on the reflector efficiency, for every individual polarization, we look at the smallest eigenvalue $|\lambda_1^{\text{TE,TM}}|$ of a bilayer transfer matrix (3.17). According to the arguments of a previous section, if $|\lambda_1^{\text{TE,TM}}| < 1$ then a reflector containing more than $N_{\text{attenuation}} = 1/\log(1/|\lambda_1^{\text{TE,TM}}|)$ bilayers will efficiently reflect an incoming radiation. In Fig. 3.5 we present regions of reflector efficiency for both polarizations. Note that for TE polarization, if $n_h > 2.24$ then a quarter-wave reflector designed for a normal incidence will actually be efficient for any angle of radiation incidence. This regime is called omnidirectional reflection. Moreover, when $n_h > 3.13$, the same reflector will be efficient for both polarizations and any angle of incidence. In the same figure we also present contour plots to indicate the $N_{\text{attenuation}}$ of the bilayers in the reflector necessary to achieve efficient reflection. Note that the reflection of TE polarization is always more efficient than that of TM polarization, thus requiring a smaller number of bilayers in the reflector, and a smaller refractive index contrast, to achieve the same reflector efficiency.

3.4   Guiding in a finite multilayer (planar dielectric waveguide)

In the problem of guiding inside a multilayer structure comprised of N dielectric layers surrounded by two semi-infinite cladding regions (see Fig. 3.2 (c)), the field coefficients corresponding to incoming waves from infinity onto a waveguide should be zero: $A_0 = 0$, $B_{N+1} = 0$. The nonzero coefficients B_0 and A_{N+1} correspond to the outgoing waves from the core region. From (3.5), it follows that:

$$\begin{pmatrix} A_{N+1} \\ 0 \end{pmatrix} = M_{N,N+1} \ldots M_{1,2} M_{0,1} \begin{pmatrix} 0 \\ B_0 \end{pmatrix} = \begin{pmatrix} a_{1,1}(k_x) & a_{1,2}(k_x) \\ a_{2,1}(k_x) & a_{2,2}(k_x) \end{pmatrix} \begin{pmatrix} 0 \\ B_0 \end{pmatrix}$$

$$\rightarrow a_{2,2}(k_x) = 0; \ A_{N+1} = a_{1,2}(k_x)B_0, \tag{3.25}$$

which represents a root-finding problem with respect to k_x, while B_0 becomes a normalization constant that can be chosen at will. The simplest example of a planar waveguide is a slab waveguide, which is considered in Problem 3.1. For every polarization, the

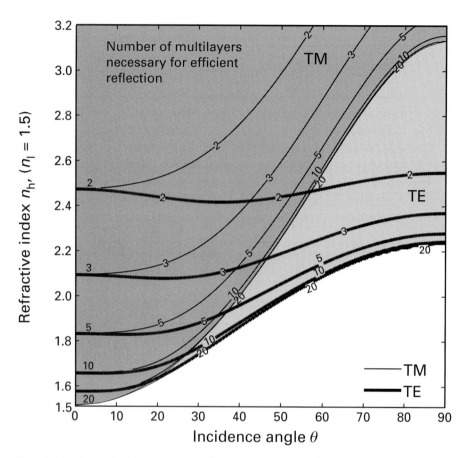

Figure 3.5 Regions of efficient operation of a quarter-wave stack designed for a normal angle of incidence $\theta_d = 0$; $n_l = 1.5$; $n_h = 2.8$. As gray regions, we present regions of reflector efficiency (where $|\lambda_1^{TE}| < 1$ or $|\lambda_1^{TM}| < 1$). For a given choice of n_h, there is always a range of incidence angles where the reflector is efficient. Moreover, for a high enough index contrast ($n_h > 3.13$) the reflector remains efficient for all the angles of incidence and any polarization: this is a region of omnidirectional TE, TM reflection. Contour plots indicate the smallest number of bilayers in the reflector necessary for efficient reflection (defined as $N_{attenuation} = 1/\log(1/|\lambda_1|)$). Note that the reflection of TE polarization is always more efficient, with a smaller index contrast necessary for the onset of omnidirectional reflectivity ($n_h > 2.24$ compared with $n_h > 3.13$ for TM polarization).

modes of a slab waveguide can be presented using a band diagram of the type shown in Fig. 2.10 (c).

3.5   Guiding in the interior of an infinitely periodic multilayer

We now consider electromagnetic modes of an infinite periodic multilayer. We are going to show that for such multilayers there exist regions of the phase space (k_x, k_z, ω) where no delocalized states are permitted. We will call such regions of phase space bandgaps.

Within a bandgap, the modes of a multilayer are evanescent, exhibiting either exponential growth or decay. The existence of bandgaps in infinite reflectors is directly related to the ability of a finite reflector of identical composition and structure to reflect light. Namely, we will demonstrate that a finite size reflector will be effective in reflecting a plane wave characterized by the propagation constant k_z and frequency ω, if (k_x, k_z, ω) finds itself within a photonic bandgap of a corresponding infinite reflector of the same geometry and composition.

From the discussion of Section 2.4.5 (see the subsection on 1D discrete translational symmetry) it follows that general form of a solution extended in space in the infinitely periodic multilayer (see Fig. 3.2 (d)) satisfies the Bloch theorem. In particular, for a discrete symmetry along $\hat{z}$ and period $a = d_l + d_h$, there is a conserved quantity k_z, such that delocalized waves have the form:

$$\begin{pmatrix} \mathbf{H} \\ \mathbf{E} \end{pmatrix}_{k_x,k_z} (\mathbf{r} + aN\hat{z}) = \exp(ik_z aN) \begin{pmatrix} \mathbf{H} \\ \mathbf{E} \end{pmatrix}_{k_x,k_z} (\mathbf{r}). \tag{3.26}$$

The explicit form of the fields can be found using (3.2), (3.7), where for TE polarization, for example, in a matrix form:

$$\begin{pmatrix} H_x^j(x,y,z) \\ E_y^j(x,y,z) \end{pmatrix} = \exp(ik_x x) \begin{pmatrix} -\frac{k_z^j}{\omega}\exp\left(ik_z^j(z-z_j)\right) & \frac{k_z^j}{\omega}\exp\left(-ik_z^j(z-z_j)\right) \\ \exp\left(ik_z^j(z-z_j)\right) & \exp\left(-ik_z^j(z-z_j)\right) \end{pmatrix} \begin{pmatrix} A_j \\ B_j \end{pmatrix}. \tag{3.27}$$

For TE waves in the form (3.27), we apply the Bloch theorem (3.26) to the H_x and E_y field components in the layers j and $j + 2N$ (layers separated by N bilayers), and taking into account that $k_z^{j+2N} = k_z^j$, we get:

$$\begin{pmatrix} H_x^{j+N}(x,y,z+aN) \\ E_y^{j+N}(x,y,z+aN) \end{pmatrix} = \exp(ik_z aN) \begin{pmatrix} H_x^j(x,y,z) \\ E_y^j(x,y,z) \end{pmatrix}. \tag{3.28}$$

Substitution of (3.27) into (3.28) leads to the equation independent of position z:

$$\begin{pmatrix} A_{j+2N} \\ B_{j+2N} \end{pmatrix} = \exp(ik_z aN) \begin{pmatrix} A_j \\ B_j \end{pmatrix}. \tag{3.29}$$

Coefficients separated by N bilayers are connected by a product of N identical bilayer transfer matrices. Assuming that layer j is of higher dielectric index, then (3.29) can be rewritten as:

$$(M_{hlh})^N \begin{pmatrix} A_j \\ B_j \end{pmatrix} = \exp(ik_z aN) \begin{pmatrix} A_j \\ B_j \end{pmatrix}. \tag{3.30}$$

Using the presentation of the transfer matrix in terms of its eigenvalues and eigenvectors $M_{hlh} = V_{hlh}\Lambda_{hlh}V_{hlh}^{-1}$, we write (3.30) as:

$$V_{hlh}\left(\Lambda_{hlh}^N - \exp(ik_z aN)\right) V_{hlh}^{-1} \begin{pmatrix} A_j \\ B_j \end{pmatrix} = 0,$$

hence:

$$\lambda_1 = \exp(ik_z a); \quad \begin{pmatrix} A_j \\ B_j \end{pmatrix} = \overline{V}_{hlh}^1 \tag{3.31}$$

or, alternatively,

$$\lambda_2 = \exp(ik_z a); \quad \begin{pmatrix} A_j \\ B_j \end{pmatrix} = \bar{V}_{\text{hlh}}^2.$$

We thus conclude that for a delocalized state to exist inside a periodic multilayer, one of its eigenvalues has to be a complex exponential in the form (3.31), while a vector of the expansion coefficients will then be one of the eigenvectors of a bilayer transfer matrix. In Section 3.5.1 we have demonstrated that for a semi-infinite periodic reflector to be effective, the smallest eigenvalue of a corresponding bilayer transfer matrix has to be less than one. In the regime described by (3.31) absolute value of an eigenvalue is one, and therefore a corresponding semi-infinite reflector is not effective, as it allows excitation of an infinitely extended radiation state in the bulk of a semi-infinite multilayer.

Exactly the same analysis can be conducted for the TM modes, where instead of H_x and E_y field components we use H_y and E_x field components. From the form of a quadratic equation (3.18) for the eigenvalues of a bilayer transfer matrix, it follows that eigenvalues have the following properties:

$$\lambda_1 \lambda_2 = 1, \ \lambda_1 + \lambda_2 = 2\cos(\phi_h)\cos(\phi_l) - \left(r_{\text{TE,TM}} + r_{\text{TE,TM}}^{-1}\right)\sin(\phi_h)\sin(\phi_l). \qquad (3.32)$$

From (3.32) it follows that if $\lambda_1 = \exp(ik_z a)$, then $\lambda_2 = \exp(-ik_z a)$, and hence $\lambda_1 + \lambda_2 = 2\cos(k_z a)$. Thus, the equation for a Bloch constant of delocalized waves becomes:

$$\cos(k_z a) = \cos(\phi_h)\cos(\phi_l) - \xi \sin(\phi_h)\sin(\phi_l), \qquad (3.33)$$

where $\xi^{\text{TE,TM}} = (r_{\text{TE,TM}} + r_{\text{TE,TM}}^{-1})/2$, $|\xi^{\text{TE,TM}}| \geq 1$ and the phases ϕ_h, ϕ_l are as in (3.16).

We now discuss in detail the dispersion relation of Bloch states presented in Fig. 3.6. We first assume that propagation is strictly perpendicular to the plane of a multilayer, namely $k_x = 0$. We use an optimal quarter-wave reflector design for a frequency $\omega_0 = 2\pi c/\lambda_0$, so that

$$\phi_{\omega_0} = \phi_{\omega_0}^h = \omega_0 n_h d_h = \phi_{\omega_0}^l = \omega_0 n_l d_l = \pi/2,$$
$$r = r_{\text{TE}} = r_{\text{TM}} = n_l/n_h.$$

As established in Section 3.1.3, a semi-infinite periodic multilayer with this chosen geometry will completely reflect radiation of frequency ω_0. Thus, one expects that there will be no propagating states at such a frequency inside an infinite periodic multilayer with the same geometry. In particular, by imposing a quarter-wave stack condition in (3.33) we get $\cos(k_z a) = -\xi$, and as $|\xi| \geq 1$, it follows that there are no real solutions of this equation and hence there are no delocalized (propagating) waves inside the multilayer. However, in the middle of a bandgap there are still complex solutions satisfying (3.33) in the form $k_z = \pi/a + i\kappa$:

$$a\kappa = \cosh^{-1}(\xi) = \pm \log(r), \qquad (3.34)$$

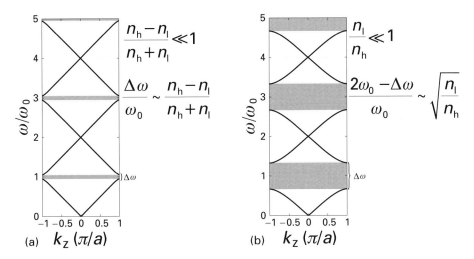

Figure 3.6 Dispersion relations of delocalized modes propagating perpendicular to the quarter-wave periodic multilayer. (a) Low refractive index contrast multilayer (example of $n_l = 1.5$, $n_h = 1.8$). The size of a bandgap is proportional to the refractive index contrast. (b) High-refractive-index contrast multilayer (example of $n_l = 1.5$, $n_h = 4.5$). The size of a bandgap is comparable to ω_0.

which defines exponentially decaying or diverging waves along the direction of propagation with a characteristic decay or divergence length l_d:

$$\kappa l_d = |\log(r)| \, l_d/a \sim 1 \rightarrow l_d = a/\log(n_h/n_l). \tag{3.35}$$

Now consider the field distribution in decaying and divergent waves. By first finding eigenvectors of a bilayer transfer matrix (3.31), we find expansion coefficients in a high-index layer. Multiplying these coefficients by a transfer matrix between high and low index layers we then find the expansion coefficients in the low index layer. In particular, at the center of a bandgap, the eigenvectors of a bilayer transfer matrix are:

$$\begin{pmatrix} A \\ B \end{pmatrix}^h_{\pi/a+i|\kappa|} = \frac{1}{2}\begin{pmatrix} -1 \\ 1 \end{pmatrix}; \begin{pmatrix} A \\ B \end{pmatrix}^h_{\pi/a-i|\kappa|} = \frac{1}{2}\begin{pmatrix} 1 \\ 1 \end{pmatrix}, \tag{3.36}$$

while the expansion coefficients in the low-refractive-index layers can be found as:

$$\begin{pmatrix} A \\ B \end{pmatrix}^l_{\pi/a+i|\kappa|} = M_{hl}\begin{pmatrix} A \\ B \end{pmatrix}^h_{\pi/a+i|\kappa|} = -\frac{i}{2}\begin{pmatrix} 1 \\ 1 \end{pmatrix}$$

$$\begin{pmatrix} A \\ B \end{pmatrix}^l_{\pi/a-i|\kappa|} = M_{hl}\begin{pmatrix} A \\ B \end{pmatrix}^h_{\pi/a-i|\kappa|} = \frac{i}{2r}\begin{pmatrix} 1 \\ -1 \end{pmatrix}.$$

From (3.27) it follows that the general expressions for the field and the field intensity are:

$$E_y^{l,h}\left(z,\frac{\pi}{a}\pm i\,|\kappa|\right) = A_{\pi/a\pm i|\kappa|}^{l,h}\,\exp(i\phi_{\omega_0}\eta^{l,h}) + B_{\pi/a\pm i|\kappa|}^{l,h}\,\exp(-i\phi_{\omega_0}\eta^{l,h})$$

$$\left|E_y^{l,h}\left(z,\frac{\pi}{a}\pm i\,|\kappa|\right)\right|^2 = \left|A_{\pi/a\pm i|\kappa|}^{l,h}\right|^2 + \left|B_{\pi/a\pm i|\kappa|}^{l,h}\right|^2$$

$$+ 2\left[\mathrm{Re}\left(A_{\pi/a\pm i|\kappa|}^{l,h}\,B_{\pi/a\pm i|\kappa|}^{*l,h}\right)\cos(2\phi_{\omega_0}\eta^{l,h}) - \mathrm{Im}\left(A_{\pi/a\pm i|\kappa|}^{l,h}\,B_{\pi/a\pm i|\kappa|}^{*l,h}\right)\sin(2\phi_{\omega_0}\eta^{l,h})\right],$$

$$\tag{3.37}$$

where $\eta^{l,h} = (z - z_{l,h})/d_{l,h}$, and $0 \le \eta^{l,h} \le 1$ for each of the layers. In the particular case of the coefficients (3.36), for a single bilayer we have:

$$E_y^h\left(z,\frac{\pi}{a}+i\,|\kappa|\right) = -i\sin\left(\frac{\pi}{2}\eta^h\right),$$

$$E_y^l\left(z,\frac{\pi}{a}+i\,|\kappa|\right) = -i\cos\left(\frac{\pi}{2}\eta^l\right),$$

$$\left|E_y^h\left(z,\frac{\pi}{a}+i\,|\kappa|\right)\right|^2 = \sin^2\left(\frac{\pi}{2}\eta^h\right),$$

$$\left|E_y^l\left(z,\frac{\pi}{a}+i\,|\kappa|\right)\right|^2 = \cos^2\left(\frac{\pi}{2}\eta^l\right),\tag{3.38}$$

and similarly:

$$E_y^h\left(z,\frac{\pi}{a}-i\,|\kappa|\right) = \cos\left(\frac{\pi}{2}\eta^h\right),$$

$$E_y^l\left(z,\frac{\pi}{a}-i\,|\kappa|\right) = -\frac{1}{r}\sin\left(\frac{\pi}{2}\eta^l\right),$$

$$\left|E_y^h\left(z,\frac{\pi}{a}-i\,|\kappa|\right)\right|^2 = \cos^2\left(\frac{\pi}{2}\eta^h\right),$$

$$\left|E_y^l\left(z,\frac{\pi}{a}-i\,|\kappa|\right)\right|^2 = \frac{1}{r^2}\sin^2\left(\frac{\pi}{2}\eta^l\right).\tag{3.39}$$

In Fig. 3.7 we present plots of the electric field intensities in the middle of a bandgap for the decaying and diverging modes over two periods. From the form of (3.33) it also follows that any two frequencies separated by a multiple of $2\omega_0$ will have the same value of propagation vector. Thus, the bandgap is periodic in the frequency direction with a period $2\omega_0$. Finally, we note that if the quarter-wave-design condition is relaxed, in addition to the wide bandgaps at $\omega_0 \cdot (2p + 1)$, where p is an integer, narrow bandgaps at $\omega_0 \cdot 2p$ will appear.

We now find the extent of a bandgap $\Delta\omega$ around ω_0 in which no delocalized states are allowed. At the edges of a bandgap $\omega_0 \pm \delta\omega$ the value of a Bloch propagation constant is $k_z = \pi/a$, while the phases can be written as $\phi_{\omega_0\pm\delta\omega}^h = \phi_{\omega_0\pm\delta\omega}^l = \pi/2 \cdot (1 \pm \delta\omega/\omega_0)$. After substitution into (3.33) we get:

$$\sin\left(\frac{\pi}{2}\frac{\delta\omega}{\omega_0}\right) = \sqrt{\frac{\xi - 1}{\xi + 1}} = \frac{n_h - n_l}{n_h + n_l}.\tag{3.40}$$

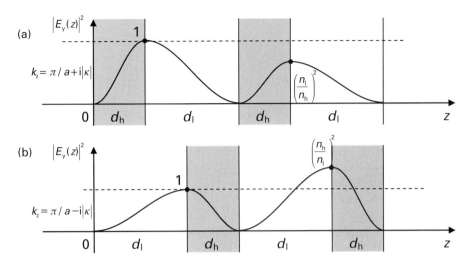

Figure 3.7 (a) Field intensity in the decaying mode at the bandgap center ($k_z = \pi/a + i|\kappa|$, ω_0). (b) Field intensity in the diverging mode at the bandgap center ($k_z = \pi/a - i|\kappa|$, ω_0). Only two periods are shown. In both examples $k_x = 0$.

The size of a bandgap is defined as $\Delta\omega = 2\delta\omega$. There are two limits when the bandgap size can be found analytically from (3.40) using Taylor expansions. In particular, in the case of a small index contrast, the relative size of a bandgap is directly proportional to the refractive index contrast:

$$\frac{\Delta\omega}{\omega_0}\bigg|_{(n_h - n_l)/n_l \ll 1} = \frac{4}{\pi}\frac{n_h - n_l}{n_h + n_l}. \tag{3.41}$$

In Fig. 3.6 (a) we present the dispersion relation of the reflector Bloch states in the limit of a small refractive index contrast. Everywhere except near the edge of a Brillouin zone band diagram is similar to one of the delocalized states of a uniform dielectric with effective refractive index $n_{avg} = (n_h d_h + n_l d_l)/a$. At the edges of a Brillouin zone $k_z = \pm\pi/a$, a bandgap appears with a size proportional to the bilayer refractive index contrast. Inside a bandgap there are no delocalized states allowed.

In the limit of high-refractive-index-contrast:

$$\frac{\Delta\omega}{\omega_0}\bigg|_{n_l/n_h \ll 1} = 2 - \frac{4\sqrt{2}}{\pi}\sqrt{\frac{n_l}{n_h}}, \tag{3.42}$$

the size of a bandgap is almost twice the value of a bandgap center frequency, and it is not sensitive to a particular value of the refractive index contrast see Fig. 3.6(b).

We now find propagation constants for the evanescent solutions with frequencies close to the bandgap edges $\omega = \omega_0 \pm \Delta\omega/2 \mp \delta\omega_e$. Near the bandgap edge, the complex-valued Bloch propagation constant will have the form $k_z = \pi/a + i\kappa$. Performing Taylor expansion near $\omega = \omega_0 \pm \Delta\omega/2$, we get:

$$a\kappa = \sqrt{2\pi \frac{n_h - n_l}{\sqrt{n_h n_l}} \frac{\delta\omega_e}{\omega_0}}. \tag{3.43}$$

This defines decay length in the multilayer ed$\sim 1/\kappa \sim 1/\sqrt{\delta\omega_e}$.

From time-reversal symmetry and the Bloch theorem, it follows that the curvature of the bands is zero at $k_z = \pm\pi/a$ (see Problem 2.1), thus signifying that the group velocity for these solutions $v_g = \partial\omega(k_z)/\partial k_z$ is also zero. Physically, this means that amplitudes of forward- and backward-propagating waves in each of the layers are equal, forming a standing wave across the multilayer. Although electromagnetic fields are excited, the forward- and backward-energy fluxes are the same, resulting in a zero total energy flux.

Finally, we consider the distribution of the electromagnetic fields inside a periodic multilayer at the upper and lower band edges of a fundamental (lowest frequency) bandgap. In fact, the eigenvectors of a bilayer transfer matrix at the bandgap edges can be found analytically. Thus, using (3.40) to define the edges of a fundamental bandgap, substituting it into the bilayer transfer matrix (3.17), and taking into account that $k_z = \pi/a$, $\phi_{\omega_0 \pm \delta\omega} = \phi^h_{\omega_0 \pm \delta\omega} = \phi^l_{\omega_0 \pm \delta\omega} = \pi/2 \cdot (1 \pm \delta\omega/\omega_0)$, $r = r_{TE} = r_{TM} = n_1/n_h$ we can derive for the upper and lower bandgap edges in the high- and low-refractive-index layers:

$$\left(\begin{array}{c} A \\ B \end{array}\right)^h_{\omega_0-\delta\omega} = \frac{1}{2}\left(\begin{array}{c} \frac{1-i\sqrt{r}}{1+i\sqrt{r}} \\ 1 \end{array}\right); \left(\begin{array}{c} A \\ B \end{array}\right)^h_{\omega_0+\delta\omega} = \frac{\sqrt{r}}{2}\left(\begin{array}{c} \frac{1+i\sqrt{r}}{1-i\sqrt{r}} \\ 1 \end{array}\right)$$

$$\left(\begin{array}{c} A \\ B \end{array}\right)^l_{\omega_0-\delta\omega} = \frac{i}{2\sqrt{r}}\left(\begin{array}{c} \frac{1-i\sqrt{r}}{1+i\sqrt{r}} \\ -1 \end{array}\right); \left(\begin{array}{c} A \\ B \end{array}\right)^l_{\omega_0+\delta\omega} = \frac{i}{2}\left(\begin{array}{c} \frac{1+i\sqrt{r}}{1-i\sqrt{r}} \\ -1 \end{array}\right). \tag{3.44}$$

From (3.40) it follows that for the lower and upper edges:

$$\exp(i\phi_{\omega_0-\delta\omega}) = \frac{(1+i\sqrt{r})^2}{1+r}; \exp(i\phi_{\omega_0+\delta\omega}) = -\frac{(1-i\sqrt{r})^2}{1+r}. \tag{3.45}$$

We now investigate the distribution of the electric field inside of a multilayer using (3.37). In Fig. 3.8 we present plots for the electric field intensities of the modes at the lower and upper edges of a bandgap. Note that at the lower edge the modes have their electric field concentrated in the high-refractive-index material, while at the upper edge the electric field is concentrated mostly in the low-refractive-index material. This trend, in general, holds for all the states in the first and second bands. Thus, one frequently calls the first band (lowest-frequency band) a high-dielectric-constant band while one calls the second band a low-dielectric-constant band. The reason why the modes in a fundamental band are concentrated in the high dielectric region has already been rationalized in Section 2.3.2 and is a direct consequence of a variational principle. Modes in a second band, however, are forced to localize in the low dielectric region to maintain their orthogonality to the modes of a fundamental band.

We now investigate electromagnetic modes propagating with a nonzero wavevector component k_x along the plane of a multilayer. First, assume that $k_z = 0$, then from (3.1),

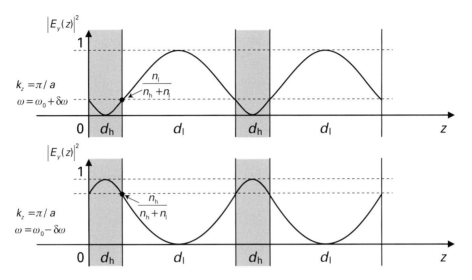

Figure 3.8 (a) Field intensity at the upper bandgap edge $k_z = \pi/2; \omega = \omega_0 + \delta\omega$. (b) Field intensity at the lower band edge $k_z = \pi/2; \omega = \omega_0 - \delta\omega$. Only two periods are shown.

(3.2) for TE polarization the fields are:

$$E_y^j(x, y, z) = \exp(ik_x x)$$
$$H_x^j(x, y, z) = 0 \tag{3.46}$$
$$H_z^j(x, y, z) = \frac{k_x}{\omega} \exp(ik_x x).$$

In the limit $k_x \to 0$, $\omega \to 0$ the only nonzero component of the electromagnetic field E_y will be essentially constant across the whole multilayer, thus defining an average displacement field across the multilayer $D_{avg} = (d_1\varepsilon_1 + d_h\varepsilon_h)/(d_1 + d_h)E_y$. This, in turn, defines an effective dielectric constant for the TE modes $\varepsilon_{avg}^{TE} = (d_1\varepsilon_1 + d_h\varepsilon_h)/(d_1 + d_h)$ in the $k_x \to 0$, $\omega \to 0$ limit, and hence a linear dispersion relation $k_x \underset{\omega \to 0}{=} \omega\sqrt{\varepsilon_{avg}^{TE}}$.

In a similar manner, for TM modes in the limit $k_x \to 0$, $\omega \to 0$:

$$H_y^j(x, y, z) = \exp(ik_x x)$$
$$E_x^j(x, y, z) = 0 \tag{3.47}$$
$$E_z^j(x, y, z) = -\frac{k_x}{\omega\varepsilon_j} \exp(ik_x x).$$

Thus, the only nonzero component of the displacement field D_z will be essentially constant throughout a multilayer, thus defining an average displacement field $E_{avg} = (d_1/\varepsilon_1 + d_h/\varepsilon_h)/(d_1 + d_h)D_z$. This relation, in turn, defines an effective dielectric constant for the TM modes $\varepsilon_{avg}^{TM} = \varepsilon_1\varepsilon_h(d_1 + d_h)/(d_1\varepsilon_h + d_h\varepsilon_1)$, and hence a linear dispersion relation $k_x \underset{\omega \to 0}{=} \omega\sqrt{\varepsilon_{avg}^{TM}}$. Note that for any choice of dielectric constants and layer thicknesses $\varepsilon_{avg}^{TE} \geq \varepsilon_{avg}^{TM}$, thus forcing TM modes to exhibit higher frequencies than TE modes (Fig. 3.9).

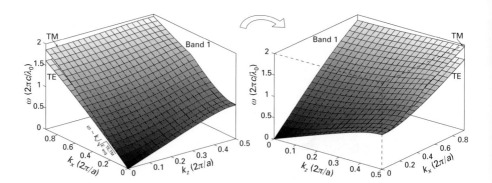

Figure 3.9 Dispersion relation of a fundamental band as a function of (k_x, k_z) for a quarter-wave stack with $n_l = 1.5; n_h = 2.8; \lambda_0 = 1; \theta_d = 0$. In the limit $k_z = 0$ and $k_x \to 0, \omega \to 0$ the dispersion relations of TE, TM modes are linear, TE modes being lower in frequency than TM modes. For higher frequencies, TE, TM dispersion relations will essentially be independent of k_z and equal to the dispersion relations of TE, TM modes of a slab waveguide formed by a single high-index layer, surrounded with a low-refractive-index cladding.

In the limit of large propagation constants $k_x > \omega n_l$ (the region of phase space below the light line of a low-index material), regardless of the value of k_z, electric fields of the modes will be confined to the layers with high refractive index. Moreover, the fields will exhibit exponential decay into the low-refractive-index layers, thus resembling a system of infinitely many identical decoupled slab waveguides. In this regime, the dispersion relation $\omega(k_x)$ will be the same for any value of k_z and equal to that of a dispersion relation of a slab waveguide.

Instead of using a 3D plot of the dispersion relation $\omega(k_x, k_z)$, it is more convenient to draw a projected band diagram where all the allowed states are plotted in a 2D plot $\omega(k_x, k_z)|_{k_z-\text{all allowed}}$ versus a single parameter k_x (see Fig. 3.10). In such a plot, bands of modes are represented by the filled regions of phase space. Note that in the inter-band region, labeled as a bandgap, there are no delocalized states allowed in the multilayer. Moreover, while the bandgap for TE polarization is continuous, the bandgap for TM polarization collapses to zero for a certain value of k_x.

Finally, we are going to discuss how the size of a bandgap changes for the optimal quarter-wave periodic reflectors designed for various propagation angles. In particular, assuming θ_d to be the modal propagation angle inside a low-refractive-index layer (see Fig. 3.11), the layer thicknesses of an optimal quarter-wave reflector are chosen according to (3.20):

$$k_z^h d_h = \frac{\pi}{2}; k_z^l d_l = \frac{\pi}{2};$$

$$k_z^h = \omega\sqrt{\varepsilon^h - \varepsilon^l \sin^2(\theta_d)}; k_z^l = \omega\sqrt{\varepsilon^l} \cos(\theta_d). \qquad (3.48)$$

If the angle of propagation θ_d is fixed, then expression (3.40) for the bandgap size is applicable, however for each of the polarizations one has to use an appropriate $\xi^{\text{TE,TM}} = (r_{\text{TE,TM}} + r_{\text{TE,TM}}^{-1})/2$ parameter. In Fig. 3.11 we plot the relative bandgap

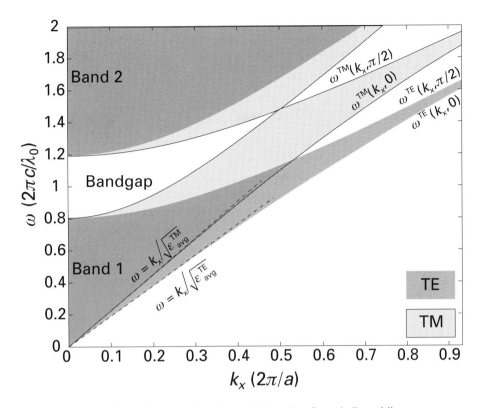

Figure 3.10 Projected band diagram of the TE and TM modes of a periodic multilayer ($n_l = 1.5; n_h = 2.8; \lambda_0 = 1; \theta_d = 0$). In a bandgap region there are no delocalized states allowed inside the multilayer. The TE bandgap is continuous, while the TM bandgap collapses for a certain value of k_x.

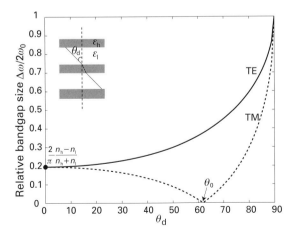

Figure 3.11 Bandwidth $\Delta\omega$ of the TE, TM fundamental gaps for different quarter-wave multilayers designed to operate at the angle of propagation θ_d, and $n_l = 1.5; n_h = 2.8; \lambda_0 = 1$. Note that for an angle of propagation $\sin(\theta_0) = \sqrt{\varepsilon_h/(\varepsilon_h + \varepsilon_l)}$, even an optimal multilayer design results in a zero bandwidth for TM polarization.

size $\Delta\omega/(2\omega_0)$ of the TE and TM fundamental gaps for various optimal quarter-wave multilayers as a function of a design angle θ_d. Note that for an angle of propagation $\sin(\theta_0) = \sqrt{\varepsilon_h/(\varepsilon_h + \varepsilon_l)}$, even an optimal multilayer design results in a zero bandgap for TM polarization.

3.5.1   Omnidirectional reflectors II

In the previous section a projected band diagram technique (Fig. 3.10) has been introduced to visualize the guided states inside an infinite periodic reflector. It was established that photonic bands (gray regions in Fig. 3.10) represent a continuum of delocalized states allowed to propagate inside a reflector. Moreover, inside the photonic bands, the absolute value of the smallest eigenvalue of a bilayer transfer matrix (3.17) equals one. From consideration of Section 3.3 it follows that a semi-infinite reflector of the same geometry will be inefficient while operating inside the photonic bands. On the other hand, we have established that inside a bandgap region of Fig. 3.10, there are no states allowed to propagate in the bulk of a periodic reflector, and, moreover, the absolute value of the smallest eigenvalue of a bilayer transfer matrix is smaller than one. From consideration of Section 3.3 it follows that a semi-infinite reflector of the same geometry will be efficient while operating inside the photonic bandgap. From this we conclude that a projected band diagram of Fig. 3.10 introduced to visualize extended states of an infinite periodic multilayer can also be used to judge the efficiency of a semi-infinite reflector of the same structure.

So far, we have described how to use a projected diagram of an infinite periodic reflector and to decide whether a corresponding semi-infinite reflector is efficient or not. However, consideration for the semi-infinite reflector has to be somewhat modified to account for the radiation incidence angle. In particular, we consider radiation to be incoming onto a reflector from a uniform cladding of refractive index n_0, with an angle of incidence $\theta \subset [0, 90°]$ (Fig. 3.2 (b)). Projection of a wave vector along $\hat{x}$ will then be:

$$k_x = \omega n_0 \sin(\theta), \tag{3.49}$$

from which it follows that, for a given frequency ω, the maximal value of k_x is limited by ωn_0. Therefore, as with Fig. 2.1, all the incoming radiation states (all the angles of radiation incidence) can be presented on a projected band diagram (k_x, ω) as a continuum of states filling the "light cone" defined by $k_x(\omega) \subset [0, \omega n_0]$ (see Fig. 3.12).

Thus, when analyzing reflection from a semi-infinite periodic multilayer, only the states within the light cone of a cladding should be considered. Within the light cone one finds bands of the delocalized reflector states as well as bandgap regions. Suppose that a certain combination (k_x, ω) within a light cone of a cladding falls into a bandgap region. This means that a plane wave of frequency ω incoming from the cladding onto a semi-infinite reflector at an angle of incidence $\theta = \sin^{-1}(k_x/\omega n_0)$ will be reflected completely. Note that in the vicinity of $\omega \sim 1$ (the dashed region in Fig. 3.12), the TE and TM bandgaps exist for all the values of k_x within the light cone of a cladding. That means that in this frequency region, the planewaves of any polarization incoming at any incidence angle will be reflected completely. This is a region of omnidirectional

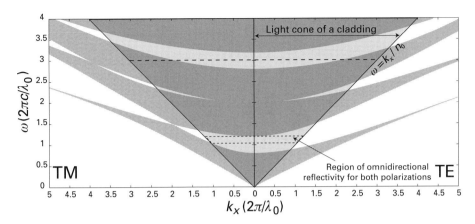

Figure 3.12 Projected band diagrams of an infinite high-index-contrast periodic multilayer $n_l = 1.5; n_h = 2.8; \lambda_0 = 1; \theta_d = 0$ for TE and TM polarizations, plus the light cone of a cladding $n_0 = 1$. Only the reflector states inside the light cone of a cladding can be excited by the radiation coming from the cladding.

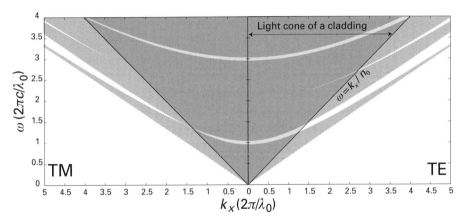

Figure 3.13 Projected band diagrams of low-index-contrast periodic multilayer $n_l = 1.4; n_h = 1.6; \lambda_0 = 1; \theta_d = 0$ for TE and TM polarizations, plus the light cone of a cladding $n_0 = 1$. As expected, photonic bandgaps are small and there is no omnidirectional reflection for either polarization.

reflectivity for both polarizations. Note that in the vicinity of $\omega \sim 3$, although there is no omnidirectional reflectivity for either polarization, the reflector is still efficient for a wide range of incidence angles. Thus, at $\omega = 3$, both TE and TM polarization will be efficiently reflected for all the incidence angles in a range $\theta \subset [0, 38°]$.

To appreciate the effect of the refractive index contrast on the reflection properties of periodic multilayers, in Fig. 3.13 we present a projected band diagram for the case of a quarter-wave reflector of low-refractive-index contrast $n_l = 1.4; n_h = 1.6$. As seen from the figure, photonic bandgaps are much smaller than in the case of a high-refractive-index-contrast reflectors in Fig. 3.12, and there is no omnidirectional reflection for either polarization. Instead, at the frequency of a fundamental bandgap $\omega = 1$, both TE and TM polarization will be efficiently reflected only for the incidence angles in a range $\theta \subset [0, 24°]$.

3.6   Defect states in a perturbed periodic multilayer: planar photonic crystal waveguides

We are going to finish this chapter by discussing guidance within a structural defect in the infinite periodic multilayer (see Fig. 3.2 (e)). In principle, the structural defect can be any geometrical or material perturbation of a system that conserves continuous translational symmetry in the multilayer plane. Examples of a typical defect would be a single layer with increased thickness, or a single layer with increased value of dielectric constant. The notion of a guided defect state in this case is understood as in-plane (of a multilayer) guidance of an electromagnetic mode localized in the $\hat{z}$ direction at a defect site. Such a defect mode would have the fields decaying into the bulk of a multilayer reflector, with most of its electromagnetic energy concentrated in the vicinity of a defect layer.

There are two design approaches to implementing field localization inside of a defect layer. The first one is by using total internal reflection guidance. In this case parameters of a defect layer are chosen such that the defect mode has an effective refractive index larger than at least one of the two refractive indices n_l or n_h. A simple way to implement a corresponding defect is by increasing the thickness of one of the high-refractive-index layers. The second approach uses the bandgap of a multilayer reflector to localize the mode spatially at a defect site. To implement this, one has to ensure that the defect mode is located inside a bandgap of a surrounding periodic reflector, resulting in exponentially fast decay of a modal field outside a defect layer. Moreover, in this case, the effective refractive index of a guided mode can be smaller than both the n_l and n_h. This opens an intriguing possibility of efficient light guidance in the low-refractive-index core (or even a hollow core), which will be discussed in detail in the next chapter.

In the following we apply transfer matrix theory to find dispersion relations of the guided defect states localized at a single-layer geometrical defect in the bulk of an infinite periodic multilayer. Note that if projection of the propagation vector along the multilayer plane k_x is nonzero, such defect states will essentially be guided modes of a waveguide with a core formed by a defect layer. If a defect layer is of low refractive index n_l, the guiding mechanism in such a waveguide is by a photonic bandgap of a surrounding periodic multilayer. In the case when the defect layer is of high refractive index n_h, both bandgap guidance and total internal reflection guidance are possible.

Figure 3.14 presents the geometry of a defect layer in a periodic multilayer stack. Without loss of generality, suppose that a defect layer of refractive index n_0 and thickness d is sandwiched between two high-index layers. Consider the field expansion coefficients in the high-index layers of the bi-layers $-N$ and N to the left and to the right of a defect layer. Defining the transfer matrix from the defect layer to the high-index layer of a first bilayer as M_{dh}, the bilayer transfer matrix connecting high-low-high index layers as M_{hlh},

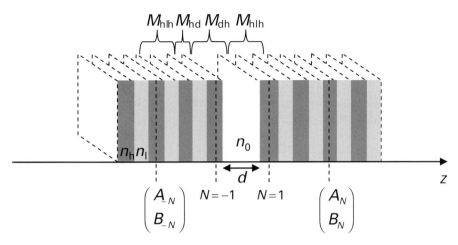

Figure 3.14 Bulk defect in a periodic multilayer.

the transfer matrix connecting the high-index layer of a minus first bilayer with a defect layer as M_{hd}, and using the transfer matrix formulation (3.13), we write:

$$\begin{pmatrix} A_N \\ B_N \end{pmatrix} = (M_{\text{hlh}})^{N-1} M_{\text{dh}} M_{\text{hd}} (M_{\text{hlh}})^{N-1} \begin{pmatrix} A_{-N} \\ B_{-N} \end{pmatrix}. \tag{3.50}$$

Using the presentation of the bilayer transfer matrix in terms of its eigenvalues and eigenvectors we get:

$$\begin{pmatrix} A_N \\ B_N \end{pmatrix} = V_{\text{hlh}} \begin{pmatrix} \lambda_{<1}^{N-1} & 0 \\ 0 & \lambda_{>1}^{N-1} \end{pmatrix} V_{\text{hlh}}^{-1} M_{\text{dh}} M_{\text{hd}} V_{\text{hlh}} \begin{pmatrix} \lambda_{<1}^{N-1} & 0 \\ 0 & \lambda_{>1}^{N-1} \end{pmatrix} V_{\text{hlh}}^{-1} \begin{pmatrix} A_{-N} \\ B_{-N} \end{pmatrix}, \tag{3.51}$$

where the eigenvalue of absolute value less than one is defined as $\lambda_{<1}$, while the one with absolute value greater than one is defined as $\lambda_{>1}$. As the field is decaying in the negative direction from the defect layer (increasing in the positive direction from the bilayer $-N$), the coefficients in the $-N$ bilayer should only excite an eigenvalue with magnitude larger than one, therefore:

$$\begin{pmatrix} A_{-N} \\ B_{-N} \end{pmatrix} = \alpha_{-N} \overline{V}_{\text{hlh}}^{>1}. \tag{3.52}$$

In the positive direction from the defect layer the fields are decaying, thus only an eigenvector with magnitude less than one is excited, therefore:

$$M_{\text{dh}} M_{\text{hd}} V_{\text{hlh}} \begin{pmatrix} \lambda_{<1}^{N-1} & 0 \\ 0 & \lambda_{>1}^{N-1} \end{pmatrix} V_{\text{hlh}}^{-1} \begin{pmatrix} A_{-N} \\ B_{-N} \end{pmatrix} = \alpha_N \overline{V}_{\text{hlh}}^{<1}, \tag{3.53}$$

where α_{-N} and α_N are some constants. Substitution of (3.52) into (3.53) leads to the following equation:

$$M_{\text{dh}} M_{\text{hd}} \overline{V}_{\text{hlh}}^{>1} = \alpha \overline{V}_{\text{hlh}}^{<1}, \tag{3.54}$$

where α is some constant. Note that the transfer matrices are the functions of a wave vector parameter k_x. Thus, to find the propagation constant k_x of a defect mode, one simply has to vary it until (3.54) is satisfied.

As an example, consider the case of a quarter-wave multilayer stack operated at the center of its fundamental bandgap, and TE polarization of an incoming light. From (3.36) it follows that $\overline{V}_{hlh}^{<1} = (-1, 1)/2; \overline{V}_{hlh}^{>1} = (1, 1)/2$, while from (3.4), (3.16), (3.20):

$$M_{hd} = \frac{i}{2} \begin{pmatrix} \left(1 + \frac{k_z^h}{k_z^0}\right) - \left(1 - \frac{k_z^h}{k_z^0}\right) \\ \left(1 - \frac{k_z^h}{k_z^0}\right) - \left(1 + \frac{k_z^h}{k_z^0}\right) \end{pmatrix};$$

$$M_{dh} = \frac{1}{2} \begin{pmatrix} \left(1 + \frac{k_z^0}{k_z^h}\right) \exp\left(ik_z^0 d\right) & \left(1 - \frac{k_z^0}{k_z^h}\right) \exp\left(-ik_z^0 d\right) \\ \left(1 - \frac{k_z^0}{k_z^h}\right) \exp\left(ik_z^0 d\right) & \left(1 + \frac{k_z^0}{k_z^h}\right) \exp\left(-ik_z^0 d\right) \end{pmatrix}, \tag{3.55}$$

where $k_z^0 = \sqrt{(\omega n_0)^2 - k_x^2}$. Substitution of (3.55) into (3.54) and elimination of a constant α leads to a simple requirement on a size of a defect layer:

$$k_z^0 d = \pi p, \quad p \subset \text{integer.} \tag{3.56}$$

A similar expression for the size of a defect layer can also be derived for TM polarization. This is a very general and useful result, which allows the design of planar photonic waveguides that support both TE and TM modes having the same desired effective refractive index $n_{eff} < n_{h,l}$. In particular, by choosing multilayer and core layer thicknesses as:

$$d_{l,h} = \frac{\lambda_0}{4\sqrt{n_{l,h}^2 - n_{eff}^2}}; d_0 = \frac{\lambda_0 p}{2\sqrt{n_0^2 - n_{eff}^2}}, \quad p \subset \text{integer,} \tag{3.57}$$

from the consideration of this section we conclude that such a waveguide will support both TE and TM polarized modes having the same propagation constant $k_x = 2\pi n_{eff}/\lambda_0$, and electromagnetic fields decaying exponentially fast away from the defect layer (waveguide core).

We conclude this chapter by considering two particular implementations of a photonic crystal slab waveguide described by (3.57). In the first case we introduce a defect in the high-refractive-index layer. Multilayer thicknesses are chosen according to (3.57), where $\lambda_0 = 1$, $n_1 = 1.5$, $n_0 = n_h = 2.8$, $n_{eff} = 1.32$, and $d_0 = 2d_h$. Figure 3.15(a) presents a band diagram of such a multilayer. In this figure, gray regions correspond to a continuum of delocalized states, which virtually coincide with the bands of a corresponding infinite periodic multilayer. Additionally, within bandgap regions, one finds several discrete states (thick lines in Fig. 3.15(a)). These are the guided-defect states. By design, one of the states (point (A) in Fig. 3.15(a)) has an effective refractive index of $n_{eff} = 1.32$ at $\lambda_0 = 1$. The inset (A) in Fig. 3.15(a) presents the $|E_y|$ field distribution across the multilayer, from which one observes that mode (A) is concentrated mostly in the defect-layer region.

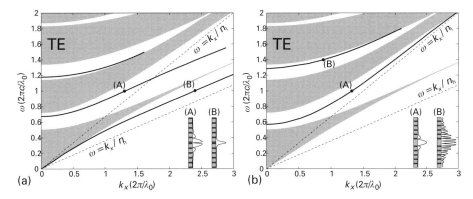

Figure 3.15 Band diagrams of guided states in a periodic multilayer with a single-layer defect. Solid thick lines – guided defect states localized in the vicinity of a defect layer. Gray regions – continuum of delocalized states in the multilayer. (a) High-refractive-index defect. Dispersion relation (A) is that of a bandgap-guided mode, while dispersion relation (B) is that of a fundamental total internal reflection guided mode. (b) Low-refractive-index defect. Both dispersion relations (A) and (B) are of the bandgap-guided modes. Judging by the field distribution, mode (A) can be considered as a fundamental defect mode, while mode (B) can be considered as a higher-order defect mode.

Moreover, corresponding modal fields are decaying exponentially into a surrounding periodic cladding, because of bandgap confinement. As discussed at the beginning of this section, a high-refractive-index defect is also expected to support a total-internal-reflection guided mode with $n_l < n_{\text{eff}} < n_h$. Indeed, a fundamental total-internal-reflection guided mode (labeled (B) in Fig. 3.15(a)) can be found below the light line of a low-refractive-index material $n_{\text{eff}} > n_l$. The inset (B) in Fig. 3.15(a) presents the $|E_y|$ field distribution across the multilayer. From the figure it is clear that the mode is mostly localized in the defect layer, with the corresponding fields decaying exponentially fast into the cladding, driven by the exponential field decay in the low-refractive-index layers. It is interesting to note that bandgap-guided mode (B) at higher frequencies, $\omega > 1.15$, crosses over the light line of a low-refractive-index material and, therefore, becomes total-internal-reflection guided.

Finally, Fig. 3.15(b) presents a band diagram of a multilayer with a low-refractive-index defect. The multilayer thicknesses are chosen according to (3.57), where $\lambda_0 = 1$, $n_0 = n_l = 1.5$, $n_h = 2.8$, $n_{\text{eff}} = 1.32$, and $d_0 = 2d_l$. In this case, the continuum of delocalized states is virtually identical with that for the multilayer with a high-refractive-index defect. Additionally, within the bandgap regions, one finds several discrete states (thick lines in Fig. 3.15(b)). These are, again, the guided-defect states. By design, one of the states (point (A) in Fig. 3.15(b)) has an effective refractive index of $n_{\text{eff}} = 1.32$ at $\lambda_0 = 1$. The inset (A) in Fig. 3.15(b) presents the $|E_y|$ field distribution across the multilayer, from which one observes that mode (A) is concentrated mostly inside a defect layer. Moreover, the corresponding modal fields are decaying exponentially into the periodic cladding, because of bandgap guidance. Inset (B) in Fig. 3.15(b) presents the $|E_y|$ field distribution across the multilayer for a higher-frequency defect mode. Although the mode is mostly localized in the defect-layer region, the modal fields exhibit a larger number of oscillations than the fields of mode (A), therefore mode (B) can be considered as a higher-order mode, while mode (A) can be considered as a fundamental defect mode. Finally, unlike the case of a multilayer with a high-refractive-index defect, there is no total-internal-reflection guided defect mode in a multilayer with a low-refractive-index defect.

3.7   Problems

3.1   Guidance in a low-refractive-index-contrast slab waveguide

Consider a slab of refractive index $n_0 + \delta n$ and thickness a surrounded by a cladding with refractive index n_0. Using (3.25) and definitions of TE, TM transfer matrices (3.4), (3.9), find the effective refractive index of a fundamental mode $n_{\text{eff}} = k_x c/\omega$ as a function of the frequency of operation ω in the long wavelength regime $\omega \ll c/(a\sqrt{2n_0\delta n})$, assuming low-refractive-index contrast $\delta n/n_0 \ll 1$. Express your answer in terms of $\delta n_{\text{eff}} = n_{\text{eff}} - n_0$. Show that the TE and TM modes are nearly degenerate in this case. In all your derivations retain only the leading-order terms in $\delta n/n_0$, $\delta n_{\text{eff}}/n_0$. Also, assume that $\delta n_{\text{eff}} \ll \delta n$ and verify whether in a long-wavelength regime this assumption holds.

3.2   Surface defect states in perturbed semi-infinite periodic multilayers

In this problem we consider TE modes of a semi-infinite quarter-wave stack with a modified first high-index layer facing the cladding. We will demonstrate that when the thickness of the first layer is varied from its quarter-wave value, a modified multilayer could support a surface state localized primarily at the cladding–multilayer interface. A periodic multilayer stack is designed as a quarter-wave stack (3.24) for the normal angle of incidence $n_l = 1.5; n_h = 2.8; \lambda_0 = 1; \theta_d = 0$, where $d_{h,l} = \lambda_0/4n_{h,l}$. The cladding is assumed to be air, $n_0 = 1.0$.

(a) We start by practicing interpretation of the band diagrams. In Fig. P3.2.1 we present a band diagram of delocalized states of an infinite periodic multilayer stack with parameters defined in the introduction of the problem. The band diagram of the photonic states of a corresponding semi-infinite multilayer will be modified by addition of the discrete defect states. The region of spatial localization of a defect state can be judged from its position in the band diagram. In Fig. P3.2.1 we mark, schematically, five different possibilities for the position of a defect mode. For every point 1–5, determine whether the corresponding mode is propagating mostly in the air, in the reflector, both in the air and in the reflector, or neither in the air nor in the reflector (mode confined to the interface). What is the difference between modes 4 and 5?

(b) Using transfer matrix theory, find an equation, the solution of which gives the propagation constant of the mode 4 localized at the multilayer–cladding interface.

(c) In Fig. P3.2.2 we present a band diagram of the modes of a semi-infinite quarter-wave stack with a high-refractive-index defect layer of thickness $d_l = 1.5d_h$. A qualitative

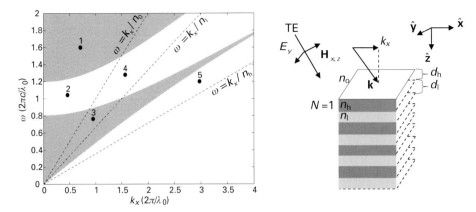

Figure P3.2.1 Schematic of a semi-infinite quarter-wave stack designed for a normal angle of incidence and a corresponding band diagram of the delocalized states in the multilayer. Also presented are the light lines of the multilayer materials and the cladding.

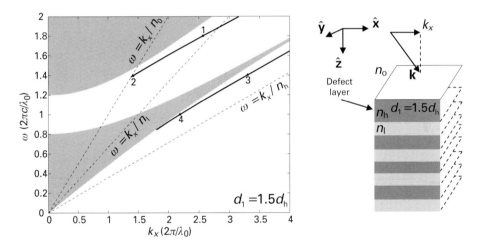

Figure P3.2.2 Band diagram of the modes of a semi-infinite multilayer stack with a high-refractive-index surface defect $d_1 = 1.5d_h$. Solid lines correspond to the dispersion relations of the guided surface (defect) states.

understanding of the field distributions in the guided surface states can be obtained directly from the form of a band diagram.

In Fig. P3.2.2, consider four points marked 1–4 corresponding to the guided surface modes.

(i) First, for the points 1 and 2, using the results of question (a), make qualitative drawings of the modal electric field amplitude $|E_y(z)|^2$ along the z direction in the vicinity of a cladding–multilayer interface. Pay special care to highlighting the difference between field distributions at the two points, and to reflecting the fact that when moving from point 1 toward point 2 along the same dispersion relation curve, one gets closer to the cladding light line. Justify your answer in terms of the value of the $k_z = \sqrt{\omega^2 n_0^2 - k_x^2}$ coefficient inside the air region and the $\exp(ik_z z)$ dependence of the modal fields.

(ii) Second, for the points 3 and 4, using the results of question (a), make qualitative drawings of the modal electric field amplitude $|E_y(z)|^2$ along the z direction in the vicinity of a cladding–multilayer interface. Pay special care to highlighting the difference between the field distributions at the two points, and to reflecting the fact that when moving from point 3 toward point 4 along the same dispersion relation curve, one gets closer to the continuum of the delocalized states in the multilayer. Justify your answer in terms of the distance from the band edge and the fact that the lowest eigenvalue of the bilayer transfer matrix behaves as $1 - |\lambda_{<1}| \sim |\omega - \omega_{\text{band edge}}|^\alpha$ near the band edge.

(d) In this final question we consider the effect of adding weak periodic perturbation of a dielectric constant in the defect layer region at the air–multilayer interface (Fig. P3.2.3). The periodic perturbation of spatial period a is assumed to be vanishingly small so that dispersion relations of the modes are not modified. As was described earlier in the chapter, in the case of a vanishingly small periodic perturbation, the band diagram of a periodic system can be constructed from the band diagram of an unperturbed system (Fig. P3.2.2) by introducing the Brillouin zone $k_x \in [-\pi/a]$ and reflecting all the dispersion curves of an unperturbed band diagram into the first Brillouin zone. For the following questions we choose six points at various locations in the band diagram, and choose the periodicity in such a way as to ensure

$$k_{1,z} - \pi/a = \pi/a - k_{2,x}, k_{3,z} - \pi/a = \pi/a - k_{4,x},$$
$$k_{5,z} - \pi/a = \pi/a - k_{6,x} \qquad \text{(see Fig. P3.2.3)}.$$

(i) Using the band diagram of an unperturbed system in Fig. P3.2.3 as a starting point, draw the band diagram of a system with vanishingly small perturbation.

(ii) In the unperturbed band diagram, mode 1 was inside the continuum of bulk states of a multilayer, while mode 2 was a defect state localized at the cladding–multilayer interface. After introduction of a periodic variation, will mode 2 still remain localized at the surface defect state? Describe in what spatial regions the fields of mode 2 will be predominantly distributed.

(iii) In the unperturbed band diagram, mode 3 was a defect state localized at the cladding–multilayer interface, while mode 4 was located inside the continuum of bulk states of a multilayer and above the light line of air. After introduction of a periodic variation will point 3 still remain a localized defect state? Describe in what spatial regions the fields of mode 3 will be predominantly distributed.

(iv) In the unperturbed band diagram, mode 5 was inside the continuum of bulk states of a multilayer and below the light line of air, while mode 4 was located inside the continuum of bulk states of a multilayer and above the air light line. After introduction of a periodic variation, describe in what spatial regions the fields of mode 5 will be predominantly distributed.

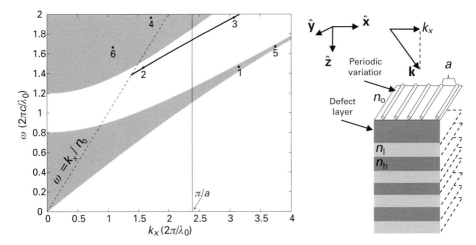

Figure P3.2.3 Schematic of a semi-infinite quarter-wave stack with a high-refractive-index layer defect and periodic perturbation at the cladding–multilayer interface. Band diagram of the modes of an unperturbed system. The position of a boundary of a first Brillouin zone is $k_x = \pi/a$.

3.3   Bandgaps of a nonquarter-wave stack multilayer

In this problem we consider electromagnetic modes propagating perpendicular to the infinite periodic-multilayer stack. As demonstrated in Section 3.5, modes of such a system can be characterized by the propagation constant k_z confined to the first Brillouin zone $k_z \subset [-\pi/a, \pi/a]$, where a is the bilayer thickness. Moreover, for the quarter-wave stack design with layer thicknesses $d_{h,l}^0 = \lambda_0/4n_{h,l}$, a band diagram of the guided modes presents an infinite collection of discrete bands separated by the bandgap regions centered on the odd harmonics of $\omega_0 = 2\pi c/\lambda_0$ (see Fig. P3.3.1(a)). In the case when the multilayer differs from a quarter-wave stack, we will demonstrate that a new set of bandgaps appears; these are centered on even harmonics of ω_0 (see Fig. P3.3.1(b)). In particular, assume that layer thicknesses are somewhat perturbed from their quarter-wave values $d_h = d_h^0(1 + \delta)$, $d_l = d_l^0(1 - \delta)$, and $\delta \ll 1$. Using (3.33), and the definition of phases ϕ_h, ϕ_l in (3.16), find the sizes of the bandgaps centered around $2p$ even harmonics of ω_0.

3.4   Dispersion relation of a core guided mode in a photonic crystal slab waveguide

In this problem, we derive the dispersion relation of a TE-polarized core-guided mode of a photonic crystal slab waveguide (see Fig. 3.14) in the vicinity of a cladding bandgap center frequency. As a waveguide, we consider a quarter-wave stack with refractive

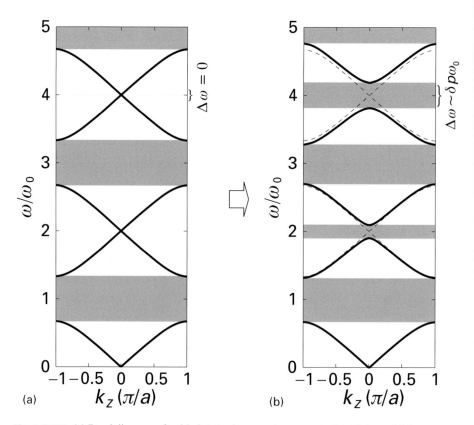

Figure P3.3.1 (a) Band diagram of guided states in a quarter-wave stack with layer thicknesses $d_{h,l}^0 = \lambda_0/4n_{h,l}$, discrete bands of guided modes are separated by the bandgap regions centered on the odd harmonics of ω_0. Example of $n_l = 1.5$; $n_h = 4.5$; $\lambda_0 = 1$. (b) When the periodic multilayer differs from a quarter-wave stack a new set of bandgaps appears, which are centered on the even harmonics of ω_0. Example of $n_l = 1.5$; $n_h = 4.5$; $\lambda_0 = 1$, and $d_h = d_h^0(1+\delta)$, $d_l = d_l^0(1-\delta)$, $\delta = 0.1$.

indices n_h, n_l and thicknesses $d_{h,l} = \lambda_0/4\sqrt{n_{h,l}^2 - n_{\text{eff}}^2}$, where $n_{\text{eff}} < n_l < n_h$. The core layer, of refractive index n_0, is chosen to have a thickness $d_0 = \lambda_0/2\sqrt{n_0^2 - n_{\text{eff}}^2}$. As was established in Section 3.6, at $\omega_0 = 2\pi c/\lambda_0$, a waveguide defined this way supports a core-guided mode with propagation constant $k_x(\omega_0) = n_{\text{eff}}\omega_0$. Find the dispersion relation of a core-guided mode in the vicinity of ω_0. Express your answer in the form $k_x(\omega_0 + \delta\omega) = n_{\text{eff}}\omega_0 + v_g^{-1}\delta\omega + O(\delta\omega^2)$, and find an expression for the core-mode group velocity v_g^{-1} at ω_0. Check the validity of your expression for a particular case $n_0 = n_1$, for which $v_g = (n_{\text{eff}}(1 + r_0^3))/(n_h^2 r_0^3 + n_l^2)$; $r_0 = \sqrt{n_l^2 - n_{\text{eff}}^2}/\sqrt{n_h^2 - n_{\text{eff}}^2}$. Finally, demonstrate that in the case $n_0 = n_1$, at ω_0 the mode group velocity is always smaller than the mode phase velocity $v_p = n_{\text{eff}}^{-1}$.

As a guide to a solution of this complex problem we suggest the following approach. In Section 3.6 it was established that the propagation constant $k_x(\omega)$ of a core guided mode of a photonic crystal slab waveguide satisfies (3.54):

$$M_{dh} M_{hd} \overline{V}_{hlh}^{>1} = \alpha \overline{V}_{hlh}^{<1},$$

where $\overline{V}_{hlh}^{>1}, \overline{V}_{hlh}^{<1}$ are the eigenvectors of a bilayer transfer matrix M_{hlh} (3.17), and α is some constant. Moreover, the $\overline{V}_{hlh}^{>1}, \overline{V}_{hlh}^{<1}$ eigenvectors correspond to the eigenvalues larger and smaller in magnitude than one, respectively. Therefore,

(a) find the eigenvalues and corresponding eigenvectors of a bilayer transfer matrix:

$$M_{hlh} = \frac{1}{4} \begin{pmatrix} (1+r)\exp(i\phi_l) & (1-r)\exp(-i\phi_l) \\ (1-r)\exp(i\phi_l) & (1+r)\exp(-i\phi_l) \end{pmatrix} \times$$

$$\begin{pmatrix} (1+r^{-1})\exp(i\phi_h) & (1-r^{-1})\exp(-i\phi_h) \\ (1-r^{-1})\exp(i\phi_h) & (1+r^{-1})\exp(-i\phi_h) \end{pmatrix} \qquad \text{(P3.4.2)}$$

for the frequency $\omega_0 + \delta\omega$, assuming $|\delta\omega| \ll \omega_0$. In your calculations, keep only the terms of the first order in $\delta\omega$. Also, use $k_x(\omega_0 + \delta\omega) = n_{eff}\omega_0 + v_g^{-1}\delta\omega + O(\delta\omega^2)$, as well as the following expansions:

$$= \frac{k_z^l(\omega)}{k_z^h(\omega)} = \frac{\sqrt{(n_l(\omega_0 + \delta\omega))^2 - k_x^2(\omega)}}{\sqrt{(n_h(\omega_0 + \delta\omega))^2 - k_x^2(\omega)}} = \frac{\sqrt{(n_l(\omega_0 + \delta\omega))^2 - \left(n_{eff}\omega_0 + v_g^{-1}\delta\omega\right)^2}}{\sqrt{(n_h(\omega_0 + \delta\omega))^2 - \left(n_{eff}\omega_0 + v_g^{-1}\delta\omega\right)^2}}$$

$$r = r_0 + C_r\delta; \delta = \frac{\delta\omega}{\omega_0}; r_0 = \frac{\sqrt{n_l^2 - n_{eff}^2}}{\sqrt{n_h^2 - n_{eff}^2}} < 1; C_r = \frac{n_{eff}\left(n_h^2 - n_l^2\right)\left(n_{eff} - v_g^{-1}\right)}{\sqrt{n_l^2 - n_{eff}^2}\left(\sqrt{n_h^2 - n_{eff}^2}\right)^3}$$

$$\phi_h = d_h k_z^h(\omega) = d_h\sqrt{(n_h(\omega_0 + \delta\omega))^2 - k_x^2(\omega)} = d_h\sqrt{(n_h(\omega_0 + \delta\omega))^2 - \left(n_{eff}\omega_0 + v_g^{-1}\delta\omega\right)^2}$$

$$\phi_h = \frac{\pi}{2} + C_h\delta; C_h = \frac{\pi}{2}\frac{n_h^2 - n_{eff}v_g^{-1}}{n_h^2 - n_{eff}^2}$$

$$\phi_l = \frac{\pi}{2} + C_l\delta; C_l = \frac{\pi}{2}\frac{n_l^2 - n_{eff}v_g^{-1}}{n_l^2 - n_{eff}^2}. \qquad \text{(P3.4.3)}$$

(b) find an explicit form of the $M_{dh} M_{hd}$ transfer matrix, keeping only the terms of the first order in $\delta\omega$, and using the following definitions:

$$M_{dh} M_{hd} = \frac{1}{4} \begin{pmatrix} (1+r_d)\exp(i\phi_d) & (1-r_d)\exp(-i\phi_d) \\ (1-r_d)\exp(i\phi_d) & (1+r_d)\exp(-i\phi_d) \end{pmatrix} \times$$

$$\begin{pmatrix} \left(1+r_d^{-1}\right)\exp(i\phi_h) & \left(1-r_d^{-1}\right)\exp(-i\phi_h) \\ \left(1-r_d^{-1}\right)\exp(i\phi_h) & \left(1+r_d^{-1}\right)\exp(-i\phi_h) \end{pmatrix} \qquad \text{(P3.4.4)}$$

$$r_{\mathrm{d}} = \frac{k_z^{\mathrm{d}}(\omega)}{k_z^{\mathrm{h}}(\omega)} = \frac{\sqrt{(n_0(\omega_0 + \delta\omega))^2 - k_x^2(\omega)}}{\sqrt{(n_{\mathrm{h}}(\omega_0 + \delta\omega))^2 - k_x^2(\omega)}} \tag{P3.4.5}$$

$$r_{\mathrm{d}} = r_{\mathrm{d}0} + C_{\mathrm{d}}\delta; \delta = \frac{\delta\omega}{\omega_0}; r_{\mathrm{d}0} = \frac{\sqrt{n_0^2 - n_{\mathrm{eff}}^2}}{\sqrt{n_{\mathrm{h}}^2 - n_{\mathrm{eff}}^2}}; C_{\mathrm{d}} = \frac{n_{\mathrm{eff}}\left(n_{\mathrm{h}}^2 - n_0^2\right)\left(n_{\mathrm{eff}} - v_{\mathrm{g}}^{-1}\right)}{\sqrt{n_0^2 - n_{\mathrm{eff}}^2}\left(\sqrt{n_{\mathrm{h}}^2 - n_{\mathrm{eff}}^2}\right)^3}$$

$$\phi_{\mathrm{d}} = d_0 k_z^0(\omega) = d_0 \sqrt{(n_0(\omega_0 + \delta\omega))^2 - k_x^2(\omega)}; \phi_{\mathrm{d}} = \pi + C_0\delta; C_0 = \pi \frac{n_0^2 - n_{\mathrm{eff}}v_{\mathrm{g}}^{-1}}{n_0^2 - n_{\mathrm{eff}}^2}.$$

(c) finally, solve (3.54) and find an expression for v_{g}^{-1}. Apply the general formula to a particular case of $n_0 = n_1$, and then demonstrate that at ω_0 mode group velocity v_{g} is always smaller than the mode phase velocity $v_{\mathrm{p}} = n_{\mathrm{eff}}^{-1}$.

4 Bandgap guidance in planar photonic crystal waveguides

In this section we discuss in more detail the guiding properties of photonic bandgap waveguides, where light is confined in a low refractive index core. We first describe guidance of TE and TM waves in a waveguide featuring an infinite periodic reflector operating at a frequency in the center of a bandgap. In this case, radiation loss from the waveguide core is completely suppressed. We then use perturbation theory to characterize modal propagation loss due to absorption losses of the constitutive materials. Finally, we characterize radiation losses when the confining reflector contains a finite number of layers.

Figure 4.1 presents a schematic of a waveguide with a low refractive index core surrounded by a periodic multilayer reflector. The analysis of guided states in such a waveguide is similar to the analysis of defect states presented at the end of Section 3. As demonstrated in that section, a core of low refractive index n_c surrounded by a quarter-wave reflector can support guided modes with propagation constants above the light line of the core refractive index $k_x \subset [0, \omega n_c]$. Note that a core state with $k_x = 0$ defines electromagnetic oscillations perpendicular to a multilayer plane, therefore such a state is a Fabry–Perot resonance rather than a guided mode. In the opposite extreme, a core state with $k_x \sim \omega n_c$ defines a mode propagating at grazing angles with respect to the walls of a waveguide core, which is typical of the lowest-loss leaky mode of a large-core photonic bandgap waveguide. Core-guided modes of the large core-diameter photonic bandgap waveguides constitute the focus of this chapter.

4.1   Design considerations of waveguides with infinitely periodic reflectors

In the case of an infinite reflector surrounding a low refractive index core, core-guided modes with frequencies inside a photonic bandgap exhibit no radiation loss. Moreover, fields of such modes decay exponentially into the periodic reflector owing to photonic bandgap confinement. According to (3.52), the vector of expansion coefficients in the high-refractive-index layer of the bilayer $-N$ is proportional to the eigenvector of a bilayer transfer matrix with corresponding eigenvalue greater than one. In particular, defining the core layer to be zero, and the high index layer in the $-N - 1$ bilayer to be $-2N - 1$ (see Fig. 4.1), then the following holds:

$$\begin{pmatrix} A_{-2N-1} \\ B_{-2N-1} \end{pmatrix} = \alpha_{-2N-1} \overline{V}_{hlh}^{>1}. \tag{4.1}$$

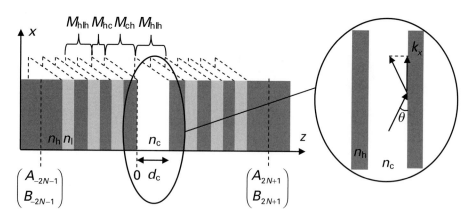

Figure 4.1 Lower-refractive-index core photonic-bandgap waveguide. The core, of size d_c and refractive index n_c, is surrounded on each side by a periodic reflector containing N bilayers of refractive indices n_1, n_h, such that $n_c \leq n_1 < n_h$. The two layers closest to the core, as well as the waveguide cladding, are of high refractive index n_h.

In the opposite direction from the core, the fields are also decaying exponentially, thus only an eigenvector with a magnitude smaller than one is excited:

$$M_{\text{ch}} M_{\text{hc}} V_{\text{hlh}} \begin{pmatrix} \lambda_{<1}^N & 0 \\ 0 & \lambda_{>1}^N \end{pmatrix} V_{\text{hlh}}^{-1} \begin{pmatrix} A_{-2N-1} \\ B_{-2N-1} \end{pmatrix} = \alpha_{2N+1} \overline{V}_{\text{hlh}}^{<1}, \qquad (4.2)$$

where α_{-2N-1} and α_{2N+1} are some constants. Substitution of (4.2) into (4.1) leads to the following equation:

$$M_{\text{ch}} M_{\text{hc}} \overline{V}_{\text{hlh}}^{>1} = \alpha \overline{V}_{\text{hlh}}^{<1}, \qquad (4.3)$$

where α is some constant. Assuming the following definitions for the components of the wave vector in terms of an angle of modal propagation inside a waveguide core (see Fig. 4.1):

$$k_x = \omega\sqrt{\varepsilon_c}\cos(\theta); \, k_z^c = \omega\sqrt{\varepsilon_c}\sin(\theta)$$

$$k_z^{n_h} = \omega\sqrt{\varepsilon_h - \varepsilon_c \cos^2(\theta)}; \, k_z^{n_1} = \omega\sqrt{\varepsilon_1 - \varepsilon_c \cos^2(\theta)}, \qquad (4.4)$$

and using the explicit form of the defect layer transfer matrices:

$$M_{\text{hc}} = \frac{1}{2}\begin{pmatrix} (1+r_c^{-1})\exp(i\phi_h) & (1-r_c^{-1})\exp(-i\phi_h) \\ (1-r_c^{-1})\exp(i\phi_h) & (1+r_c^{-1})\exp(-i\phi_h) \end{pmatrix}$$

$$M_{\text{ch}} = \frac{1}{2}\begin{pmatrix} (1+r_c)\exp(i\phi_c) & (1-r_c)\exp(-i\phi_c) \\ (1-r_c)\exp(i\phi_c) & (1+r_c)\exp(-i\phi_c) \end{pmatrix}$$

$$r_c = r_c^{\text{TE}}, r_c^{\text{TM}}; \quad r_c^{\text{TE}} = \frac{k_z^c}{k_z^h}; \quad r_c^{\text{TM}} = \frac{\varepsilon_h k_z^c}{\varepsilon_c k_z^h}$$

$$\phi_c = (k_z^c d_c); \quad \phi_h = (k_z^h d_h); \quad \phi_1 = (k_z^1 d_1), \qquad (4.5)$$

(4.3) can be resolved analytically for the case of a quarter-wave reflector.

In particular, assuming that reflector geometry is chosen to satisfy the quarter-wave condition $\phi_h = \phi_l = \pi/2$ at frequency $\omega = 2\pi/\lambda$, then from (3.36) $\overline{V}_{hlh}^{<1} = (-1, 1)/2; \overline{V}_{hlh}^{>1} = (1, 1)/2$, and, similarly to (3.56), a simple requirement on a core size follows from (4.3):

$$\phi_c = k_z^c d_c = \pi p; \quad \text{where } p \text{ is an integer,} \qquad (4.6)$$

which is valid for any polarization of a core guided mode. In terms of individual layer thicknesses, this constitutes:

$$d_{l,h} = \frac{\lambda}{4\sqrt{n_{l,h}^2 - n_c^2 \cos^2(\theta_c)}}; \quad d_c = \frac{\lambda p}{2 n_c \sin(\theta_c)}. \qquad (4.7)$$

In the following, we consider in more detail the case of $p = 1$. From (4.6) we can conclude that for a given core size $d_c \geq \lambda/2n_c$ there always exists a bandgap-guided core mode with an effective propagation angle θ_c:

$$\sin(\theta_c) = \frac{\lambda}{2 n_c d_c} \xrightarrow[d_c \gg \lambda]{} \theta_c \simeq \frac{\lambda}{2 n_c d_c}, \qquad (4.8)$$

and effective refractive index $n_{\text{eff}} = n_c \cos(\theta_c)$, given that the surrounding multilayer reflector is quarter-wave.

Note that when the core size is much larger than the wavelength of light, the angle of propagation of a core-guided fundamental mode approaches zero inversely proportional to the core size (grazing angles of incidence onto a reflector). In this regime, the core mode can be envisioned as a plane wave propagating almost parallel to the waveguide center line, while exhibiting infrequent bounces from the reflector walls every $L_1 \underset{d_c \gg \lambda}{=} d_c/\theta_c = 2 n_c d_c^2/\lambda$ units of length (see Fig. 4.2). Practically, there are no perfect reflectors, causing the guided mode to lose its energy at every bounce. Such energy loss is typically a result of either absorption loss of the reflector materials, or scattering from imperfections at the multilayer interfaces. Assuming that the power loss of a core mode per single reflection is t, then, after each reflection only $1 - t$ of modal power will continue propagating. The modal propagation loss after a distance L can, therefore, be written as $P(L)/P(0) = (1 - t)^{L/L_1}$. One frequently defines propagation loss α in units of [dB/m] as:

$$\alpha[\text{dB/m}] = -\frac{10}{L} \log_{10}\left(\frac{P(L)}{P(0)}\right). \qquad (4.9)$$

If the loss per single reflection is small $t \ll 1$, the expression for the propagation loss of a core guided mode of a photonic-bandgap waveguide can be approximated as:

$$\alpha[\text{dB/m}] = -\frac{10}{L_1} \log_{10}(1 - t) \underset{\substack{t \ll 1 \\ \lambda/d_c \ll 1}}{=} \frac{5}{\ln(10)} \frac{\lambda t}{n_c d_c^2}. \qquad (4.10)$$

Moreover, for grazing angles $\theta_c \sim 0$ of modal incidence on the reflector, from the expressions of Fresnel's reflection coefficients (3.12), and from Fig. 3.3, one finds that $t \sim \theta_c$, and finally, $\alpha[\text{dB/m}] \underset{\lambda/d_c \ll 1}{\sim} \lambda^2/d_c^3$.

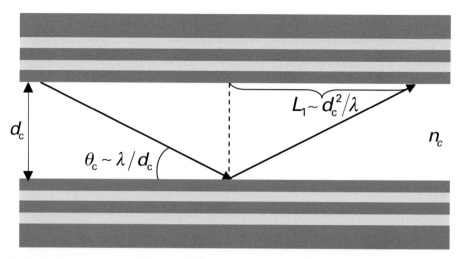

Figure 4.2 When the waveguide core size is much larger than the wavelength of guided light, the angle of incidence of a fundamental mode onto the reflector approaches zero. In this regime, light propagation in the core can be envisioned as infrequent bouncing of a plane wave from the reflector walls every L_1 units of length.

4.2   Fundamental TE mode of a waveguide with infinitely periodic reflector

In this section, we study the field distribution in the fundamental TE mode of a lower-refractive-index core waveguide. We start by finding field expansion coefficients in the waveguide layers. Assuming a quarter-wave reflector $\phi_h = \phi_l = \pi/2$, $\phi_c = \pi$ condition for the fundamental core mode, and the fact that for TE polarization $|r_{TE}|$, as defined in (3.16), is always less than or equal to one, from (3.17), (3.21), (3.36), (4.1), and (4.5) we can find the coefficients in each of the layers (see Fig. 4.3):

$$\begin{pmatrix} A \\ B \end{pmatrix}_0^c = \frac{1}{2} \begin{pmatrix} 1 \\ -1 \end{pmatrix}$$

$$\begin{pmatrix} A \\ B \end{pmatrix}_{2n+1}^h = \begin{pmatrix} A \\ B \end{pmatrix}_{-2n-2}^l = \frac{r_c \left(-r_{TE}\right)^n}{2} \begin{pmatrix} -1 \\ 1 \end{pmatrix}$$

$$\begin{pmatrix} A \\ B \end{pmatrix}_{-2n-1}^h = \begin{pmatrix} A \\ B \end{pmatrix}_{2n+2}^l = \frac{r_c \left(-r_{TE}\right)^n}{2i} \begin{pmatrix} 1 \\ 1 \end{pmatrix}$$

$$n \subset \text{integer} \geq 0; \quad r_c = r_c^{TE} = \frac{k_z^c}{k_z^h}; \quad r^{TE} = \frac{k_z^l}{k_z^h}. \tag{4.11}$$

The expansion coefficients are chosen in such a way as to normalize to unity the maximum value of the electric field: $\max |E_y(z)| = 1$.

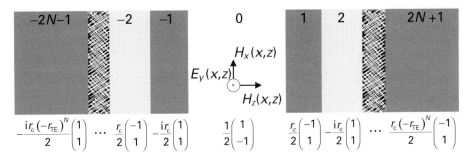

Figure 4.3 Expansion coefficients in the reflector layers for the TE-polarized fundamental mode.

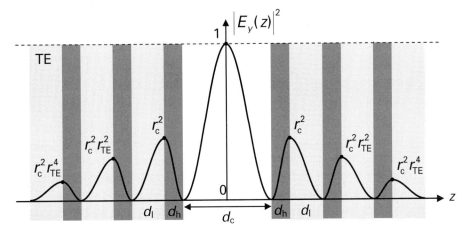

Figure 4.4 Schematic of the $|E_y(z)|^2$ field intensity distribution of the TE-polarized core mode at the center frequency of a quarter-wave reflector bandgap.

In Fig. 4.4, in each layer, we plot intensities of the only nonzero component of the electric field $|E_y(z)|^2$ as calculated by (3.37). In each of the reflector layers the field intensities $|E_y(z)|^2$ are proportional to $r_c^2 \cos^2(\pi \eta_{h,l}/2)$ or $r_c^2 \sin^2(\pi \eta_{h,l}/2)$, where $\eta_{h,l} = (z - z_{\text{left interface}})/d_{h,l}$, while in the core layer the field intensity is $\cos^2(\pi \eta_c)$, where $\eta_c = z/d_c$. Note that the factor

$$r_c = \frac{\sqrt{\varepsilon_c}\sin(\theta_c)}{\sqrt{\varepsilon_h - \varepsilon_c \cos^2(\theta_c)}} \underset{d_c \gg \lambda}{\simeq} \frac{1}{\sqrt{\varepsilon_h - \varepsilon_c}}\left(\frac{\lambda}{2d_c}\right), \qquad (4.12)$$

thus amplitudes of the electric field at the core boundaries are exactly zero, while inside the reflector they are inversely proportional to the core size.

We now calculate the amount of the electric energy in the reflector and its fraction compared to the total electric energy of the mode:

$$E_{total}^{el} = \frac{1}{2} \int\limits_{-\infty}^{+\infty} dz \varepsilon(z) |E_y(z)|^2$$

$$= \frac{1}{2} \int\limits_{-d_c/2}^{d_c/2} dz \varepsilon_c \cos^2(\pi \eta_c) + \sum_{j=0}^{+\infty} r_c^2 r_{TE}^{2j} \left(\int\limits_{0}^{d_h} dz \varepsilon_h \sin^2\left(\frac{\pi}{2}\eta_h\right) + \int\limits_{0}^{d_l} dz \varepsilon_l \cos^2\left(\frac{\pi}{2}\eta_l\right) \right)$$

$$= \frac{1}{2}\left[\frac{\varepsilon_c d_c}{2} + (\varepsilon_h d_h + \varepsilon_l d_l) \frac{r_c^2}{1 - r_{TE}^2} \right], \tag{4.13}$$

and thus:

$$\frac{E_{reflector}^{el}}{E_{total}^{el}} = \frac{(\varepsilon_h d_h + \varepsilon_l d_l) \dfrac{r_c^2}{1 - r_{TE}^2}}{\dfrac{\varepsilon_c d_c}{2} + (\varepsilon_h d_h + \varepsilon_l d_l) \dfrac{r_c^2}{1 - r_{TE}^2}}$$

$$\simeq \begin{cases} \left(\dfrac{\lambda}{2d_c}\right)^3 \left(\dfrac{\varepsilon_h/\varepsilon_c}{\sqrt{\varepsilon_h - \varepsilon_c}} + \dfrac{\varepsilon_l/\varepsilon_c}{\sqrt{\varepsilon_l - \varepsilon_c}}\right) \dfrac{1}{\varepsilon_h - \varepsilon_l}; \begin{pmatrix} \theta_c \to 0 \\ d_c \gg \lambda \end{pmatrix} \\ \\ \dfrac{n_c}{n_c + n_h - n_l}; \begin{pmatrix} \theta_c \to \pi/2 \\ d_c \to \lambda/2n_c \end{pmatrix} \end{cases}, \tag{4.14}$$

from which we conclude that, with increasing core size, the fraction of modal energy in the reflector decreases very fast, and is inversely proportional to the third power of the core size. If the materials of the reflector exhibit absorption losses, for example, and assuming that a core material is lossless, the propagation loss of the core mode will be proportional to the electric energy fraction in the reflector, thus decreasing as the inverse third power of the core size.

4.3   Infinitely periodic reflectors, field distribution in TM modes

Depending upon the choice of the core dielectric constant, there are two distinct guiding regimes for the TM polarized fundamental core mode. In particular, if the core refractive index is low enough, the intensity distribution of the magnetic and electric fields in the waveguide core are Gaussian-like, similar to the field intensity distribution of the fundamental TE polarized core mode. In the case when the core refractive index is not low enough, the intensity distributions of the magnetic and electric fields in the waveguide core show minima in the core center, reaching their maxima at the core–reflector boundary.

4.3.1   Case of the core dielectric constant $\varepsilon_c < \varepsilon_h \varepsilon_l/(\varepsilon_h + \varepsilon_l)$

In this regime, as demonstrated at the beginning of Section 3.3, for any angle of radiation incidence from the uniform half space with dielectric constant ε_c, an

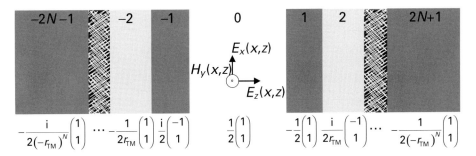

Figure 4.5 Expansion coefficients in the reflector layers for the TM-polarized fundamental mode, case of $\varepsilon_c < \varepsilon_h \varepsilon_1/(\varepsilon_h + \varepsilon_1)$.

infinite quarter-wave reflector designed for that incidence angle will completely reflect the incoming radiation. Proceeding as in Section 4.2, we first find the field expansion coefficients in the waveguide layers. Assuming a quarter-wave reflector $\phi_h = \phi_1 = \pi/2$, $\phi_c = \pi$ condition for the fundamental core mode, and the fact that for TM polarization $|r_{TM}|$, as defined in (3.16), is always greater than or equal to one, from (3.17), (3.22), (3.36), (4.1), and (4.5) we can find the coefficients in each of the layers (see Fig. 4.5):

$$\begin{pmatrix} A \\ B \end{pmatrix}_0^c = \frac{1}{2}\begin{pmatrix} 1 \\ 1 \end{pmatrix}$$

$$\begin{pmatrix} A \\ B \end{pmatrix}_{2n+1}^h = r_{TM}\begin{pmatrix} A \\ B \end{pmatrix}_{-2n-2}^1 = \frac{-1}{2\left(-r_{TM}\right)^n}\begin{pmatrix} 1 \\ 1 \end{pmatrix}$$

$$\begin{pmatrix} A \\ B \end{pmatrix}_{-2n-1}^h = r_{TM}\begin{pmatrix} A \\ B \end{pmatrix}_{2n+2}^1 = \frac{i}{2\left(-r_{TM}\right)^n}\begin{pmatrix} -1 \\ 1 \end{pmatrix}$$

$$n \subset \text{integer} \geq 0; \quad r^{TM} = \frac{\varepsilon_h k_z^1}{\varepsilon_1 k_z^h}. \tag{4.15}$$

For TM polarization and grazing angles of incidence $\theta_c \to 0$, from (3.6), (3.7), (4.4) it follows that in the core:

$$\left|E_x^{core}(z)\right| \underset{\theta_c \to 0}{\ll} \left|E_z^{core}(z)\right| \simeq \left|H_y^{core}(z)\right|/n_c. \tag{4.16}$$

As the electric field component $E_z(z)$ is discontinuous across the dielectric boundaries, and as $E_x(z)$ is not negligible in the reflector layers, it is more convenient to plot the continuous $H_y(z)$ field instead. In Fig. 4.6, a solid line shows the intensity of the only nonzero component of the magnetic field $|H_y(z)|^2$ as calculated by (3.37).

The dotted line plots the distribution of the electric field intensity $\varepsilon_c |E_z(z)|^2$ in the core region in the limit as $\theta_c \to 0$. Note that, unlike the case of the TE polarized mode, the electric field does not vanish at the core boundary; on the contrary, it reaches its maximum. For grazing angles of mode propagation, from (4.16) it follows that magnetic and electric energies in the core are close to each other. Using the fact that for harmonic

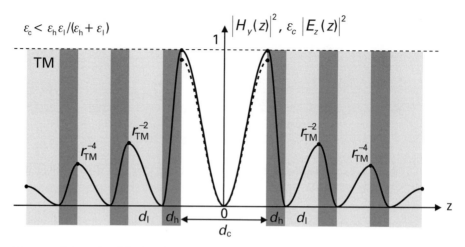

Figure 4.6 Schematic of the field intensity distribution of the TM-polarized mode at the center frequency of a quarter-wave reflector bandgap; $|H_y(z)|^2$, the field intensity of the fundamental core mode (solid); $\varepsilon_c|E_z(z)|^2$, the field intensity in the core, $\theta_c \to 0$ (dotted).

fields the total magnetic energy equals the total electric energy, we derive:

$$E_{total}^{el} = E_{total}^{mg} = \frac{1}{2} \int\limits_{-\infty}^{+\infty} dz \, |H_y(z)|^2$$

$$= \frac{1}{2} \int\limits_{-d_c/2}^{d_c/2} dz \, \sin^2(\pi \eta_c) + \int\limits_0^{d_h} dz \, \cos^2\left(\frac{\pi}{2}\eta_h\right)$$

$$+ \sum_{j=1}^{+\infty} r_{TM}^{-2j} \left(\int\limits_0^{d_l} dz \, \sin^2\left(\frac{\pi}{2}\eta_l\right) + \int\limits_0^{d_h} dz \, \cos^2\left(\frac{\pi}{2}\eta_h\right) \right)$$

$$= \frac{1}{2}\left[\frac{d_c}{2} + d_h + \frac{(d_l + d_h)}{r_{TM}^2 - 1} \right]. \tag{4.17}$$

Finally, the fraction of the electric energy in the waveguide reflector is then:

$$\frac{E_{reflector}^{el}}{E_{total}^{el}} = \frac{\left(d_h + \dfrac{d_l + d_h}{r_{TM}^2 - 1} \right)}{\dfrac{d_c}{2} + \left(d_h + \dfrac{d_l + d_h}{r_{TM}^2 - 1} \right)}$$

$$\simeq \left(\frac{\lambda}{2d_c} \right) \frac{\dfrac{\varepsilon_h^2(\varepsilon_l - \varepsilon_c)}{\sqrt{\varepsilon_h - \varepsilon_c}} + \dfrac{\varepsilon_l^2(\varepsilon_h - \varepsilon_c)}{\sqrt{\varepsilon_l - \varepsilon_c}}}{\left(\varepsilon_h^2 - \varepsilon_l^2 \right) \left(\dfrac{\varepsilon_h \varepsilon_l}{\varepsilon_h + \varepsilon_l} - \varepsilon_c \right)}; \left(\begin{matrix} \theta_c \to 0 \\ d_c \gg \lambda \end{matrix} \right), \tag{4.18}$$

signifying that the fraction of the electric energy of a TM-polarized mode in the reflector is only inversely proportional with the core size. Therefore, if the materials of the reflector exhibit absorption losses, the modal propagation loss due to modal energy absorption will also decrease as the inverse of core size.

4.3.2   Case of the core dielectric constant $\varepsilon_l \geq \varepsilon_c > \varepsilon_h \varepsilon_l/(\varepsilon_h + \varepsilon_l)$

In this regime, as demonstrated at the beginning of Section 3.3, there is an angle of modal propagation in the core, $\cos(\theta_0) = \varepsilon_h \, \varepsilon_l/(\varepsilon_h + \varepsilon_l)/\varepsilon_c$, for which even an infinite quarter-wave stack designed for such an angle will not be efficient. Proceeding as in Section 4.3.1, and taking into consideration that in this regime $|r_{TM}|$ as defined in (3.16) is smaller than one for grazing angles of incidence $\theta_0 > \theta_c \to 0$, we arrive at a similar set of expansion coefficients as in (4.11):

$$\begin{pmatrix} A \\ B \end{pmatrix}_0^c = \frac{1}{2} \begin{pmatrix} 1 \\ -1 \end{pmatrix}$$

$$\begin{pmatrix} A \\ B \end{pmatrix}_{2n+1}^h = \begin{pmatrix} A \\ B \end{pmatrix}_{-2n-2}^1 = \frac{r_c \, (-r_{TM})^n}{2} \begin{pmatrix} -1 \\ 1 \end{pmatrix}$$

$$\begin{pmatrix} A \\ B \end{pmatrix}_{-2n-1}^h = \begin{pmatrix} A \\ B \end{pmatrix}_{2n+2}^1 = \frac{r_c \, (-r_{TM})^n}{2i} \begin{pmatrix} 1 \\ 1 \end{pmatrix}$$

$$n \subset \text{integer} \geq 0; \quad r_c = r_c^{TM} = \frac{\varepsilon_h k_z^c}{\varepsilon_c k_z^h}; \quad r^{TM} = \frac{\varepsilon_h k_z^l}{\varepsilon_l k_z^h}. \tag{4.19}$$

Moreover, from (4.16) it follows again that $|E_x^{core}(z)| \underset{\theta_c \to 0}{\ll} |E_z^{core}(z)| \simeq |H_y^{core}(z)|/n_c$. As the electric field component $E_z(z)$ is discontinuous across the dielectric boundaries, and as $E_x(z)$ is not negligible in the reflector layers, it is more convenient to plot a continuous $H_y(z)$ field instead. In Fig. 4.7 we present as a solid line the intensity of the only nonzero component of the magnetic field $|H_y(z)|^2$ as calculated by (3.37).

The dotted line plots the distribution of the electric field intensity $\varepsilon_c|E_z(z)|^2$ in the core region in the limit as $\theta_c \to 0$. Note that, like the case of the TE-polarized mode, the electric field vanishes at the core boundary. For grazing angles of mode propagation, from (4.19) it follows that magnetic and electric energies in the core are close to each other. Moreover, using the fact that for harmonic fields the total magnetic energy equals the total electric energy, we derive:

$$E_{total}^{el} = E_{total}^{mg} = \frac{1}{2} \int_{-\infty}^{+\infty} dz \, |H_y(z)|^2$$

$$= \frac{1}{2} \int_{-d_c/2}^{d_c/2} dz \cos^2(\pi \eta_c) \; + \; \sum_{j=0}^{+\infty} r_c^2 r_{TM}^{2j} \left(\int_0^{d_h} dz \sin^2\left(\frac{\pi}{2}\eta_h\right) + \int_0^{d_l} dz \cos^2\left(\frac{\pi}{2}\eta_l\right) \right)$$

$$= \frac{1}{2} \left[\frac{d_c}{2} + (d_h + d_l) \frac{r_c^2}{1 - r_{TM}^2} \right], \tag{4.20}$$

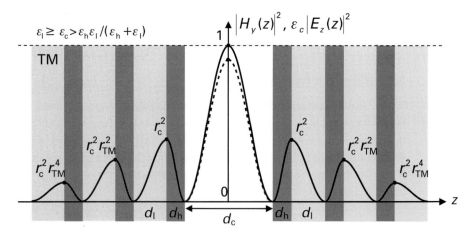

Figure 4.7 Schematic of the field intensity distribution of TM polarized mode at the center frequency of a quarter-wave reflector bandgap; $|H_y(z)|^2$, the field intensity of the fundamental core mode (solid); $\varepsilon_c|E_z z)|^2$ the field intensity in the core, $\theta_c \to 0$ (dotted).

and thus:

$$
\frac{E_{\text{reflector}}^{\text{el}}}{E_{\text{total}}^{\text{el}}} = \frac{(d_h + d_l)\dfrac{r_c^2}{1 - r_{\text{TM}}^2}}{\dfrac{d_c}{2} + (d_h + d_l)\dfrac{r_c^2}{1 - r_{\text{TM}}^2}}
$$

$$
\simeq \left(\frac{\lambda}{2d_c}\right)^3 \frac{\varepsilon_l^2 \varepsilon_h}{\varepsilon_c} \frac{\dfrac{1}{\sqrt{\varepsilon_h - \varepsilon_c}} + \dfrac{1}{\sqrt{\varepsilon_l - \varepsilon_c}}}{\left(\varepsilon_h^2 - \varepsilon_l^2\right)\left(\varepsilon_c - \dfrac{\varepsilon_h\varepsilon_l}{\varepsilon_h + \varepsilon_l}\right)} ; \left(\begin{array}{c} \theta_0 > \theta_c \to 0 \\ d_c \gg \lambda \end{array}\right), \quad (4.21)
$$

signifying that, in this regime, the fraction of the electric energy of a TM-polarized mode in the reflector decreases with the core size as fast as in the case of a TE-polarized mode, namely, inversely to the third power of the core size. If the reflector materials exhibit absorption losses, the modal propagation loss due to modal energy absorption will also decrease as the inverse third power of the core size.

Comparing expressions for the electric energies in the multilayer reflector for the cases $\varepsilon_l \geq \varepsilon_c > \varepsilon_h\varepsilon_l/(\varepsilon_h + \varepsilon_l)$ (4.21) and $\varepsilon_c < \varepsilon_h\varepsilon_l/(\varepsilon_h + \varepsilon_l)$ (4.18), we conclude that the fraction of the electric energy in the reflector decreases significantly faster with increase in the core size in the former case than in the latter. This is an interesting conclusion. It signifies that to minimize the propagation loss of a TM-polarized fundamental core mode of a low index core waveguide it is beneficial to operate in the regime when $\varepsilon_l \geq \varepsilon_c > \varepsilon_h\varepsilon_l/(\varepsilon_h + \varepsilon_l)$. Moreover, in this regime, the modal field distribution is Gaussian-like, resembling that of a TE-polarized mode.

4.4   Perturbation theory for Maxwell's equations, frequency formulation

Up until now we have discussed finding guided modes of geometrically perfect periodic systems with strictly real dielectric profiles (zero material absorption). In practice, realistic photonic crystals exhibit small variations in the geometry of their constituent features. Moreover, realistic dielectrics show material absorption losses, and, therefore, their dielectric constants typically contain small imaginary contributions. The influence of such small deviations from ideal on the propagation properties of eigenmodes can be efficiently studied in the framework of perturbation theory (PT) or coupled-mode theory (CMT). In this section, we present perturbation theory for the case when contours of the dielectric interfaces are fixed, while dielectric profile is weakly perturbed.

At the basis of PT and CMT is an assumption that for small variations in the dielectric profile, changes in the modal field distribution and modal frequency (or propagation constant) are also small. One then assumes that the perturbed mode can be expressed as a linear combination of the unperturbed modes, where expansion coefficients can be found either analytically (PT) or by solving a certain eigenvalue problem (CMT). We start with an unperturbed Hermitian Maxwell Hamiltonian $\hat{H}_0$ having eigenmodes $|H_j^0\rangle$, $j \subset$ integer ≥ 1 and eigenvalues $\omega_{0,j}^2$ that satisfy:

$$\hat{H}_0 |H_j^0\rangle = \omega_{0,j}^2 |H_j^0\rangle. \tag{4.22}$$

The eigenmode $|H\rangle$ of a perturbed Hamiltonian $\hat{H} = \hat{H}_0 + \delta\hat{H}$ (which can be nonHermitian) is then expressed as a linear combination of the unperturbed modes with a vector of expansion coefficients $\overline{C}$ as:

$$|H\rangle = \sum_{k=1}^{+\infty} C^k |H_k^0\rangle. \tag{4.23}$$

The eigenvalue equation for perturbed modes will then be

$$(\hat{H}_0 + \delta\hat{H})|H\rangle = \omega^2 |H\rangle. \tag{4.24}$$

Truncating (4.23) to a finite sum, substituting (4.23) into (4.24), multiplying the left and right parts of (4.24) by $\langle H_i^0|$, and taking into account (4.22), we arrive at a generalized coupled mode eigenproblem with respect to the new frequency ω:

$$(ND + \delta H)\overline{C} = \omega^2 N\overline{C}. \tag{4.25}$$

Here, N is a normalization matrix with elements $N_{i,j} = \langle H_i^0|H_j^0\rangle$ (N is diagonal for the modes of a Hermitian Hamiltonian as follows from (2.82)), D is a diagonal matrix of unperturbed eigenvalues $D_{i,i} = \omega_{0,i}^2$, and $\delta H_{i,j} = \langle H_i^0|\delta\hat{H}|H_j^0\rangle$ is a generally nondiagonal matrix of coupling elements.

In perturbation theory formulation, analytical corrections to a particular eigenstate and eigenfield can be found semi-analytically by assuming that perturbed eigenstates can be Taylor expanded around the point of zero perturbation with respect to some small parameter δ. In particular, defining corrections of various orders as:

$$|H_k\rangle = \left|H_k^0\right\rangle + \delta \left|H_k^1\right\rangle + \delta^2 \left|H_k^2\right\rangle + \cdots$$
$$\omega_k = \omega_{0,k} + \delta\omega_{1,k} + \delta^2\omega_{2,k} + \cdots \tag{4.26}$$

Substituting (4.26) into (4.24) and equating the terms with the same order of a small parameter δ we arrive at the following equations:

0th order PT : $\qquad\qquad\qquad \hat{H}_0 \left|H_k^0\right\rangle = \omega_{0,k}^2 \left|H_k^0\right\rangle$

1st order PT : $\hat{H}_0\delta \left|H_k^1\right\rangle + \delta\hat{H} \left|H_k^0\right\rangle = \omega_{0,k}^2\delta \left|H_k^1\right\rangle + 2\delta\omega_{1,k}\omega_{0,k} \left|H_k^0\right\rangle. \tag{4.27}$

2nd order PT : $\qquad\qquad\qquad\qquad \cdots$

As in (4.23) we express the first-order correction $|H_k^1\rangle$ in terms of a linear combination of the unperturbed modes. Multiplying the second equation by $\langle H_k^0|$ and using orthogonality of the eigenmodes of a Hermitian Hamiltonian (2.82), we find:

$$\omega_k = \omega_{0,k} + \frac{1}{2\omega_{0,k}} \frac{\left\langle H_k^0|\delta\hat{H}|H_k^0\right\rangle}{\left\langle H_k^0|H_k^0\right\rangle} + O(\delta^2)$$

$$|H_k\rangle = \left|H_k^0\right\rangle + \sum_{j\neq k} \frac{\left\langle H_j^0|\delta\hat{H}|H_k^0\right\rangle}{\omega_{0,k}^2 - \omega_{0,j}^2} \left|H_j^0\right\rangle + O(\delta^2). \tag{4.28}$$

4.4.1   Accounting for the absorption losses of the waveguide materials: calculation of the modal lifetime and decay length

As an example of the application of perturbation theory, we calculate the finite lifetime of a fundamental core mode of a planar photonic-crystal waveguide, considered in Section 4.1, in the case when reflector materials exhibit absorption losses. As mentioned earlier, in the case of nonzero absorption losses, the material dielectric constant includes a small imaginary part. Thus, the dielectric profile of an imperfect waveguide can be expressed as $\varepsilon(\mathbf{r}) = \varepsilon_0(\mathbf{r}) + i\alpha(\mathbf{r})$, where $i\alpha(\mathbf{r})$ defines the material absorption loss. The impact of the imaginary part of a dielectric constant is especially easy to demonstrate on the example of the modes of an infinite uniform dielectric characterized by the complex dielectric constant $\varepsilon = \varepsilon_0 + i\alpha$. As established in Section 2.1, eigenmodes of a uniform dielectric are plane waves $\mathbf{H}_0 \exp(i\mathbf{kr} - i\omega t)$, with a corresponding dispersion relation $|\mathbf{k}|^2 = \varepsilon\omega^2$.

Assuming that the propagation wave vector is purely real, for the dispersion relation to be satisfied in the material with a complex dielectric constant, the value of a modal frequency has to be complex. In particular, in the limit $\alpha \ll \varepsilon_0$ one can rewrite the dispersion relation as $\omega \simeq \omega_0 - i\alpha\omega_0/2\varepsilon_0$, where $\omega_0 = |\mathbf{k}|/\sqrt{\varepsilon_0}$. Finally, substitution of the complex frequency into the functional form of a plane-wave solution gives $\mathbf{H}_0 \exp(i\mathbf{kr} - i\omega_0 t) \exp(-t\,\alpha\omega_0/2\varepsilon_0)$. This result suggests power decay in the electromagnetic wave due to material absorption with a characteristic time:

$$\tau_0^{\mathrm{abs}} = \varepsilon_0/\alpha\omega_0. \tag{4.29}$$

Alternatively, the complex dielectric constant can be interpreted as defining a finite decay length over which modal energy is absorbed by the lossy materials. In particular, assuming that the modal frequency is purely real, for the dispersion relation to be satisfied in the material with a complex dielectric constant, the modal wave vector has to be complex. For example, for a plane wave propagating in the $\hat{x}$ direction, the corresponding wave vector in the limit $\alpha \ll \varepsilon_0$ will be $k_x \simeq k_0 + i\alpha k_0/2\varepsilon_0$, where $k_0 = \omega_0\sqrt{\varepsilon_0}$. Finally, substitution of a complex wave vector into the functional form of a plane-wave solution gives $H_0 \exp(ik_0 x - i\omega_0 t)\exp(-x\alpha k_0/2\varepsilon_0)$. This result suggests the power decay of the electromagnetic wave due to material absorption with a characteristic length:

$$L_0^{\mathrm{abs}} = \varepsilon_0/\alpha k_0 = \tau_0^{\mathrm{abs}} v_g^0, \tag{4.30}$$

where $v_g^0 = \omega_0/k_0$ is the group velocity of a plane wave.

We now use the perturbation theory expression (4.28) to characterize the lifetime of a guided mode propagating in a waveguide incorporating absorbing materials in its structure. In particular, using the explicit form of the Maxwell Hamiltonian (2.71) we find:

$$\hat{H} = \nabla \times \left(\frac{1}{\varepsilon(\mathbf{r})}\nabla\times\right) = \hat{H}_0 + \delta\hat{H}$$

$$\hat{H}_0 = \nabla \times \left(\frac{1}{\varepsilon_0(\mathbf{r})}\nabla\times\right); \quad \delta\hat{H}\underset{\alpha\ll\varepsilon_0}{=} -i\nabla \times \left(\frac{\alpha(\mathbf{r})}{\varepsilon_0^2(\mathbf{r})}\nabla\times\right). \tag{4.31}$$

Using integration by parts and following the same steps as in (2.86) we can express the first-order correction to the modal frequency as:

$$\frac{\omega_k - \omega_{0,k}}{\omega_{0,k}} = \frac{1}{2\omega_{0,k}^2}\frac{\left\langle H_k^0|\delta\hat{H}|H_k^0\right\rangle}{\left\langle H_k^0 \mid H_k^0\right\rangle} = -\frac{i}{2}\frac{\displaystyle\int_V d\mathbf{r}\alpha(\mathbf{r})\left|E_k^0(\mathbf{r})\right|^2}{\displaystyle\int_V d\mathbf{r}\varepsilon(\mathbf{r})\left|E_k^0(\mathbf{r})\right|^2}, \tag{4.32}$$

where we used the fact that the total electric and magnetic energies are equal to each other, and integration is performed over the volume (or system over the waveguide cross-section).

We now apply (4.32) to find the lifetime of the fundamental core-guided mode of a hollow photonic-bandgap waveguide. Assuming that there are no losses in the core and that the losses of the reflector materials are $\varepsilon(\mathbf{r}) = \varepsilon_0(\mathbf{r})(1 + i\delta)$, (4.32) transforms into:

$$\frac{\omega_k - \omega_{0,k}}{\omega_{0,k}} = \frac{1}{2\omega_{0,k}^2}\frac{\left\langle H_k^0|\delta\hat{H}|H_k^0\right\rangle}{\left\langle H_k^0 \mid H_k^0\right\rangle} = -i\frac{\delta}{2}\frac{E_{\mathrm{reflector}}^{\mathrm{el}}}{E_{\mathrm{total}}^{\mathrm{el}}}. \tag{4.33}$$

Because of the $\exp(-i\omega t)$ time dependence of the modal electromagnetic fields, the imaginary contribution to the eigenfrequency (4.33) defines the finite lifetime for the waveguide guided mode as:

$$\frac{1}{\tau_{\mathrm{wg}}^{\mathrm{abs}}} = \delta\omega_{0,k}\frac{E_{\mathrm{reflector}}^{\mathrm{el}}}{E_{\mathrm{total}}^{\mathrm{el}}}. \tag{4.34}$$

Note that compared with the lifetime of a plane wave in a uniform dielectric (4.29), the lifetime of a mode guided in the hollow core of a bandgap waveguide can be considerably increased by reducing the fraction of light in the absorbing reflector. Depending on the polarization of the core mode and the value of the core dielectic constant compared with those of the reflector, the lifetime of the mode can vary significantly. Thus, using expressions for the fraction of the electric energy in the reflector (4.14), (4.18) and (4.21) of Sections 4.2 and 4.3, we derive the following scaling relations with the waveguide core size:

$$
\frac{1}{\tau_{\text{wg}}^{\text{abs}}} = \delta\omega_{0,k} \frac{E_{\text{reflector}}^{\text{el}}}{E_{\text{total}}^{\text{el}}} \sim
\begin{bmatrix}
\delta\dfrac{\lambda^2}{d_{\text{c}}^3} \text{ for } \begin{bmatrix} \text{TE}, \theta_{\text{c}} \to 0 \\ \text{TM}, \theta_0 > \theta_{\text{c}} \to 0, \varepsilon_1 \geq \varepsilon_{\text{c}} > \varepsilon_{\text{h}}\varepsilon_1/(\varepsilon_{\text{h}} + \varepsilon_1) \end{bmatrix} \\
\delta\dfrac{1}{d_{\text{c}}} \text{ for TM}, \theta_{\text{c}} \to 0, \varepsilon_{\text{c}} < \varepsilon_{\text{h}}\varepsilon_1/(\varepsilon_{\text{h}} + \varepsilon_1).
\end{bmatrix}
\tag{4.35}
$$

Note that for both polarizations, larger core sizes lead to smaller losses and, therefore, larger lifetimes of the guided modes. Note also that the finite lifetime of a guided mode also defines a characteristic decay length $L^{\text{abs}} = \tau^{\text{abs}} v_{\text{g}}$ over which modal power loss is incurred, where v_{g} is the modal group velocity.

4.5   Perturbative calculation of the modal radiation loss in a photonic bandgap waveguide with a finite reflector

In this section, we derive a perturbative expression for the radiation loss of a fundamental core mode of a photonic-bandgap waveguide having a finite number of layers in the reflector. The estimation of radiation losses, however, requires a formulation different from the one presented in Section 4.4.

4.5.1   Physical approach

From physical considerations, modal propagation loss is caused by the nonzero electromagnetic energy flux outgoing from the waveguide volume. From Maxwell's equations it can be demonstrated that in the absence of free currents, the rate of electromagnetic energy change inside a finite volume V is related to the outgoing energy flux through the volume boundary A as:

$$
\oint_A \mathbf{da} \cdot \mathbf{S} = \frac{\partial}{\partial t} \int_V dV \rho_{\text{E}},
\tag{4.36}
$$

where energy flux (Poynting vector) and energy density are defined as:

$$
\mathbf{S} = (\mathbf{E} \times \mathbf{H})
$$
$$
\rho = \frac{1}{2}(\mathbf{ED} + \mathbf{BH}).
\tag{4.37}
$$

A characteristic decay time of radiation from volume V can, thus, be defined as:

$$
\frac{1}{\tau^{\text{rad}}} = \frac{\dfrac{\partial}{\partial t} \int_V dV \rho}{\int_V dV \rho} = \frac{\oint_A \mathbf{da} \cdot \mathbf{S}}{\int_V dV \rho}.
\tag{4.38}
$$

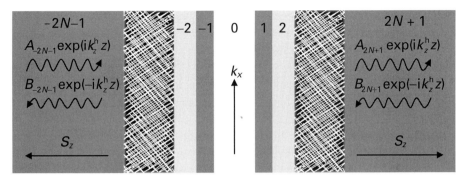

Figure 4.8 Radiation losses in photonic-crystal waveguides.

In the case of the guided mode of a waveguide, (4.38) becomes:

$$\frac{1}{\tau_{wg}^{rad}} = \frac{S_z}{E_{total}^{el}},$$ (4.39)

where S_z is a transverse component of the energy flux (see Fig. 4.8), while E_{total}^{el} is the electric energy (which, for harmonic modes, is also equal to magnetic energy) in a waveguide cross-section.

We now estimate the lifetime of a core-guided mode of a photonic-bandgap waveguide with a finite reflector using (4.39). For the TE mode, the transverse flux is $S_z = -\operatorname{Re}(E_y)\operatorname{Re}(H_x)$, while for the TM mode, it is $S_z = \operatorname{Re}(E_x)\operatorname{Re}(H_y)$. Note that to evaluate the flux correctly, one has to use real fields; thus, for the harmonic fields:

$$S_z = \operatorname{Re}\left(\mathbf{E}_\parallel(\mathbf{r})\exp(i\omega t)\right) \times \operatorname{Re}\left(\mathbf{H}_\parallel(\mathbf{r})\exp(i\omega t)\right)$$

$$= \frac{1}{4}\left(\mathbf{E}_\parallel(\mathbf{r})\exp(-i\omega t) + \mathbf{E}_\parallel^*(\mathbf{r})\exp(i\omega t)\right) \times \left(\mathbf{H}_\parallel(\mathbf{r})\exp(-i\omega t) + \mathbf{H}_\parallel^*(\mathbf{r})\exp(i\omega t)\right)$$

$$\langle S_z \rangle = \frac{1}{2}\operatorname{Re}(\mathbf{E}_\parallel(\mathbf{r}) \times \mathbf{H}_\parallel^*(\mathbf{r})),$$ (4.40)

where subscript $\parallel$ signifies "in-plane component of a vector", while $\langle\rangle$ signifies "time average". In a similar manner for the energy:

$$\rho = \frac{1}{2}\left[\varepsilon(r)\left(\operatorname{Re}\left(\mathbf{E}(\mathbf{r})\exp(i\omega t)\right)\right)^2 + \mu(r)\left(\operatorname{Re}\left(\mathbf{H}(\mathbf{r})\exp(i\omega t)\right)\right)^2\right]$$

$$= \frac{1}{8}\left[\varepsilon(r)\left(\mathbf{E}(\mathbf{r})\exp(-i\omega t) + \mathbf{E}^*(\mathbf{r})\exp(i\omega t)\right)^2\right.$$

$$\left. + \mu(r)\left(\mathbf{H}(\mathbf{r})\exp(-i\omega t) + \mathbf{H}^*(\mathbf{r})\exp(i\omega t)\right)^2\right]$$

$$\langle \rho \rangle = \frac{1}{4}\left[\varepsilon(r)|\mathbf{E}(\mathbf{r})|^2 + \mu(r)|\mathbf{H}(\mathbf{r})|^2\right].$$ (4.41)

From the expressions (3.1), (3.2), (3.6), (3.7) for the electromagnetic fields in a general multilayer system, it follows that in a waveguide cladding to the right of the core (see Fig. 4.8) the total flux is:

$$\langle S_z \rangle^{TE} = \varepsilon_h \langle S_z \rangle^{TM} = \frac{k_z^h}{2\omega}\left(|A_{2N+1}|^2 - |B_{2N+1}|^2\right).$$ (4.42)

Expression (4.42) suggests that the total flux through the waveguide boundary equals the difference in the fluxes of the outgoing wave with coefficient A_{2N+1} and an incoming wave with coefficient B_{2N+1}.

In Sections 4.2 and 4.3, we have established several solutions for the TE and TM modes of the photonic-bandgap waveguides with infinite reflectors (expansion coefficients (4.11) and (4.15)). Exactly the same expansion coefficients with indexes in the range $[-2N - 1, 2N + 1]$ also define a solution in a finite-size photonic crystal waveguide with N reflector bilayers, terminated by the high-refractive-index cladding. From the form of expansion coefficients (4.11) and (4.15), it follows that $|A_{2N+1}| = |B_{2N+1}|$, thus making the total flux (4.42) traversing the waveguide boundary zero. To estimate the lifetime of a guided mode due to radiation from the waveguide core one has to use only an outgoing flux in (4.42). Thus, using $|A_{2N+1}|$ coefficients in (4.11) and (4.15), as well as expressions for the energy in a waveguide cross-section (4.13), (4.17), and (4.20), we finally get:

$$\frac{1}{\tau_{TE}^{rad}}, \quad \frac{1}{\tau_{TM}^{rad}} \sim \frac{|A_{2N+1}|^2}{E_{total}^{el}} \sim \frac{\left(r_c^{TE,TM}\right)^2 r_{TE,TM}^{2N}}{d_c} \sim \frac{\lambda^2}{d_c^3} r_{TE,TM}^{2N}$$
$$\theta_c \to 0 \quad \varepsilon_c > \varepsilon_h \varepsilon_l/(\varepsilon_h+\varepsilon_l)$$
$$\theta_0 > \theta_c \to 0$$

$$\frac{1}{\tau_{TM}^{rad}} \sim \frac{|A_{2N+1}|^2}{E_{total}^{el}} \sim \frac{1}{d_c r_{TM}^{2N}} . \tag{4.43}$$
$$\varepsilon_c < \varepsilon_h \varepsilon_l/(\varepsilon_h+\varepsilon_l)$$
$$\theta_c \to 0$$

Note for (4.43) that the lifetime of a bandgap-guided mode increases algebraically with increasing core size, while it increases exponentially with the number of reflector bilayers. Finally, note that in the case of a waveguide, the finite lifetime of a guided mode also defines a characteristic decay length, $L^{rad} = \tau^{rad} v_g$, over which modal power loss is incurred, where v_g is the modal group velocity.

4.5.2   Mathematical approach

While the qualitative expressions (4.43) for the rates of radiation decay capture correctly the underlying scaling laws, to get a quantitative answer one has to solve an exact scattering problem with boundary conditions of zero incoming flux from infinity; $|B_{2N+1}| = |A_{-2N-1}| = 0$. In the transfer matrix formalism, the propagation constant k_x of a mode of a photonic-crystal waveguide with a finite periodic reflector (Fig. 4.8) satisfies:

$$\begin{pmatrix} A_{2N+1} \\ 0 \end{pmatrix} = (M_{hlh})^N M_{ch} M_{hc} (M_{hlh})^N \begin{pmatrix} 0 \\ B_{-2N-1} \end{pmatrix}. \tag{4.44}$$

In the following analysis we assume that the fundamental core mode is described by a real frequency and a complex propagation constant $k_x = \omega\sqrt{\varepsilon_c}\cos(\theta_c) + i\alpha^{rad}/2$, where α^{rad} is assumed to be small compared with the real part of the propagation constant.

Because of the $|\exp(ik_x z)|^2 \sim \exp(-\alpha^{\mathrm{rad}} z)$ dependence of the energy flux, $1/\alpha^{\mathrm{rad}}$ will define a characteristic modal power decay length L^{rad} along the direction of propagation. Starting with a photonic-bandgap waveguide with an infinite quarter-wave reflector designed for an angle of incidence θ_c, and truncating the number of bilayers to N on both sides of a core, the solution of (4.44) in terms of propagation loss α^{rad} can be found analytically. In particular, using individual layer thicknesses (4.7), and assuming that $\alpha^{\mathrm{rad}} \ll \omega\sqrt{\varepsilon_c}$, we use the Taylor expansion of all the matrices in (4.44) around the point of zero losses $\alpha^{\mathrm{rad}} = 0$:

$$k_x = \omega\sqrt{\varepsilon_c}\cos(\theta_c) + i\frac{\alpha^{\mathrm{rad}}}{2};$$

$$k_z^c = \omega\sqrt{\varepsilon_c}\sin(\theta_c) - i\frac{\alpha^{\mathrm{rad}}}{2}\frac{\cos(\theta_c)}{\sin^2(\theta_c)}$$

$$k_z^{h,1} = \omega\sqrt{\varepsilon_{h,1} - \varepsilon_c\cos^2(\theta_c)} - i\frac{\alpha^{\mathrm{rad}}}{2}\frac{\sqrt{\varepsilon_c}\cos(\theta_c)}{\sqrt{\varepsilon_{h,1} - \varepsilon_c\cos^2(\theta_c)}}$$

$$\phi_c = \pi - i\frac{\alpha^{\mathrm{rad}}}{2}\phi_c^1 = \pi - i\frac{\alpha^{\mathrm{rad}}}{2}\frac{\lambda\cos(\theta_c)}{2\sqrt{\varepsilon_c}\sin^2(\theta_c)}$$

$$\phi_{h,1} = \frac{\pi}{2} - i\frac{\alpha^{\mathrm{rad}}}{2}\phi_{h,1}^1 = \frac{\pi}{2} - i\frac{\alpha^{\mathrm{rad}}}{2}\frac{\lambda\sqrt{\varepsilon_c}\cos(\theta_c)}{4(\varepsilon_{h,1} - \varepsilon_c\cos^2(\theta_c))}. \qquad (4.45)$$

Keeping only the first-order terms in α^{rad}, we can then solve (4.44) analytically to find:

$$\alpha^{\mathrm{rad}} = \frac{4r^{2N}(1 - r^2)r_c}{[\phi_c^1(1 - r^2) + 2r_c(\phi_h^1 + r\phi_1^1)] - r^{4N}r_c[r_c\phi_c^1(r^2 - 1) + 2r(r\phi_h^1 + \phi_1^1)]}, \qquad (4.46)$$

where, depending upon modal polarization, $r_c = r_c^{\mathrm{TE}}, r_c^{\mathrm{TM}}, r = r_{\mathrm{TE}}, r_{\mathrm{TM}}$ as defined in (4.11) and (4.19).

Assuming a large number of bilayers, so that $r_{\mathrm{TE}}^{2N} \ll 1$, $r_{\mathrm{TM}}^{2N} \underset{\substack{\varepsilon_c > \varepsilon_h\varepsilon_1/(\varepsilon_h+\varepsilon_1)\\\theta_0 > \theta_c \to 0}}{\ll} 1$, and

$r_{\mathrm{TM}}^{2N} \underset{\substack{\varepsilon_0 < \varepsilon_h\varepsilon_1/(\varepsilon_h+\varepsilon_1)\\\theta_c \to 0}}{\gg} 1$, and a large size of a waveguide core $d_c \gg \lambda$, for the radiation losses

of the fundamental core-guided mode, we find:

$$\alpha^{\mathrm{rad}} = \frac{1}{L^{\mathrm{rad}}} \simeq \begin{cases} \dfrac{\lambda^2}{d_c^3}\dfrac{r_{\mathrm{TE,TM}}^{2N}}{\sqrt{\varepsilon_c}(\varepsilon_h - \varepsilon_c)} & \text{for} \begin{array}{l} \text{TE}: \theta_c \to 0 \\ \text{TM}: \varepsilon_c > \varepsilon_h\varepsilon_1/(\varepsilon_h + \varepsilon_1) \\ \theta_0 > \theta_c \to 0 \end{array} \\[4mm] \dfrac{4}{d_c}\dfrac{r_{\mathrm{TM}}^{-2N}}{\sqrt{\varepsilon_c}(\varepsilon_h - \varepsilon_c)} & \text{for TM}: \varepsilon_c < \varepsilon_h\varepsilon_1/(\varepsilon_h + \varepsilon_1); \theta_c \to 0 \end{cases} \qquad (4.47)$$

5 Hamiltonian formulation of Maxwell's equations for waveguides (propagation-constant consideration)

In Chapter 4 we have derived perturbation theory for Maxwell's equations to find corrections to the electromagnetic state eigenfrequency ω due to small changes in the material dielectric constant. Applied to the case of systems incorporating absorbing materials, we have concluded that absorption introduces an imaginary part to the modal frequency, thus resulting in decay of the modal power in time. While this result is intuitive for resonator states localized in all spatial directions, it is somewhat not straightforward to interpret for the case of waveguides in which energy travels freely along the waveguide direction. In the case of waveguides, a more natural description of the phenomenon of energy dissipation would be in terms of a characteristic modal decay length, or, in other words, in terms of the imaginary contribution to the modal propagation constant. The Hamiltonian formulation of Maxwell's equations in the form (2.70) is an eigenvalue problem with respect to ω^2, thus perturbation theory formalism based on this Hamiltonian form is most naturally applicable for finding frequency corrections. In the following sections we develop the Hamiltonian formulation of Maxwell's equations in terms of the modal propagation constant, which allows, naturally, perturbative formulation with respect to the modal propagation constant.

5.1 Eigenstates of a waveguide in Hamiltonian formulation

In what follows, we introduce the Hamiltonian formulation of Maxwell's equations for waveguides, [1] which is an eigenvalue problem with respect to the modal propagation constant β. A waveguide is considered to possess continuous translational symmetry in the longitudinal $\hat{z}$ direction. The waveguide dielectric profile is then a function of the transverse coordinates only $\varepsilon = \varepsilon(x, y)$. Assuming that the harmonic time dependence of the electromagnetic fields $\mathbf{F}(x, y, z, t) = \mathbf{F}(x, y, z) \exp(-i\omega t)$ ($\mathbf{F}$ denotes the electric or magnetic field vector), we introduce transverse and longitudinal components of the fields as $\mathbf{F} = \mathbf{F}_t + F_z\hat{z}$, where $\mathbf{F}_t = (\hat{z} \times \mathbf{F}) \times \hat{z}$. Maxwell's equations (2.11), (2.12), (2.13), (2.14) can then be written in terms of the transverse and longitudinal field components as:

$$\frac{\partial \mathbf{E}_t}{\partial z} + i\omega\mu\,(\hat{z} \times \mathbf{H}_t) = \nabla_t E_z; \quad \frac{\partial \mathbf{H}_t}{\partial z} - i\varepsilon\mu\,(\hat{z} \times \mathbf{E}_t) = \nabla_t H_z, \tag{5.1}$$

$$\hat{z}\nabla_t \times \mathbf{E}_t = i\omega\mu H_z; \quad \hat{z}\nabla_t \times \mathbf{H}_t = -i\omega\varepsilon E_z, \tag{5.2}$$

$$\nabla_t\,(\varepsilon\mathbf{E}_t) = -\frac{\partial\,(\varepsilon E_z)}{\partial z}; \quad \nabla_t\,(\mu\mathbf{H}_t) = -\frac{\partial\,(\mu H_z)}{\partial z}. \tag{5.3}$$

Eliminating the longitudinal E_z and H_z components from (5.1) by using (5.2) and after some rearrangement using the identity $\hat{\mathbf{z}} \times (\nabla_t F) = -\nabla_t \times (\hat{\mathbf{z}} F)$ we arrive at the following:

$$-\mathrm{i}\frac{\partial}{\partial z} \hat{B} \begin{pmatrix} \mathbf{E}_t(x,y,z) \\ \mathbf{H}_t(x,y,z) \end{pmatrix} = \hat{H} \begin{pmatrix} \mathbf{E}_t(x,y,z) \\ \mathbf{H}_t(x,y,z) \end{pmatrix}, \tag{5.4}$$

where we define the normalization operator $\hat{B}$ and a waveguide Hamiltonian $\hat{H}$ as:

$$\hat{B} = \begin{pmatrix} 0 & -\hat{\mathbf{z}}\times \\ \hat{\mathbf{z}}\times & 0 \end{pmatrix}; \tag{5.5}$$

$$\hat{H} = \begin{pmatrix} \omega\varepsilon - \omega^{-1}\nabla_t \times \left(\hat{\mathbf{z}}\left(\mu^{-1}\hat{\mathbf{z}}\left(\nabla_t\times\right)\right)\right) & 0 \\ 0 & \omega\mu - \omega^{-1}\nabla_t \times \left(\hat{\mathbf{z}}\left(\varepsilon^{-1}\hat{\mathbf{z}}\left(\nabla_t\times\right)\right)\right) \end{pmatrix},$$

Finally, from consideration of Chapter 2, the general form of a solution in a system with continuous translational symmetry along the $\hat{\mathbf{z}}$ direction can be written as:

$$\mathbf{F}(x,y,z,t) = \mathbf{F}_\beta(x,y)\exp(\mathrm{i}(\beta z - \omega t)). \tag{5.6}$$

After substitution of this form into (5.4) we arrive at the following eigenvalue problem with respect to the modal propagation constant, and the transverse components of the electromagnetic fields:

$$\hat{H} \begin{pmatrix} \mathbf{E}_t(x,y) \\ \mathbf{H}_t(x,y) \end{pmatrix}_\beta = \beta\hat{B} \begin{pmatrix} \mathbf{E}_t(x,y) \\ \mathbf{H}_t(x,y) \end{pmatrix}_\beta, \tag{5.7}$$

In Dirac notation, (5.7) can be written as:

$$\hat{H}\left|\mathbf{F}_\beta\right\rangle = \beta\hat{B}\left|\mathbf{F}_\beta\right\rangle; \quad \left|\mathbf{F}_\beta\right\rangle = \begin{pmatrix} \mathbf{E}_t(x,y) \\ \mathbf{H}_t(x,y) \end{pmatrix}_\beta. \tag{5.8}$$

5.1.1 Orthogonality relation between the modes of a waveguide made of lossless dielectrics

In the case of a waveguide with a purely real dielectric profile $\varepsilon = \varepsilon^*$, $\mu = \mu^*$ (lossless dielectrics), the generalized eigenvalue problem (5.7) is Hermitian. This allows us to define an orthogonality relation between the distinct waveguide modes, as well as to derive several integral expressions for the values of phase and group velocities.

To demonstrate that (5.7) is Hermitian, we have to show that both the normalization operator $\hat{B}$ and the Hamiltonian operator $\hat{H}$ are Hermitian. We start by demonstrating that the normalization operator $\hat{B}$ is Hermitian. Namely, for the dot product between the two modes the following holds:

$$\langle \mathbf{F}_{\beta'}|\hat{B}|\mathbf{F}_\beta\rangle = \int\limits_{\substack{\text{waveguide}\\\text{cross-section}}} dxdy \begin{pmatrix} \mathbf{E}_t(x,y) \\ \mathbf{H}_t(x,y) \end{pmatrix}_{\beta'}^\dagger \begin{pmatrix} 0 & -\hat{\mathbf{z}}\times \\ \hat{\mathbf{z}}\times & 0 \end{pmatrix} \begin{pmatrix} \mathbf{E}_t(x,y) \\ \mathbf{H}_t(x,y) \end{pmatrix}_\beta$$

$$= \int\limits_{\text{wc}} dxdy\hat{\mathbf{z}}\left(\mathbf{E}_{t\beta'}^*(x,y) \times \mathbf{H}_{t\beta}(x,y) + \mathbf{E}_{t\beta}(x,y) \times \mathbf{H}_{t\beta'}^*(x,y)\right)$$

$$= \langle \mathbf{F}_\beta|\hat{B}|\mathbf{F}_{\beta'}\rangle^*, \tag{5.9}$$

where † signifies complex transpose, and all the integrations are performed across a two-dimensional waveguide cross-section. We now demonstrate that Hamiltonian $\hat{H}$ is also Hermitian. Particularly:

$$\langle \mathbf{F}_{\beta'} | \hat{H} | \mathbf{F}_\beta \rangle = \int_{wc} dxdy\, \mathbf{E}^*_{t\beta'}(x, y) \left[\omega\varepsilon - \omega^{-1} \nabla_t \times \left(\hat{z} \left(\mu^{-1} \hat{z} \left(\nabla_t \times \right) \right) \right) \right] \mathbf{E}_{t\beta}(x, y)$$

$$+ \int_{wc} dxdy\, \mathbf{H}^*_{t\beta'}(x, y) \left[\omega\mu - \omega^{-1} \nabla_t \times \left(\hat{z} \left(\varepsilon^{-1} \hat{z} \left(\nabla_t \times \right) \right) \right) \right] \mathbf{H}_{t\beta}(x, y).$$

$$(5.10)$$

As an example, consider one of the complex terms in (5.10), which can be simplified by using a 2D vector identity $\mathbf{b} \cdot (\nabla_t \times \mathbf{a}) = \mathbf{a} \cdot (\nabla_t \times \mathbf{b}) + \nabla_t \cdot (\mathbf{a} \times \mathbf{b})$:

$$\int_{wc} dxdy\, \omega^{-1} \mathbf{E}^*_{t\beta'}(x, y) \nabla_t \times \left(\hat{z} \left(\mu^{-1} \hat{z} \left(\nabla_t \times \mathbf{E}_{t\beta}(x, y) \right) \right) \right)$$

$$= \int_{wc} dxdy\, \mu^{-1} \left[\hat{z} \left(\nabla_t \times \mathbf{E}_{t\beta}(x, y) \right) \right] \cdot \left[\hat{z} \left(\nabla_t \times \mathbf{E}^*_{t\beta'}(x, y) \right) \right]$$

$$+ \int_{wc} dxdy\, \nabla_t \left(\mathbf{E}^*_{t\beta'}(x, y) \times \hat{z} \left(\mu^{-1} \hat{z} \left(\nabla_t \times \mathbf{E}_{t\beta}(x, y) \right) \right) \right). \qquad (5.11)$$

The last integral in (5.11) is over a 2D waveguide cross-section. This integral can be transformed into a 1D contour integral along the curve encircling the waveguide cross-section by using $\int dA \nabla_t \cdot \mathbf{a} = \oint d\mathbf{l} \cdot \mathbf{a}$. Assuming localized states with vanishing fields at distances far from the waveguide core, the last integral is zero. Finally, using (5.2), (5.11) transforms into:

$$\int_{wc} dxdy\, \omega^{-1} \mathbf{E}^*_{t\beta'}(x, y) \nabla_t \times \left(\hat{z} \left(\mu^{-1} \hat{z} \left(\nabla_t \times \mathbf{E}_{t\beta}(x, y) \right) \right) \right)$$

$$= \omega \int_{wc} dxdy\, \mu^* H^*_{z\beta'}(x, y) H_{z\beta}(x, y). \qquad (5.12)$$

Treating the other terms in (5.10) in the same manner, we arrive at the following expression for the matrix element:

$$\langle \mathbf{F}_{\beta'} | \hat{H} | \mathbf{F}_\beta \rangle = \omega \int_{wc} dxdy\, \left[\varepsilon \mathbf{E}^*_{t\beta'}(x, y) \mathbf{E}_{t\beta}(x, y) + \mu \mathbf{H}^*_{t\beta'}(x, y) \mathbf{H}_{t\beta}(x, y) \right]$$

$$- \omega \int_{wc} dxdy\, \left[\varepsilon^* E^*_{z\beta'}(x, y) E_{z\beta}(x, y) + \mu^* H^*_{z\beta'}(x, y) H_{z\beta}(x, y) \right].$$

$$(5.13)$$

From the explicit form of (5.13) it also follows that for the real dielectric profiles $\langle \mathbf{F}_{\beta'} | \hat{H} | \mathbf{F}_\beta \rangle = \langle \mathbf{F}_\beta | \hat{H} | \mathbf{F}_{\beta'} \rangle^*$, and, thus, operator $\hat{H}$ is Hermitian.

Now, using the fact that both operators in (5.7) are Hermitian, we can demonstrate orthogonality between the two distinct waveguide modes. In particular, for the two modes with propagation constants β and β' we can write:

$$\langle \mathbf{F}_{\beta'} | \hat{H} | \mathbf{F}_\beta \rangle = \beta \langle \mathbf{F}_{\beta'} | \hat{B} | \mathbf{F}_\beta \rangle$$
$$\langle \mathbf{F}_\beta | \hat{H} | \mathbf{F}_{\beta'} \rangle = \beta' \langle \mathbf{F}_\beta | \hat{B} | \mathbf{F}_{\beta'} \rangle.$$

(5.14)

Using the fact that operators $\hat{H}$ and $\hat{B}$ are Hermitian, after complex conjugation of the first equation in (5.14) and its subtraction from the second one we get:

$$(\beta' - \beta^*) \langle \mathbf{F}_\beta | \hat{B} | \mathbf{F}_{\beta'} \rangle = 0.$$

(5.15)

Therefore, if $\beta^* \neq \beta'$, the only way to satisfy (5.15) is through modal orthogonality in the sense:

$$\langle \mathbf{F}_{\beta'} | \hat{B} | \mathbf{F}_\beta \rangle = 0; \quad \beta^* \neq \beta'.$$

(5.16)

Note that (5.16) is a somewhat unusual orthogonality condition. In fact, even a waveguide with a purely real dielectric profile can have modes with imaginary propagation constants. These physical modes are the evanescent waves that decay exponentially fast along the direction of their propagation. For an evanescent wave, the overlap integral in (5.16) is not zero only between the mode itself and another evanescent mode with a complex-conjugate propagation constant. Generally speaking, the dot product (5.9), although supporting an orthogonality condition (5.16), does not constitute a strict norm and, therefore, can take any complex value.

Finally, for an evanescent wave with a complex propagation constant β, the existence of an evanescent wave with a complex conjugate value of the propagation constant β^* is assured for purely real dielectric profiles. Namely, starting with an evanescent wave satisfying:

$$\hat{H} | \mathbf{F}_\beta \rangle = \beta \hat{B} | \mathbf{F}_\beta \rangle,$$

(5.17)

and after complex conjugation of (5.17) we get:

$$\hat{H} | \mathbf{F}_\beta \rangle^* = \beta^* \hat{B} | \mathbf{F}_\beta \rangle^*,$$

(5.18)

where from the explicit form (5.5) of the waveguide Hamiltonian it follows that $\hat{H}^* = \hat{H}$ for the real dielectric profiles. Therefore, we conclude that $| \mathbf{F}_\beta \rangle^*$ is also an eigenstate of the Hamiltonian (5.5) with propagation constant β^*.

In what follows we assume that the mode in question is a true guided wave that has a purely real propagation constant β, which is the case for the most problems of interest. All the results derived in the following sections are also applicable for the evanescent waves with complex propagation constants; however, in all the expressions, matrix elements of the form $\langle \mathbf{F}_\beta | \ldots | \mathbf{F}_{\beta'} \rangle$ have to be substituted by matrix elements of the form $\langle \mathbf{F}_{\beta^*} | \ldots | \mathbf{F}_{\beta'} \rangle$.

5.1.2   Expressions for the modal phase velocity

In this section we derive an integral expression for the phase velocity of a mode of a waveguide with a purely real dielectric profile. Applying (5.13) to the same mode gives:

$$\langle \mathbf{F}_\beta | \hat{H} | \mathbf{F}_\beta \rangle = 4\omega \left(E_{\mathrm{wc}}^{\mathrm{t}} - E_{\mathrm{wc}}^{z} \right), \tag{5.19}$$

where we have used (4.41) to define transverse and longitudinal time-averaged electromagnetic energy densities:

$$E_{\mathrm{wc}}^{\mathrm{t}} = \frac{1}{4} \int_{\mathrm{wc}} dxdy \left[\varepsilon \left| \mathbf{E}_{\mathrm{t}\beta}(x,y) \right|^2 + \mu \left| \mathbf{H}_{\mathrm{t}\beta}(x,y) \right|^2 \right]$$

$$E_{\mathrm{wc}}^{z} = \frac{1}{4} \int_{\mathrm{wc}} dxdy \left[\varepsilon \left| E_{z\beta}(x,y) \right|^2 + \mu \left| H_{z\beta}(x,y) \right|^2 \right]. \tag{5.20}$$

Applying (5.9) to the same mode gives:

$$\langle \mathbf{F}_\beta | \hat{B} | \mathbf{F}_\beta \rangle = \int_{\mathrm{wc}} dxdy\, \hat{z} \left(\mathbf{E}_{\mathrm{t}\beta}^*(x,y) \times \mathbf{H}_{\mathrm{t}\beta}(x,y) + \mathbf{E}_{\mathrm{t}\beta}(x,y) \times \mathbf{H}_{\mathrm{t}\beta}^*(x,y) \right) = 4 S_{\mathrm{wc}}^z, \tag{5.21}$$

where S_{wc}^z is a time-averaged longitudinal component of the energy flux (see (4.40)).

Finally, from (5.9) and (5.13) we derive a useful identity for the propagation constant of a waveguide-guided mode, as well as for the modal phase velocity:

$$\frac{1}{v_{\mathrm{p}}} = \frac{\beta}{\omega} = \frac{1}{\omega} \frac{\langle \mathbf{F}_\beta | \hat{H} | \mathbf{F}_\beta \rangle}{\langle \mathbf{F}_\beta | \hat{B} | \mathbf{F}_\beta \rangle} = \frac{E_{\mathrm{wc}}^{\mathrm{t}} - E_{\mathrm{wc}}^z}{S_{\mathrm{wc}}^z}. \tag{5.22}$$

5.1.3   Expressions for the modal group velocity

In this section, we derive an expression for the group velocity of a mode of a waveguide with a purely real dielectric profile. As a theoretical approach we use a so-called Hellmann–Feynman theorem, according to which, derivative of the eigenvalue of a Hermitian operator can be calculated as a mean of the operator derivative.

We start with a mode having propagation constant β and satisfying the Hamiltonian equation:

$$\hat{H} \left| \mathbf{F}_\beta \right\rangle = \beta \hat{B} \left| \mathbf{F}_\beta \right\rangle. \tag{5.23}$$

To find the modal group velocity, defined as $v_{\mathrm{g}}^{-1} = \partial\beta/\partial\omega$, we first differentiate (5.23) with respect to ω:

$$\frac{\partial \hat{H}}{\partial \omega} \left| \mathbf{F}_\beta \right\rangle + \hat{H} \left| \frac{\partial \mathbf{F}_\beta}{\partial \omega} \right\rangle = \frac{\partial \beta}{\partial \omega} \hat{B} \left| \mathbf{F}_\beta \right\rangle + \beta \hat{B} \left| \frac{\partial \mathbf{F}_\beta}{\partial \omega} \right\rangle, \tag{5.24}$$

and then multiply the resultant expression from the left by $\langle \mathbf{F}_\beta |$:

$$\langle \mathbf{F}_\beta | \frac{\partial \hat{H}}{\partial \omega} \left| \mathbf{F}_\beta \right\rangle + \langle \mathbf{F}_\beta | \hat{H} \left| \frac{\partial \mathbf{F}_\beta}{\partial \omega} \right\rangle = \frac{\partial \beta}{\partial \omega} \langle \mathbf{F}_\beta | \hat{B} \left| \mathbf{F}_\beta \right\rangle + \beta \langle \mathbf{F}_\beta | \hat{B} \left| \frac{\partial \mathbf{F}_\beta}{\partial \omega} \right\rangle. \tag{5.25}$$

Using the Hermitian property of the Maxwell Hamiltonian (5.5) for the waveguides with purely real dielectric profiles, we find:

$$\left\langle \mathbf{F}_\beta \left| \hat{H} \right| \frac{\partial \mathbf{F}_\beta}{\partial \omega} \right\rangle = \left\langle \frac{\partial \mathbf{F}_\beta}{\partial \omega} \right| \hat{H} \left| \mathbf{F}_\beta \right\rangle^* = \beta \left\langle \mathbf{F}_\beta \left| \hat{B} \right| \frac{\partial \mathbf{F}_\beta}{\partial \omega} \right\rangle. \tag{5.26}$$

From (5.25) it then follows that:

$$\frac{1}{v_g} = \frac{\partial \beta}{\partial \omega} = \frac{\left\langle \mathbf{F}_\beta \left| \dfrac{\partial \hat{H}}{\partial \omega} \right| \mathbf{F}_\beta \right\rangle}{\left\langle \mathbf{F}_\beta \left| \hat{B} \right| \mathbf{F}_\beta \right\rangle}. \tag{5.27}$$

To calculate the matrix element in the denominator of (5.27) we first differentiate with respect to ω the explicit form (5.5) of the Hamiltonian $\hat{H}$ to find:

$$\frac{\partial \hat{H}}{\partial \omega} = \begin{pmatrix} \varepsilon + \omega^{-2} \nabla_t \times \left(\hat{\mathbf{z}} \left(\mu^{-1} \hat{\mathbf{z}} (\nabla_t \times) \right) \right) & 0 \\ 0 & \mu + \omega^{-2} \nabla_t \times \left(\hat{\mathbf{z}} \left(\varepsilon^{-1} \hat{\mathbf{z}} (\nabla_t \times) \right) \right) \end{pmatrix}, \tag{5.28}$$

and then use integration by parts, similarly to (5.11), to arrive at the following form of the matrix element:

$$\left\langle \mathbf{F}_\beta \left| \frac{\partial \hat{H}}{\partial \omega} \right| \mathbf{F}_\beta \right\rangle = \int_{wc} dxdy \left[\varepsilon \left| \mathbf{E}_\beta(x, y) \right|^2 + \mu \left| \mathbf{H}_\beta(x, y) \right|^2 \right] = 4 E_{wc}^{total}. \tag{5.29}$$

Substituting (5.29) into (5.27), and using definitions (5.20) and (5.21) for the modal electromagnetic energy flux, we finally get:

$$\frac{1}{v_g} = \frac{\partial \beta}{\partial \omega} = \frac{E_{wc}^{total}}{S_{wc}^z} = \frac{E_{wc}^t + E_{wc}^z}{S_{wc}^z}, \tag{5.30}$$

which can also be compared to the expression for the modal phase velocity (5.22)

$$v_p^{-1} = \beta / \omega = \left(E_{wc}^t - E_{wc}^z \right) / S_{wc}^z.$$

5.1.4 Orthogonality relation between the modes of a waveguide made of lossy dielectrics

In the case of a waveguide with complex dielectric profile $\varepsilon \neq \varepsilon^*$ or $\mu \neq \mu^*$ (absorbing dielectrics), the generalized eigenvalue problem (5.7) is no longer Hermitian, therefore the results of the previous section, including the orthogonality relation (5.15), are no longer valid. Interestingly, by modifying the definition of a dot product (5.9) between the modes, one can still derive a modified orthogonality relation, as well as several useful identities involving the value of the propagation constant.

In particular, we modify the dot product between the two modes to be:

$$\langle \mathbf{F}_{\beta'} | \hat{B} | \mathbf{F}_\beta \rangle = \langle \mathbf{F}_\beta | \hat{B} | \mathbf{F}_{\beta'} \rangle = \int_{\substack{waveguide \\ cross\text{-}section}} dxdy \begin{pmatrix} \mathbf{E}_t(x, y) \\ \mathbf{H}_t(x, y) \end{pmatrix}_{\beta'}^T \begin{pmatrix} 0 & -\hat{\mathbf{z}} \times \\ \hat{\mathbf{z}} \times & 0 \end{pmatrix} \begin{pmatrix} \mathbf{E}_t(x, y) \\ \mathbf{H}_t(x, y) \end{pmatrix}_\beta$$

$$= \int_{wc} dxdy \hat{\mathbf{z}} \left(\mathbf{E}_{t\beta'}(x, y) \times \mathbf{H}_{t\beta}(x, y) + \mathbf{E}_{t\beta}(x, y) \times \mathbf{H}_{t\beta'}(x, y) \right), \tag{5.31}$$

where superscript ^{T} signifies vector transposition. We now introduce the modified matrix elements consistent with the dot product (5.31):

$$\langle \mathbf{F}_{\beta'}|\hat{H}|\mathbf{F}_{\beta}\rangle = \int_{wc} dxdy\, \mathbf{E}_{t\beta'}(x,y)\left[\omega\varepsilon - \omega^{-1}\nabla_t \times \left(\hat{\mathbf{z}}\left(\mu^{-1}\hat{\mathbf{z}}(\nabla_t\times)\right)\right)\right]\mathbf{E}_{t\beta}(x,y)$$

$$+ \int_{wc} dxdy\, \mathbf{H}_{t\beta'}(x,y)\left[\omega\mu - \omega^{-1}\nabla_t \times \left(\hat{\mathbf{z}}\left(\varepsilon^{-1}\hat{\mathbf{z}}(\nabla_t\times)\right)\right)\right]\mathbf{H}_{t\beta}(x,y).$$

$$(5.32)$$

Proceeding as in (5.11), one can show that:

$$\langle \mathbf{F}_{\beta'}|\hat{H}|\mathbf{F}_{\beta}\rangle = \langle \mathbf{F}_{\beta}|\hat{H}|\mathbf{F}_{\beta'}\rangle \qquad (5.33)$$

$$= \omega \int_{wc} dxdy\left[\varepsilon \mathbf{E}_{\beta'}(x,y)\mathbf{E}_{\beta}(x,y) + \mu \mathbf{H}_{\beta'}(x,y)\mathbf{H}_{\beta}(x,y)\right],$$

where $\mathbf{E}$ and $\mathbf{H}$ are the full vectors (all three components) of the electromagnetic fields. Applying (5.31) and (5.33) to the same mode, and using (5.8), gives:

$$\beta = \frac{\langle \mathbf{F}_{\beta}|\hat{H}|\mathbf{F}_{\beta}\rangle}{\langle \mathbf{F}_{\beta}|\hat{B}|\mathbf{F}_{\beta}\rangle} = \omega \frac{\displaystyle\int_{wc} dxdy\left[\varepsilon \mathbf{E}_{\beta}^2(x,y) + \mu \mathbf{H}_{\beta}^2(x,y)\right]}{2\displaystyle\int_{wc} dxdy\hat{\mathbf{z}}\left(\mathbf{E}_{t\beta}(x,y) \times \mathbf{H}_{t\beta}(x,y)\right)}. \qquad (5.34)$$

Finally, we demonstrate the modified orthogonality relation between the two distinct waveguide modes. In particular, for the modes with propagation constants β and β' from (5.8) it follows that:

$$\langle \mathbf{F}_{\beta'}|\,\hat{H}\,|\mathbf{F}_{\beta}\rangle = \beta\,\langle \mathbf{F}_{\beta'}|\,\hat{B}\,|\mathbf{F}_{\beta}\rangle$$

$$\langle \mathbf{F}_{\beta}|\,\hat{H}\,|\mathbf{F}_{\beta'}\rangle = \beta'\,\langle \mathbf{F}_{\beta}|\,\hat{B}\,|\mathbf{F}_{\beta'}\rangle. \qquad (5.35)$$

Using the equalities in (5.31) and (5.33) and subtracting the first equation in (5.35) from the second one we get:

$$(\beta' - \beta)\langle \mathbf{F}_{\beta}|\,\hat{B}\,|\mathbf{F}_{\beta'}\rangle = 0. \qquad (5.36)$$

Therefore, if $\beta \neq \beta'$, the only way to satisfy (5.36) is through modal orthogonality in the sense:

$$\langle \mathbf{F}_{\beta'}|\,\hat{B}\,|\mathbf{F}_{\beta}\rangle = 0; \quad \beta \neq \beta'. \qquad (5.37)$$

5.2   Perturbation theory for uniform variations in a waveguide dielectric profile

In this section, we derive corrections to the modal propagation constants and eigenfields, assuming that the waveguide dielectric profile is weakly perturbed in such a manner as

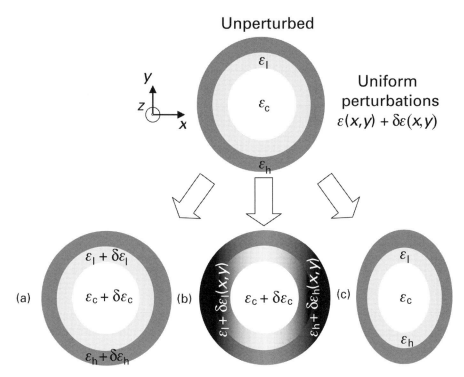

Figure 5.1 Possible uniform variations of a waveguide profile. (a) Perturbations that do not change the positions of the dielectric interfaces and do not change the original symmetry. For example, addition of material absorption losses. (b) Perturbations that do not change the positions of the dielectric interfaces but do break the original symmetry. For example, variations in dielectric profile that break the circular symmetry of a fiber, such as stress-induced birefringence. (c) Perturbations that do change the positions of the dielectric interfaces. For example, fiber ellipticity.

to maintain its uniformity along the direction of modal propagation. In this case, the perturbed waveguide still maintains continuous translational symmetry, and, therefore, allows harmonic eigenmodes labeled by the conserved propagation constants. An example of such uniform perturbation can be adding a small imaginary part to the dielectric constant of the underlying materials, thus introducing material absorption losses, as shown in Fig. 5.1(a). Another example is modal birefringence induced by trasversely nonuniform variation in the dielectric profile (Fig. 5.1(b)) without the actual change in the position of dielectric interfaces. Such perturbation can be induced, for example, by stress accumulation during imperfect fiber drawing.

The perturbation theory formulation presented in this section is valid for any uniform variation of a waveguide dielectric profile, regardless of the index contrast, as long as this variation does not change the position of the dielectric interfaces. Perturbative expressions similar to the ones derived in this section can also be found in earlier works. [2,3] Moreover, if the index contrast in a waveguide dielectric profile is small (typically, less than 10% in the ratio of the index contrast to the average index of refraction [1]), then the perturbation theory developed in this section tends to be also valid even when material boundaries are shifted (the case of fiber ellipticity shown in Fig. 5.1(c), for example). In general, the development of perturbation theory for the case of high-index-contrast waveguides with shifting material boundaries is nontrivial and is beyond the scope of this book. [4,5,6] One of the few cases when it can be done relatively simply is in the case of isotropic scaling of a dielectric profile, which is presented at the end of this section.

Finally, in what follows we assume that the mode in question is a true guided wave that has a purely real propagation constant β. However, all the results derived in the following sections are also applicable for evanescent waves with complex propagation constants, but, in all the expressions, matrix elements of the form $\langle \mathbf{F}_\beta | \ldots | \mathbf{F}_{\beta'} \rangle$ have to be substituted by matrix elements of the form $\langle \mathbf{F}_{\beta^*} | \ldots | \mathbf{F}_{\beta'} \rangle$.

5.2.1   Perturbation theory for the nondegenerate modes: example of material absorption

In this section we derive expressions for the first- and second-order perturbation-theory corrections to the modal propagation constant and modal fields of a perturbed Hamiltonian. We then apply these expressions to characterize modal propagation losses in a waveguide that incorporates absorbing dielectrics. In this section we suppose that the waveguide mode is either nondegenerate (TE or TM mode of a planar waveguide, for example), or that the perturbation in question does not lead to coupling between the degenerate modes (for example, in the case of a doubly degenerate linear-polarized mode of a circular fiber under a circular symmetric perturbation of a dielectric constant, like material absorption).

Defining $\delta \hat{H}$ to be the correction to an unperturbed Hamiltonian $\hat{H}_0$, assuming purely real unperturbed dielectric profile and dot product (5.9), and using eigenmodes of Hermitian Hamiltonian $\hat{H}_0$ satisfying:

$$\hat{H}_0 \left| \mathbf{F}_{\beta_0}^0 \right\rangle = \beta_0 \hat{B} \left| \mathbf{F}_{\beta_0}^0 \right\rangle, \tag{5.38}$$

similarly to (4.26), (4.27), (4.28), one can demonstrate that the propagation constants and fields of the perturbed guided modes are related to those of the unperturbed modes as:

$$\beta = \beta_0 + \underbrace{\frac{\left\langle \mathbf{F}_{\beta_0}^0 \left| \delta \hat{H} \right| \mathbf{F}_{\beta_0}^0 \right\rangle}{\left\langle \mathbf{F}_{\beta_0}^0 \left| \hat{B} \right| \mathbf{F}_{\beta_0}^0 \right\rangle}}_{\text{first-order correction}} + \underbrace{\sum_{\substack{\beta_0' \neq \beta_0 \\ \beta_0' \subset \text{Real}}} \frac{\left\langle \mathbf{F}_{\beta_0}^0 \left| \delta \hat{H} \right| \mathbf{F}_{\beta_0'}^0 \right\rangle \left\langle \mathbf{F}_{\beta_0'}^0 \left| \delta \hat{H} \right| \mathbf{F}_{\beta_0}^0 \right\rangle}{\left\langle \mathbf{F}_{\beta_0}^0 \left| \hat{B} \right| \mathbf{F}_{\beta_0}^0 \right\rangle \left\langle \mathbf{F}_{\beta_0'}^0 \left| \hat{B} \right| \mathbf{F}_{\beta_0'}^0 \right\rangle} \frac{1}{\beta_0 - \beta_0'}}_{\text{second-order correction due to coupling to true guided modes}}$$

$$+ \underbrace{\sum_{\beta_0' \subset \text{complex}} \frac{\left\langle \mathbf{F}_{\beta_0}^0 \left| \delta \hat{H} \right| \mathbf{F}_{\beta_0'}^0 \right\rangle \left\langle \mathbf{F}_{\beta_0'^*}^0 \left| \delta \hat{H} \right| \mathbf{F}_{\beta_0}^0 \right\rangle}{\left\langle \mathbf{F}_{\beta_0}^0 \left| \hat{B} \right| \mathbf{F}_{\beta_0}^0 \right\rangle \left\langle \mathbf{F}_{\beta_0'^*}^0 \left| \hat{B} \right| \mathbf{F}_{\beta_0}^0 \right\rangle} \frac{1}{\beta_0 - \beta_0'}}_{\text{second-order correction due to coupling to evanescent modes}} + O(\delta^3), \tag{5.39}$$

and

$$\left|\mathbf{F}_{\beta}\right\rangle = \left|\mathbf{F}_{\beta_0}^0\right\rangle + \sum_{\substack{\beta_0' \neq \beta_0 \\ \beta_0' \subset \text{real}}} \frac{\left\langle \mathbf{F}_{\beta_0'}^0 \left|\delta\hat{H}\right| \mathbf{F}_{\beta_0}^0\right\rangle}{\left\langle \mathbf{F}_{\beta_0'}^0 \left|\hat{B}\right| \mathbf{F}_{\beta_0'}^0\right\rangle} \frac{\left|\mathbf{F}_{\beta_0'}^0\right\rangle}{\beta_0 - \beta_0'}$$

$$\underbrace{\phantom{\sum_{\substack{\beta_0' \neq \beta_0 \\ \beta_0' \subset \text{real}}} \frac{\left\langle \mathbf{F}_{\beta_0'}^0 \left|\delta\hat{H}\right| \mathbf{F}_{\beta_0}^0\right\rangle}{\left\langle \mathbf{F}_{\beta_0'}^0 \left|\hat{B}\right| \mathbf{F}_{\beta_0'}^0\right\rangle} \frac{\left|\mathbf{F}_{\beta_0'}^0\right\rangle}{\beta_0 - \beta_0'}}}_{\text{first-order correction due to coupling to true guided modes}}$$

$$+ \underbrace{\sum_{\beta_0' \subset \text{complex}} \frac{\left\langle \mathbf{F}_{\beta_0'^*}^0 \left|\delta\hat{H}\right| \mathbf{F}_{\beta_0}^0\right\rangle}{\left\langle \mathbf{F}_{\beta_0'^*}^0 \left|\hat{B}\right| \mathbf{F}_{\beta_0'}^0\right\rangle} \frac{\left|\mathbf{F}_{\beta_0'}^0\right\rangle}{\beta_0 - \beta_0'}}_{\text{first-order correction due to coupling to evanescent modes}} + O(\delta^2). \tag{5.40}$$

Moreover, the modified propagation constants (5.39) and fields (5.40) are the true eigenvalues and eigenfunctions of the perturbed Hamiltonian up to the second order in the perturbation strength:

$$\left(\hat{H}_0 + \delta\hat{H}\right)\left|\mathbf{F}_{\beta}\right\rangle = \beta\hat{B}\left|\mathbf{F}_{\beta}\right\rangle + O(\delta^2). \tag{5.41}$$

We now compute the first-order correction (5.39) to the modal propagation constant due to a small perturbation in a purely real dielectric profile $\varepsilon(x, y)$ of a waveguide. Assuming perturbation in the form $\varepsilon(x, y) \to \varepsilon(x, y) + \delta\varepsilon(x, y)$, $\mu = 1$ we derive for the perturbation correction to the Hamiltonian:

$$\delta\hat{H} = \hat{H}(\varepsilon + \delta\varepsilon) - \hat{H}(\varepsilon) \underset{\delta\varepsilon \ll \varepsilon}{\simeq} \begin{pmatrix} \omega\delta\varepsilon & 0 \\ 0 & \omega^{-1}\nabla_t \times \left(\hat{\mathbf{z}}\left(\frac{\delta\varepsilon}{\varepsilon^2}\hat{\mathbf{z}}(\nabla_t\times)\right)\right) \end{pmatrix}. \tag{5.42}$$

The matrix element for the first-order correction in (5.39) can be simplified by using integration by parts and equations (5.1), (5.2), and (5.3). Following the same steps as in (5.10) one can demonstrate that:

$$\left\langle \mathbf{F}_{\beta}^0 \left|\delta\hat{H}\right| \mathbf{F}_{\beta'}^0\right\rangle = \omega \int_{\text{wc}} dx dy\, \delta\varepsilon(x, y) \left[E_{t\beta}^{*0}(x, y)E_{t\beta'}^0(x, y) + E_{z\beta}^{*0}(x, y)E_{z\beta'}^0(x, y)\right], \tag{5.43}$$

and, finally, to the first order:

$$\beta - \beta_0 = \frac{\left\langle \mathbf{F}_{\beta_0}^0 \left|\delta\hat{H}\right| \mathbf{F}_{\beta_0}^0\right\rangle}{\left\langle \mathbf{F}_{\beta_0}^0 \left|\hat{B}\right| \mathbf{F}_{\beta_0}^0\right\rangle} \tag{5.44}$$

$$= \omega \frac{\displaystyle\int_{\text{wc}} dx dy\, \delta\varepsilon(x, y)\left|\mathbf{E}_{\beta_0}^0(x, y)\right|^2}{\displaystyle\int_{\text{wc}} dx dy\, \hat{\mathbf{z}}\left(\mathbf{E}_{t\beta_0}^{0*}(x, y) \times \mathbf{H}_{t\beta_0}^0(x, y) + \mathbf{E}_{t\beta_0}^0(x, y) \times \mathbf{H}_{t\beta_0}^{0*}(x, y)\right)}.$$

Modal propagation loss due to waveguide material absorption

We now apply (5.44) to characterize modal propagation losses in a waveguide made of absorbing materials. Material absorption can be characterized by a small imaginary contribution to the waveguide refractive index:

$$\varepsilon(x, y) = (\sqrt{\varepsilon_0(x, y)} + in_i(x, y))^2 \tag{5.45}$$

$$\rightarrow \delta\varepsilon(x, y) \underset{n_i \ll \varepsilon_0}{=} 2in_i(x, y)\sqrt{\varepsilon_0(x, y)}.$$

According to (5.44), this defines the imaginary contribution to the propagation constant $\beta - \beta_0 \propto i\langle n_i \rangle$, leading to the modal field decay along the direction of propagation due to the $\exp(i\beta z)$ dependence of the fields. If the losses of all the absorbing materials are the same $n_i(x, y) = n_i$, (5.44) can be further simplified. Namely:

$$\beta - \beta_0 = in_i\omega f$$

$$f = \frac{2\displaystyle\int_{\substack{\text{absorbing region}}} dxdy\,\sqrt{\varepsilon_0(x, y)}\,\left|E^0_{\beta_0}(x, y)\right|^2}{\displaystyle\int_{wc} dxdy\,\hat{z}\left(E^{0*}_{t\beta_0}(x, y) \times H^0_{t\beta_0}(x, y) + E^0_{t\beta_0}(x, y) \times H^{0*}_{t\beta_0}(x, y)\right)}, \tag{5.46}$$

where integration in the denominator is performed only over the spatial regions containing absorbing materials, while f defines the modal field fraction in the absorbing regions. After substitution of a complex propagation constant (5.46) into the functional form of modal fields (5.6), for the modal energy flux we get $S_z \sim \exp(-2zn_i\omega f)$. This defines the propagation loss of a waveguide mode as $\alpha^{wg}_{abs}[1/m] = 2n_i\omega f$. In engineering, it is also customary to express the modal power loss in units of [dB/m], which for exponentially decaying functions is defined as:

$$\alpha^{wg}_{abs}[dB/m] = -\frac{10}{z[m]}\log_{10}\left(\frac{S_z(z)}{S_z(0)}\right) = \frac{20}{\log(10)}n_i\omega f. \tag{5.47}$$

To understand (5.47) better, we compare it with the absorption loss of a plane wave propagating along the $\hat{z}$ direction in a uniform absorbing dielectric described by $\varepsilon = (\sqrt{\varepsilon_0} + in_i)^2$. From the dispersion relation of a plane-wave solution $k_z^2 = \omega^2\varepsilon$, it follows that $k_z = \omega_0\sqrt{\varepsilon_0} + in_i\omega$, and, therefore, the bulk material absorption loss can be defined as $\alpha^{material}_{abs}[1/m] = 2n_i\omega$. Thus, the propagation loss of a waveguide mode is related to the bulk absorption loss of a constituent material as:

$$\alpha^{wg}_{abs} = \alpha^{material}_{abs} f, \tag{5.48}$$

where f, defined in (5.46), is a modal field fraction in the absorbing region.

5.2.2 Perturbation theory for the degenerate modes coupled by perturbation: example of polarization-mode dispersion

In this section, we deal with modes that are degenerate (having the same propagation constant) in an unperturbed waveguide, while later coupled by the perturbation in a waveguide dielectric profile. As a result, the propagation constants of the perturbed modes will become different from each other, with the difference proportional to the perturbation strength. An example of such modes might be the doubly degenerate circular-polarized modes of a circular symmetric fiber. Under a noncircular symmetric perturbation of a dielectric profile, such as stress-induced change in the refractive index of a fiber under pressure (see Fig. 5.1(b)), two originally degenerate modes will form two properly symmetrized supermodes with distinct propagation constants β^+ and β^-. The goal of this section is to derive first-order perturbation theory corrections for the modal propagation constants and modal fields of a perturbed Hamiltonian.

We define $|\mathbf{F}_{\beta_0}^{0+}\rangle$ and $|\mathbf{F}_{\beta_0}^{0-}\rangle$ to be the orthogonal degenerate eigenmodes with a real propagation constant β_0. In the unperturbed system, any linear combination of such modes:

$$\left|\mathbf{F}_{\beta_0}\right\rangle = C^+ \left|\mathbf{F}_{\beta_0}^{0+}\right\rangle + C^- \left|\mathbf{F}_{\beta_0}^{0-}\right\rangle, \tag{5.49}$$

is also an eigenstate of a waveguide Hamiltonian with the same value of a propagation constant. When perturbation is introduced, (5.49) no longer remains an eigenstate of a perturbed Hamiltonian, except for a specific choice of the expansion coefficients. In particular, consider the eigenequation for the new eigenstates under the presence of a perturbation:

$$\left(\hat{H}_0 + \delta\hat{H}\right)\left(C^+\left|\mathbf{F}_{\beta_0}^{0+}\right\rangle + C^-\left|\mathbf{F}_{\beta_0}^{0-}\right\rangle\right)$$
$$= (\beta_0 + \delta\beta)\,\hat{B}\left(C^+\left|\mathbf{F}_{\beta_0}^{0+}\right\rangle + C^-\left|\mathbf{F}_{\beta_0}^{0-}\right\rangle\right) + O(\delta^2). \tag{5.50}$$

Multiplying the left and right sides of the equations by $\langle\mathbf{F}_{\beta_0}^{0+}|$ and $\langle\mathbf{F}_{\beta_0}^{0-}|$, assuming orthogonality of the degenerate states $\langle\mathbf{F}_{\beta_0}^{0\pm}|\hat{B}|\mathbf{F}_{\beta_0}^{0\mp}\rangle = 0$, $\langle\mathbf{F}_{\beta_0}^{0+}|\hat{B}|\mathbf{F}_{\beta_0}^{0+}\rangle = \langle\mathbf{F}_{\beta_0}^{0-}|\hat{B}|\mathbf{F}_{\beta_0}^{0-}\rangle \neq 0$, and keeping all the terms up to the first order we arrive at the following eigenvalue problem with respect to the correction in the propagation constant $(\beta - \beta_0)$:

$$\begin{pmatrix} \left\langle\mathbf{F}_{\beta_0}^{0+}\left|\delta\hat{H}\right|\mathbf{F}_{\beta_0}^{0+}\right\rangle & \left\langle\mathbf{F}_{\beta_0}^{0+}\left|\delta\hat{H}\right|\mathbf{F}_{\beta_0}^{0-}\right\rangle \\ \left\langle\mathbf{F}_{\beta_0}^{0-}\left|\delta\hat{H}\right|\mathbf{F}_{\beta_0}^{0+}\right\rangle & \left\langle\mathbf{F}_{\beta_0}^{0-}\left|\delta\hat{H}\right|\mathbf{F}_{\beta_0}^{0-}\right\rangle \end{pmatrix}\begin{pmatrix} C^+ \\ C^- \end{pmatrix}$$
$$= (\beta - \beta_0)\begin{pmatrix} \left\langle\mathbf{F}_{\beta_0}^{0+}\left|\hat{B}\right|\mathbf{F}_{\beta_0}^{0+}\right\rangle & 0 \\ 0 & \left\langle\mathbf{F}_{\beta_0}^{0-}\left|\hat{B}\right|\mathbf{F}_{\beta_0}^{0-}\right\rangle \end{pmatrix}\begin{pmatrix} C^+ \\ C^- \end{pmatrix}. \tag{5.51}$$

Frequently, one encounters the case when diagonal elements are zero $\langle\mathbf{F}_{\beta_0}^{0\pm}|\delta\hat{H}|\mathbf{F}_{\beta_0}^{0\pm}\rangle = 0$, while $\langle\mathbf{F}_{\beta_0}^{0+}|\delta\hat{H}|\mathbf{F}_{\beta_0}^{0-}\rangle = \langle\mathbf{F}_{\beta_0}^{0-}|\delta\hat{H}|\mathbf{F}_{\beta_0}^{0+}\rangle^*$. The solution of (5.51) is then particularly simple:

$$\beta^\pm = \beta_0 \pm \frac{\left|\left\langle\mathbf{F}_{\beta_0}^{0+}\left|\delta\hat{H}\right|\mathbf{F}_{\beta_0}^{0-}\right\rangle\right|}{\sqrt{\left\langle\mathbf{F}_{\beta_0}^{0-}\left|\hat{B}\right|\mathbf{F}_{\beta_0}^{0-}\right\rangle}}, \tag{5.52}$$

while fields of waveguide supermodes are:

$$|\mathbf{F}_{\beta\pm}\rangle = \frac{1}{2}\left(\left|\mathbf{F}_{\beta_0}^{0+}\right\rangle \pm \left|\mathbf{F}_{\beta_0}^{0-}\right\rangle\right).\tag{5.53}$$

Modal birefringence and polarization-mode dispersion induced by the elliptical variations in a circularly symmetric fiber

As an example, consider mode birefringence in a circularly symmetric fiber induced by a noncircularly symmetric perturbation of a dielectric profile of the form:

$$\varepsilon(r) \rightarrow \varepsilon(r)(1 + \delta\cos(2\theta)), \mu = 1.\tag{5.54}$$

Such a perturbation can arise from a uniaxial compression or heating of the fiber. Unperturbed circular symmetric fibers possess both continuous translational and continuous rotational symmetries, defining a general form of a solution (in cylindrical coordinates):

$$|\mathbf{F}_{m,\beta_0}\rangle = \exp(im\theta)\mathbf{F}_{m,\beta_0}(\rho).\tag{5.55}$$

For any $m \geq 1$, the modes with angular momenta $\pm m$ are degenerate with the same value of propagation constant β_0. Moreover, it can be demonstrated directly from Maxwell's equations that the fields in the degenerate modes (m, β_0), $(-m, \beta_0)$ can be chosen to be related by the following transformations:

$$E_z^{-m}(\rho) = E_z^m(\rho), \quad E_\rho^{-m}(\rho) = E_\rho^m(\rho), \quad E_\theta^{-m}(\rho) = -E_\theta^m(\rho)$$
$$H_z^{-m}(\rho) = -H_z^m(\rho), \quad H_\rho^{-m}(\rho) = -H_\rho^m(\rho), \quad E_\theta^{-m}(\rho) = H_\theta^m(\rho).\tag{5.56}$$

We now apply (5.52) to estimate the modal birefringence due to a noncircularly symmetric perturbation of a dielectric profile of the form (5.54). For the fundamental mode $m = 1$, the coupling element $\langle \mathbf{F}_{1,\beta_0}^0|\delta\hat{H}|\mathbf{F}_{-1,\beta_0}^0\rangle$ can be found easily from (5.43):

$$\left\langle \mathbf{F}_{1,\beta_0}^0 \left|\delta\hat{H}\right| \mathbf{F}_{-1,\beta_0}^0 \right\rangle$$

$$= \delta\omega \int_0^{+\infty} \rho\mathrm{d}\rho \int_0^{2\pi} \mathrm{d}\theta \cos(2\theta)\varepsilon(\rho) \left[\begin{array}{l} E_{\rho(1,\beta_0)}^{*0}(\rho,\theta)E_{\rho(-1,\beta_0)}^0(\rho,\theta) \\ + E_{\theta(1,\beta_0)}^{*0}(\rho,\theta)E_{\theta(-1,\beta_0)}^0(\rho,\theta) \\ + E_{z(1,\beta_0)}^{*0}(x,y)E_{z(-1,\beta_0)}^0(x,y) \end{array} \right]$$

$$= \delta\omega \int_0^{+\infty} \rho\mathrm{d}\rho \int_0^{2\pi} \mathrm{d}\theta \cos(2\theta)\exp(-\mathrm{i}2\theta)\varepsilon(\rho)$$

$$\times \left[\left|E_{\rho(1,\beta_0)}^0(\rho)\right|^2 - \left|E_{\theta(1,\beta_0)}^0(\rho)\right|^2 + \left|E_{z(1,\beta_0)}^0(\rho)\right|^2 \right]$$

$$= \frac{\delta}{2}\omega \int_0^{+\infty} 2\pi\rho\mathrm{d}\rho\varepsilon(\rho) \left[\left|E_{\rho(1,\beta_0)}^0(\rho)\right|^2 - \left|E_{\theta(1,\beta_0)}^0(\rho)\right|^2 + \left|E_{z(1,\beta_0)}^0(\rho)\right|^2 \right]$$

$$= 2\omega\delta \left[E_\rho^{\mathrm{el}} - E_\theta^{\mathrm{el}} + E_z^{\mathrm{el}} \right],\tag{5.57}$$

where $E^{el}_{\rho,\theta,z}$ are the electric energies of the various field components in a waveguide cross-section. Finally, from (5.52) it follows that the difference in the propagation constants of the newly formed supermodes (perturbation induced birefringence) for $m = 1$ is:

$$\beta^+ - \beta^- = \delta \cdot \omega \frac{E^{el}_\rho - E^{el}_\theta + E^{el}_z}{S^z_{wc}}, \qquad (5.58)$$

while the supermode fields are linearly polarized modes of the form:

$$\left|\mathbf{F}_{\beta\pm}\right\rangle = \frac{1}{2}\left(\left|\mathbf{F}^{+1}_{\beta_0}\right\rangle \pm \left|\mathbf{F}^{-1}_{\beta_0}\right\rangle\right)$$

$$\left|\mathbf{F}_{\beta+}\right\rangle = \begin{pmatrix} E^1_z(\rho)\cos(\theta),\ E^1_\rho(\rho)\cos(\theta),\ iE^1_\theta(\rho)\sin(\theta), \\ iH^1_z(\rho)\sin(\theta),\ iH^1_\rho(\rho)\sin(\theta),\ H^1_\theta(\rho)\cos(\theta) \end{pmatrix} \qquad (5.59)$$

$$\left|\mathbf{F}_{\beta-}\right\rangle = \begin{pmatrix} iE^1_z(\rho)\sin(\theta),\ iE^1_\rho(\rho)\sin(\theta),\ E^1_\theta(\rho)\cos(\theta), \\ H^1_z(\rho)\cos(\theta),\ H^1_\rho(\rho)\cos(\theta),\ iH^1_\theta(\rho)\sin(\theta) \end{pmatrix}.$$

The splitting of a doubly degenerate fundamental $m = 1$ mode under external perturbation into two supermodes with somewhat different propagation constants can have serious implications on the information capacity carried by the fiber link. In particular, after an initial launch into a doubly degenerate mode, in the region of perturbation the signal will be split between the two supermodes propagating with somewhat different group velocities $v^+_g = \partial\omega/\partial\beta^+ \neq v^-_g = \partial\omega/\partial\beta^-$. Assuming that the same perturbation persists over the whole fiber link, after the propagation distance $L((v^+_g)^{-1} - (v^-_g)^{-1}) \propto B^{-1}$, where B is a signal bit rate, the signal will effectively be scrambled. The parameter $\tau = (v^+_g)^{-1} - (v^-_g)^{-1}$ is called an inter-mode dispersion parameter and can be expressed through the frequency derivative of a modal birefringence (5.58) as:

$$\tau = \left(v^+_g\right)^{-1} - \left(v^-_g\right)^{-1} = \frac{\partial(\beta^+ - \beta^-)}{\partial\omega} \sim \delta. \qquad (5.60)$$

5.2.3 Perturbations that change the positions of dielectric interfaces

In the case of high-index-contrast waveguides and variations leading to the changes in the position of dielectric interfaces, the correct formulation of perturbation theory is not trivial. [4,5] The conventional approach to the evaluation of the matrix elements described in 5.2.1 and 5.2.2 proceeds by first expanding the perturbed modal fields into the modal fields of an unperturbed system, and then, given the explicit form of a perturbation operator, by computing the required matrix elements. Unfortunately, this approach encounters difficulties when applied to the problem of finding perturbed electromagnetic modes of the waveguides with shifted high-index-contrast dielectric interfaces. In this case, the expansion of the perturbed modes into an increasing number of the modes of an unperturbed system does not converge to a correct solution when the standard form of the matrix elements (5.43) is used. [4] The mathematical reasons for such a failure lie in the incompleteness of the basis of the eigenmodes of an unperturbed waveguide in the domain of the eigenmodes of a perturbed waveguide, as well as in the fact that the mode-orthogonality condition (5.16) does not constitute a strict norm.

We would like to point out that coupled-mode theory using standard matrix elements (5.43) can still be used even in the problem of perturbations in high-index-contrast waveguides with shifting dielectric interfaces. [6] However, as an expansion basis for the fields of a perturbed mode, one has to use modes of a waveguide with a continuous dielectric profile (graded-index waveguide, for example), rather than modes of an unperturbed waveguide. Unfortunately, in this case, the convergence of such a method with respect to the number of modes in the basis is slow (at most linear). A perturbation formulation within this approach is also problematic and, even for a small perturbation, a complete matrix of the coupling elements has to be recomputed.

Other methods developed to deal with shifting metallic boundaries and dielectric interfaces originate primarily from the works on metallic waveguides and microwave circuits. [6] There, however, the Hermitian nature of Maxwell's equations in the problem of radiation propagation along the waveguides is not emphasized, and consequent development of perturbation expansions is usually omitted. Moreover, dealing with nonuniform waveguides, these formulations usually employ an expansion basis of instantaneous modes. Such modes have to be recalculated at each different waveguide cross-section, thus leading to computationally demanding propagation schemes.

Recently, the method of perturbation matching [4] was developed to allow computation of the correct matrix elements using, as an expansion basis, the modes of an unperturbed waveguide. This method is valid for a general case of any analytical variation of the waveguide geometry. To derive a correct form of the matrix elements one starts by defining an analytical function that describes the variation of the waveguide dielectric profile. One then constructs a novel expansion basis using spatially stretched modes of an unperturbed waveguide. The stretching is performed in such a way as to match the regions of the field discontinuities in the expansion modes with the positions of the perturbed dielectric interfaces. By substituting such expansions into Maxwell's equations, one then finds the required expansion coefficients. It becomes more convenient to perform further algebraic manipulations in a coordinate system where stretched expansion modes become again unperturbed modes of an original waveguide. Thus, the final steps of evaluation of the coupling elements involve transforming and manipulating Maxwell's equations in the perturbation-matched curvilinear coordinates. Although powerful, this method is best suited to deal with the geometrical variations described analytically.

To demonstrate the method of perturbation matching, we consider a uniform scaling perturbation, where all the transverse coordinates are scaled by the same factor:

$$x' = x(1 + \delta), \, y' = y(1 + \delta), \, z' = z. \tag{5.61}$$

As in the rest of the chapter, we develop perturbation theory in β keeping the frequency ω fixed. We start by defining a perturbed dielectric profile using the dielectric profile of an unperturbed waveguide through coordinate transformation (5.61) as:

$$\varepsilon_{\text{perturbed}} = \varepsilon_{\text{unperturbed}} \, (x'(x, y), y'(x, y)). \tag{5.62}$$

Thus defined, the stretching corresponds to a uniform increase or decrease of all the waveguide dimensions. Given the modes of an unperturbed waveguide $\mathbf{F}^0_{\beta_0}(x, y)$, we define an expansion basis of the stretched modes as:

$$\mathbf{F}^0_{\beta_0(\omega)}(x'(x, y), y'(x, y)). \tag{5.63}$$

Note that the positions of the dielectric interfaces in a perturbed dielectric profile (5.62) will coincide, by definition, with positions of the field discontinuities in the perturbation-matched expansion basis (5.63). Finally, the fields $\mathbf{F}_{\beta(\omega)}(x, y)$ of a perturbed mode are expanded in terms of a linear combination of the basis functions (5.63):

$$\mathbf{F}_{\beta(\omega)}(x, y) = \sum_{\beta'_0} C^{\beta'_0} \mathbf{F}^0_{\beta'_0(\omega)}(x'(x, y), y'(x, y)). \tag{5.64}$$

In the case of a uniform variation (5.61), from (5.5) and (5.7) it follows that $\mathbf{F}^0_{\beta_0}(x(1 + \delta), y(1 + \delta))$ is an eigenfunction of the same waveguide Hamiltonian (5.5), however, with a different propagation constant $\beta = \beta_0(1 + \delta)$, and a different frequency $\tilde{\omega} = \omega(1 + \delta)$. Thus:

$$\mathbf{F}_{\beta_0(1+\delta),\omega(1+\delta)}(x, y) = \mathbf{F}_{\beta_0,\omega}(x(1 + \delta), y(1 + \delta)). \tag{5.65}$$

Note that even though the perturbation-matched basis function $\mathbf{F}^0_{\beta_0}(x(1 + \delta), y(1 + \delta))$ is an eigenstate at $\tilde{\omega}$, it is, however, not an eigenstate at ω. Therefore, even for a simple scaling perturbation (5.61), the perturbed eigenstate $\mathbf{F}_\beta(x, y)$ is still described by an infinite linear combination of basis functions (5.64). Finally, by substitution of (5.64) into the perturbed Hamiltonian, after making a coordinate transformation (5.61), keeping only the terms of the first order, and using orthogonality relations between the modes of an unperturbed Hamiltonian, one can derive the first-order perturbative corrections to the values of the propagation constants.

Fortunately, in the particular case of scaling perturbation (5.61), the same result can be achieved much more simply. Thus, from (5.5), (5.7) it follows that given an unperturbed waveguide dispersion relation $\beta(\omega)$, for any scaling factor $(1 + \delta)$ the following holds for the dispersion relation $\tilde{\beta}(\omega)$ of a scaled waveguide:

$$\tilde{\beta}(\omega(1 + \delta)) = \beta(\omega) \cdot (1 + \delta) \rightarrow \tilde{\beta}(\omega) = \beta\left(\frac{\omega}{1 + \delta}\right) \cdot (1 + \delta). \tag{5.66}$$

Using Taylor expansion of (5.66) to the first order in a small parameter δ, we then find the first-order correction to the modal propagation constant due to uniform scaling:

$$\tilde{\beta}(\omega) - \beta(\omega) = \delta\left(\beta(\omega) - \omega\frac{\partial \beta}{\partial \omega}\right)$$

$$\tilde{\beta}(\omega) - \beta(\omega) = \delta \cdot \omega\left(\frac{1}{v_p} - \frac{1}{v_g}\right) = -\delta \cdot 2\omega\frac{E^z_{wc}}{S^z_{wc}}, \tag{5.67}$$

where we have used integral expressions (5.22), (5.30) for the phase and group velocities.

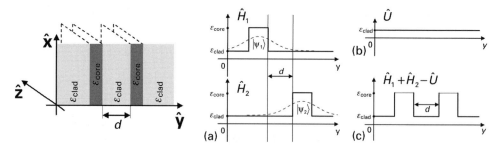

Figure P5.1.1 Schematic of a two-waveguide coupler. (a) Dielectric profiles of the two constituent slab waveguides. (b) Dielectric profile of a uniform cladding. (c) Dielectric profile of a two-slab waveguide coupler.

5.3   Problems

5.1   Supermodes of a two-waveguide coupler

In this problem we use perturbation theory to find electromagnetic eigenstates of a coupler made of the two phase-matched, weakly coupled slab waveguides (see schematic in Fig. P5.1.1). Propagation is assumed to be along the $\hat{z}$ direction.

Consider $\hat{H}_1$ and $\hat{H}_2$ to be the Hamiltonians of the two constituent slab waveguides (not necessarily identical) with step-like dielectric profiles, shown in Fig. P5.1.1(a). Let $\hat{U}$ be the Hamiltonian of a uniform cladding (see Fig. P5.1.1(b)). We define $|\psi_1\rangle$, $|\psi_2\rangle$ to be the modes of the two individual slab waveguides, which are phase-matched at a given frequency ω. In this case, the waveguide modes have the same value of a propagation constant k_z^0, and they satisfy the following equations:

$$\hat{H}_1|\psi_1\rangle = k_z^0|\psi_1\rangle$$
$$\hat{H}_2|\psi_2\rangle = k_z^0|\psi_2\rangle. \tag{P5.1.1}$$

For the step-index dielectric profile shown in Fig. P5.1.1(c), it can be confirmed directly from the explicit form of a Hamiltonian operator (5.5) that the Hamiltonian for a two-waveguide coupler can be written as:

$$\hat{H} = \hat{H}_1 + \hat{H}_2 - \hat{U}. \tag{P5.1.2}$$

We now presume that the coupler eigenstate $|\psi_c\rangle$, satisfying:

$$(\hat{H}_1 + \hat{H}_2 - \hat{U})|\psi_c\rangle = k_z|\psi_c\rangle, \tag{P5.1.3}$$

can be approximated by the linear combination of the eigenstates of two constituent waveguides:

$$|\psi_c\rangle = a_1|\psi_1\rangle + a_2|\psi_2\rangle, \tag{P5.1.4}$$

with the new eigenvalues:

$$k_z = k_z^0 + \delta k_z. \tag{P5.1.5}$$

Substituting (P5.1.4) and (P5.1.5) into the coupler eigenequation (P5.1.3), find the first-order accurate expressions for the values of the supermode propagation constants (P5.1.5), as well as the corresponding field combinations (P5.1.4), assuming:

$$\langle \psi_1 | \hat{H}_2 - \hat{U} | \psi_2 \rangle = \langle \psi_2 | \hat{H}_1 - \hat{U} | \psi_1 \rangle = \delta \quad \subset \text{Real}$$

$$\langle \psi_1 | \psi_1 \rangle = 1; \quad \langle \psi_2 | \psi_2 \rangle = 1. \tag{P5.1.6}$$

Hint: as a small parameter for perturbative expansions use the value of the intermodal coupling strength δ. Terms of the form $\langle \psi_1 | \psi_2 \rangle$ and $\langle \psi_2 | \psi_1 \rangle$ are of the first order in coupling strength, while the terms $\langle \psi_1 | \hat{H}_2 - \hat{U} | \psi_1 \rangle$ and $\langle \psi_2 | \hat{H}_1 - \hat{U} | \psi_2 \rangle$ are of the second order in the coupling strength and can be omitted.

5.2   Suppression of the propagation losses for the core modes of a hollow photonic crystal waveguide

In this problem we compute absorption losses of the fundamental TE-polarized mode propagating in the hollow core of a photonic bandgap waveguide. We assume that the hollow core is filled with a weakly absorbing gas of refractive index $n_g = n_g^r + i n_g^i$. We also assume that the periodic reflector is made of much stronger absorbing dielectrics with refractive indices $n_l = n_l^r + i n_l^i$, $n_h = n_h^r + i n_h^i$, such that $n_g^i \ll n_l^i, n_h^i$. For the TE-polarized mode guided by the bandgap of a quarter-wave reflector, the electric field distribution is presented in Fig. 4.4. In particular, in each of the reflector layers, the field intensities $|E_y(z)|^2$ are proportional to either $\cos^2(\pi \eta_{h,l}/2)$ (low refractive index layers) or $\sin^2(\pi \eta_{h,l}/2)$ (high refractive index layers), where $\eta_{h,l} = (z - z_{\text{left interface}})/d_{h,l}$. In the core layer, the field intensity is $\cos^2(\pi \eta_c)$, where $\eta_c = z/d_c$. The amplitude coefficients in front of the cosines and sines are also indicated in Fig. 4.4.

Using the perturbation theory expression (5.44), find the modal propagation loss $\alpha_{\text{abs}}^{\text{wg}}[1/m] = 2|\beta_i|$ of a fundamental TE-polarized core-guided mode of a hollow-core photonic crystal waveguide, where $\beta_i = \beta - \beta_0$ is an imaginary correction to the modal propagation constant.

In the limit of a waveguide with a large core diameter $d_c \gg \lambda$, simplify the expression for the modal propagation loss and confirm that the contributions of the bulk absorption losses of the reflector materials are suppressed by the factor $(\lambda/d_c)^3$. Establish the minimal diameter of a hollow core above which modal propagation losses are dominated only by the absorption loss of a gas filling the hollow core.

References

[1] M. Skorobogatiy, M. Ibanescu, S. G. Johnson, *et al.* Analysis of general geometric scaling perturbations in a transmitting waveguide. The fundamental connection between polarization mode dispersion and group-velocity dispersion, *J. Opt. Soc. Am. B* **19** (2002), 2867–2875.

[2] D. Marcuse. *Theory of Dielectric Optical Waveguides*. Quantum Electronics – Principles and Applications Series (New York: Academic Press, 1974).

[3] A. W. Snyder and, J. Love. *Optical Waveguide Theory*. Science Paperbacks, 190 (London: Chapman and Hall, 1983).

[4] M. Skorobogatiy, S. A. Jacobs, S. G. Johnson, and Y. Fink. Geometric variations in high index-contrast waveguides, coupled mode theory in curvilinear coordinates, *Opt. Express* **10** (2002), 1227–1243.

[5] S. G. Johnson, M. Ibanescu, M. Skorobogatiy, *et al.* Perturbation theory for Maxwell's equations with shifting material boundaries, *Phys. Rev. E* **65** (2002), 66611.

[6] B. Z. Katsenelenbaum, L. M. Del Rio, M. Pereyaslavets, M. S. Ayza, and M. Thumm. *Theory of Nonuniform Waveguides: The Cross-Section Method*. Electromagnetic Waves Series (London: IEE, 1998).

6   Two-dimensional photonic crystals

In this section we investigate photonic bandgaps in two-dimensional photonic crystal lattices. We start by plotting a band diagram for a periodic lattice with negligible refractive-index-contrast. We then introduce a plane-wave expansion method for calculating the eigenmodes of a general 2D photonic crystal, and then develop a perturbation approach to describe bandgap formation in the case of photonic crystal lattices with small refractive index contrast. Next, we introduce a modified plane-wave expansion method to treat line and point defects in photonic crystal lattices. [1,2] Finally, we introduce perturbation formulation to describe bifurcation of the defect states from the bandgap edges in lattices with weak defects.

The two-dimensional dielectric profiles considered in this section exhibit discrete translational symmetry in the plane of a photonic crystal, and continuous translational symmetry perpendicular to the photonic crystal plane direction (Fig. 6.1). The mirror symmetry described in Section 2.4.7 suggests that the eigenmodes propagating strictly in the plane of a crystal can be classified as either TE or TM, depending on whether the vector of a modal magnetic or electric field is directed along the $\hat{\mathbf{z}}$ axis.

6.1   Two-dimensional photonic crystals with diminishingly small index contrast

In the case of a 2D discrete translational symmetry, the dielectric profile transforms into itself $\varepsilon(\mathbf{r} + \delta\mathbf{r}) = \varepsilon(\mathbf{r})$ for any translation along the lattice vector $\delta\mathbf{r}$ defined as $\delta\mathbf{r} = \bar{a}_1 N_1 + \bar{a}_2 N_2, (N_1, N_2) \subset$ integer. Lattice basis vectors $(\bar{a}_1, \bar{a}_2) \perp \hat{\mathbf{z}}$ are said to define a Bravais lattice, and it is presumed that they are noncollinear. As described in Section 2.4.5 (see the 2D discrete translational symmetry subsection), the general form of an electromagnetic solution reflecting 2D discrete translational symmetry is a Bloch form:

$$\mathbf{F}_{\mathbf{k}_t}(\mathbf{r}) = \exp(i\mathbf{k}_t \mathbf{r}_t)\mathbf{U}_{\mathbf{k}_t}^{\mathrm{F}}(\mathbf{r}_t)$$
$$\mathbf{U}_{\mathbf{k}_t}^{\mathrm{F}}(\mathbf{r}_t + \bar{a}_1 N_1 + \bar{a}_2 N_2) = \mathbf{U}_{\mathbf{k}_t}^{\mathrm{F}}(\mathbf{r}_t), \qquad (6.1)$$
$$(N_1, N_2) \subset \text{integer},$$

where $\mathbf{F}$ denotes either the electric or magnetic field, and t denotes transverse component of a vector confined to the plane of photonic crystal.

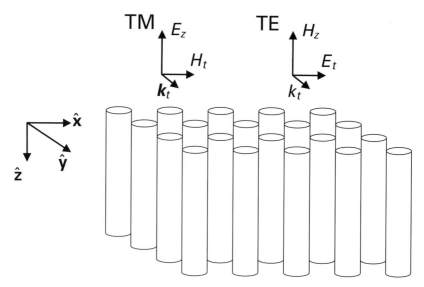

Figure 6.1 Two-dimensional photonic crystal exhibiting discrete translational symmetry in the xy plane and continuous translational symmetry along the $\hat{z}$ direction. The symmetry consideration suggests two possible polarizations, which are TE- and TM-polarized modes with the directions of the electromagnetic field vectors as demonstrated above.

We now define the basis vectors of a reciprocal lattice as:

$$\bar{b}_1 = 2\pi \frac{\bar{a}_2 \times \hat{\mathbf{z}}}{\bar{a}_1 \cdot (\bar{a}_2 \times \hat{\mathbf{z}})};$$

$$\bar{b}_2 = 2\pi \frac{\hat{\mathbf{z}} \times \bar{a}_1}{\bar{a}_1 \cdot (\bar{a}_2 \times \hat{\mathbf{z}})}. \qquad (6.2)$$

Bloch states with $\mathbf{k}_t$ and $\mathbf{k}_t + \mathbf{G}$ are identical for any $\mathbf{G} = \bar{b}_1 P_1 + \bar{b}_2 P_2, (P_1, P_2) \subset$ integer, thus only a small volume of the reciprocal phase space can be used to label the modes (first Brillouin zone). Moreover, if discrete rotational symmetries are present, then only a part of the first Brillouin zone (the so-called irreducible Brillouin zone (IBZ)) is required to label all the states (2.139). To describe eigenstates of a 2D photonic crystal one frequently plots a dispersion relation along the edge of an irreducible Brillouin zone. The reason for such a choice is an observation that all the frequencies corresponding to the interior points of a first Brillouin zone typically fall in between the lowest and the highest frequencies at the Brillouin zone edge. This is especially useful for observation of the photonic crystal bandgap structure.

We now use Bloch theorem to understand the general structure of a 2D photonic crystal dispersion relation. The photonic crystal under consideration is a square lattice of dielectric rods with dielectric constant ε_a and radius r_a placed in a uniform background of dielectric constant ε_b, and separated by a lattice constant a (see Fig. 6.2). We start with a photonic crystal with vanishingly small refractive index contrast $\varepsilon_a \rightarrow \varepsilon_b$. In the

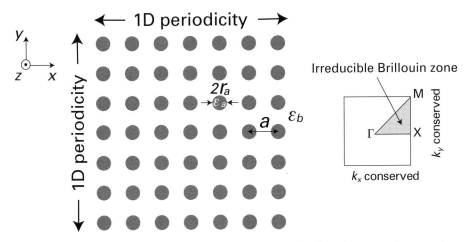

Figure 6.2 According to the Bloch theorem, modes of a periodic photonic system (for example a square array of dielectric rods in the air) can be labeled by the Bloch wave vectors confined to the first Brillouin zone of the reciprocal lattice. If additional rotational symmetries are present, then only a fraction of a Brillouin zone is needed to label all the modes. For a square lattice, high symmetry points are Γ, X, M.

uniform dielectric, the dispersion relation of the photonic states is simply:

$$\omega\sqrt{\varepsilon_b} = \sqrt{|\mathbf{k}_t|^2}, \tag{6.3}$$

where $\mathbf{k}_t$ is any 2D vector. However, according to the Bloch theorem, even the smallest periodic variation maps all the states into the first Brillouin zone, thus introducing a modified dispersion relation in the form:

$$\omega\sqrt{\varepsilon_b} = \sqrt{\left|\mathbf{k}_t + \bar{b}_1 P_1 + \bar{b}_2 P_2\right|^2}, \tag{6.4}$$

where $\mathbf{k}_t$ is now a vector confined to the first Brillouin zone, reciprocal basis vectors $(\bar{b}_1, \bar{b}_2)$ are defined in (6.2), and P_1, P_2 are any integers. In the case of a square lattice with lattice constant a, the reciprocal basis vectors are $\bar{b}_1 = (2\pi/a, 0); \bar{b}_2 = (0, 2\pi/a)$. In this case (6.4) transforms into:

$$\omega_{P_1, P_2}\sqrt{\varepsilon_b} = \sqrt{\left(k_x + \frac{2\pi}{a}P_1\right)^2 + \left(k_y + \frac{2\pi}{a}P_2\right)^2}. \tag{6.5}$$

For a square lattice of circular rods, the irreducible Brillouin zone is shown in Fig. 6.2. It is triangular with vertices marked as Γ, X, M. In Fig. 6.3 we demonstrate several lowest bands evaluated for the Bloch wave vectors positioned along the edge of an irreducible Brillouin zone. In the following numerical examples we assume that $\varepsilon_b = 2.25$. Every band is marked by the pair of coefficients P_1, P_2 that generate such a band using (6.5). Some bands are degenerate, meaning that different pairs of coefficients generate the same band of states.

In what follows we aim at understanding what happens to the photonic band structure when a small, but sizable, periodic index contrast is introduced into a system. As

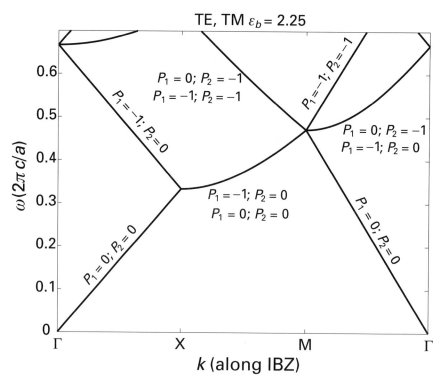

Figure 6.3 Dispersion relations of the lowest frequency bands of a 2D photonic crystal (shown in Fig. 6.2) with vanishingly small refractive index contrast. The dispersion relation is presented along the Γ-X-M-Γ edge of an irreducible Brillouin zone. Bands are marked by their proper generating coefficients according to (6.5).

established in Chapter 5 using perturbation theory, whenever a Maxwell Hamiltonian allows a degenerate state, after the introduction of a perturbation, such a state might be split into several closely spaced states. Treating a small index contrast as a perturbation, we demonstrate that such a perturbation, indeed, lifts the degeneracy for the bands shown in Fig. 6.3, ultimately resulting in opening of the photonic bandgaps. Our further derivations are made in the framework of the plane-wave expansion method, which is presented next.

6.2   Plane-wave expansion method

From Bloch theorem (6.1) it follows that the modal field in a periodic system can be presented in the form of a product of a complex exponential and a periodic function in space. From the theorems of Fourier analysis it also follows that a periodic function can be expanded in terms of an infinite discrete sum of spatial harmonics. Thus, electromagnetic fields in a periodic medium (of any dimension) can be written as:

$$\mathbf{E_k(r)} = \sum_{\mathbf{G}} \mathbf{E_k(G)} \exp(i(\mathbf{k+G})\mathbf{r})$$
$$\mathbf{H_k(r)} = \sum_{\mathbf{G}} \mathbf{H_k(G)} \exp(i(\mathbf{k+G})\mathbf{r}), \qquad (6.6)$$
$$\mathbf{G} = \bar{b}_1 P_1 + \bar{b}_2 P_2 + \bar{b}_3 P_3$$

where reciprocal basis vectors $\bar{b}_1, \bar{b}_2, \bar{b}_3$ are defined as in (2.129), and we chose an example of a 3D periodic structure. Similarly, a periodic dielectric profile can be expanded as:

$$\frac{1}{\varepsilon(\mathbf{r})} = \sum_{\mathbf{G}} \kappa(\mathbf{G}) \exp(i\mathbf{Gr}). \qquad (6.7)$$

Recalling Maxwell's equations written in terms of only the electric (2.15), (2.16) or magnetic (2.17), (2.18) fields:

$$\omega^2 \mathbf{E} = \frac{1}{\varepsilon(\mathbf{r})} \nabla \times (\nabla \times \mathbf{E}), \qquad (6.8)$$

$$\nabla \cdot \varepsilon(\mathbf{r})\mathbf{E} = 0, \qquad (6.9)$$

$$\omega^2 \mathbf{H} = \nabla \times \left(\frac{1}{\varepsilon(\mathbf{r})} \nabla \times \mathbf{H} \right), \qquad (6.10)$$

$$\nabla \cdot \mathbf{H} = 0, \qquad (6.11)$$

after substituting (6.6) into (6.8), (6.10), and by using orthogonality of the plane waves in a sense $\int d\mathbf{r} \exp(i\mathbf{Gr}) \sim \delta(\mathbf{G})$ (delta function), we arrive at the following equations in terms of the Fourier components of the electric or magnetic fields:

$$-\sum_{\mathbf{G'}} \kappa(\mathbf{G-G'})(\mathbf{k+G'}) \times \left[(\mathbf{k+G'}) \times \mathbf{E_k(G')} \right] = \omega_{\mathbf{k}}^2 \mathbf{E_k(G)}, \qquad (6.12)$$

$$-\sum_{\mathbf{G'}} \kappa(\mathbf{G-G'})(\mathbf{k+G}) \times \left[(\mathbf{k+G'}) \times \mathbf{H_k(G')} \right] = \omega_{\mathbf{k}}^2 \mathbf{H_k(G)}. \qquad (6.13)$$

For a given value of a Bloch wave vector $\mathbf{k}$, these equations present a linear eigenvalue problem with respect to the value of the modal frequency $\omega_{\mathbf{k}}^2$. In practice, instead of an infinite number of Fourier coefficients, one uses a finite number N of them. As electromagnetic fields are, in general, 3D vectors there will be $3N$ unknown coefficients to solve for in (6.12) or in (6.13). To avoid spurious solutions one has to make sure that the eigensolution of (6.12) or (6.13) also satisfies (6.9) or (6.11), respectively. It turns out that when using the formulation in terms of the magnetic fields (6.13), imposing (6.11) is trivial. Indeed, by substitution of (6.6) into (6.11) one finds that condition (6.11) amounts to the transversality of the Fourier components to their corresponding wavevectors:

$$\nabla \cdot \mathbf{H} = \sum_{\mathbf{G}} \mathbf{H_k(G)} \nabla \cdot \exp(i(\mathbf{k+G})\mathbf{r})$$

$$= i \sum_{\mathbf{G}} \mathbf{H_k(G)}(\mathbf{k+G}) \exp(i(\mathbf{k+G})\mathbf{r}) \underset{\mathbf{H_k(G)(k+G)}=0}{=} 0. \qquad (6.14)$$

The transversality condition (6.14) reduces the number of unknown coefficients to $2N$, and is trivial to implement in practice. Moreover, because of the Hermitian nature of the Maxwell equation formulation in terms of the magnetic fields (6.10), the resultant matrices in (6.13) are also Hermitian, thus allowing for efficient numerical methods to be used to solve for the matrix eigenvalues. These considerations make (6.13) in combination with (6.14) a method of choice for the computation of the modes in periodic dielectric media.

6.2.1   Calculation of the modal group velocity

Given the solution for the magnetic field of a photonic crystal eigenstate (6.13), using a Hellman–Feynman theorem similar to considerations of section 5.1.3, we can easily compute the photonic state group velocity. In particular, using the fact that eigenvalue formulation (6.13) in terms of the magnetic field results in Hermitian matrices, the eigenstate group velocity can be expressed as:

$$\mathbf{v}_g = \frac{\partial \omega_{\mathbf{k}}}{\partial \mathbf{k}} = \frac{1}{2\omega_{\mathbf{k}}} \frac{\langle H_{\mathbf{k}}| \, \partial \hat{H}/\partial \mathbf{k} \, |H_{\mathbf{k}}\rangle}{\langle H_{\mathbf{k}} \mid H_{\mathbf{k}}\rangle}, \tag{6.15}$$

where, from (6.13), we have the following definitions:

$$\langle H_{\mathbf{k}} \mid H_{\mathbf{k}}\rangle = \sum_{\mathbf{G}} |\mathbf{H}_{\mathbf{k}}(\mathbf{G})|^2, \tag{6.16}$$

$$\langle H_{\mathbf{k}}| \, \partial \hat{H}/\partial \mathbf{k} \, |H_{\mathbf{k}}\rangle = -\sum_{\mathbf{G},\mathbf{G}'} \mathbf{H}_{\mathbf{k}}^*(\mathbf{G}) \kappa(\mathbf{G}-\mathbf{G}') \frac{\partial [(\mathbf{k}+\mathbf{G}) \times [(\mathbf{k}+\mathbf{G}') \times]]}{\partial \mathbf{k}} \mathbf{H}_{\mathbf{k}}(\mathbf{G}'). \tag{6.17}$$

We now simplify the vector products in (6.17) by using the vector identity $\mathbf{a} \times (\mathbf{b} \times \mathbf{c}) = \mathbf{b}(\mathbf{ac}) - \mathbf{c}(\mathbf{ab})$:

$$(\mathbf{k}+\mathbf{G}) \times \left[(\mathbf{k}+\mathbf{G}') \times \mathbf{H}_{\mathbf{k}}(\mathbf{G}')\right]$$
$$= (\mathbf{k}+\mathbf{G}')((\mathbf{k}+\mathbf{G})\,\mathbf{H}_{\mathbf{k}}(\mathbf{G}')) - \mathbf{H}_{\mathbf{k}}(\mathbf{G})((\mathbf{k}+\mathbf{G})\,(\mathbf{k}+\mathbf{G}')), \tag{6.18}$$

which after substitution into (6.17) results in:

$$\langle H_{\mathbf{k}}| \, \partial \hat{H}/\partial \mathbf{k} \, |H_{\mathbf{k}}\rangle$$
$$= \sum_{\mathbf{G},\mathbf{G}'} \kappa(\mathbf{G}-\mathbf{G}') \left[\begin{array}{l} (\mathbf{H}_{\mathbf{k}}^*(\mathbf{G})\mathbf{H}_{\mathbf{k}}(\mathbf{G}'))(2\mathbf{k}+\mathbf{G}+\mathbf{G}') - \mathbf{H}_{\mathbf{k}}^*(\mathbf{G})((\mathbf{k}+\mathbf{G})\,\mathbf{H}_{\mathbf{k}}(\mathbf{G}')) \\ - \mathbf{H}_{\mathbf{k}}(\mathbf{G}')((\mathbf{k}+\mathbf{G}')\mathbf{H}_{\mathbf{k}}^*(\mathbf{G})) \end{array} \right].$$
$$\tag{6.19}$$

All the terms in (6.15) can now be computed in terms of the sums (6.16) and (6.19).

6.2.2   Plane-wave method in 2D

In application to the two-dimensional photonic crystals, the plane-wave expansion method can be greatly simplified. This is related to the fact that eigenstates in 2D systems can be classified as either TE- or TM-polarized (see Fig. 6.1), with the following choices of the electromagnetic field vectors:

$$\text{TE:} \quad (0, 0, H_z(x, y, z)); \quad (E_x(x, y, z), E_y(x, y, z), 0)$$
$$\text{TM:} \quad (H_x(x, y, z), H_y(x, y, z), 0); \quad (0, 0, E_z(x, y, z)) \tag{6.20}$$

Using Fourier expansion of the $\hat{\mathbf{z}}$ vector components:

$$\text{TM:} \quad E_{z,\mathbf{k}_t}(\mathbf{r}_t) = \sum_{\mathbf{G}} E_{z,\mathbf{k}_t}(\mathbf{G}) \exp(\mathrm{i}\,(\mathbf{k}_t + \mathbf{G})\,\mathbf{r}_t)$$

$$\text{TE:} \quad H_{z,\mathbf{k}_t}(\mathbf{r}_t) = \sum_{\mathbf{G}} H_{z,\mathbf{k}_t}(\mathbf{G}) \exp(\mathrm{i}\,(\mathbf{k}_t + \mathbf{G})\,\mathbf{r}_t), \tag{6.21}$$

$$\mathbf{G} = \bar{b}_1 P_1 + \bar{b}_2 P_2$$

and after their substitution into (6.12) and (6.13) we find that, in 2D, the plane-wave expansion method leads to the following equations with respect to the scalar Fourier components:

$$\text{TM:} \quad \sum_{\mathbf{G}'} \kappa(\mathbf{G} - \mathbf{G}')|\mathbf{k}_t + \mathbf{G}'|^2 E_{z,\mathbf{k}_t}(\mathbf{G}') = \omega_{\mathbf{k}_t}^2 E_{z,\mathbf{k}_t}(\mathbf{G}), \tag{6.22}$$

$$\text{TE:} \quad \sum_{\mathbf{G}'} \kappa(\mathbf{G} - \mathbf{G}')(\mathbf{k}_t + \mathbf{G}')\,(\mathbf{k}_t + \mathbf{G})\, H_{z,\mathbf{k}_t}(\mathbf{G}') = \omega_{\mathbf{k}_t}^2 H_{z,\mathbf{k}_t}(\mathbf{G}), \tag{6.23}$$

$$\kappa(\mathbf{G}) = \frac{1}{S_{\text{unit cell}}} \int_{\text{unit cell}} d\mathbf{r}_t \frac{1}{\varepsilon(\mathbf{r}_t)} \exp(-\mathrm{i}\mathbf{G}\mathbf{r}_t). \tag{6.24}$$

The coupling coefficients (6.24) can be computed analytically for simple geometries. For example, for a square lattice of dielectric rods shown in Fig. 6.2, unit cell is a square of area $S_{\text{unit cell}} = a^2$. Integration in (6.24) can be performed analytically to give:

$$\kappa(0) = f\varepsilon_a^{-1} + (1 - f)\varepsilon_b^{-1}$$

$$\kappa(|\mathbf{G}|)_{\mathbf{G}\neq 0} = 2f\left(\varepsilon_a^{-1} - \varepsilon_b^{-1}\right) \frac{J_1(|\mathbf{G}|\,r_a)}{|\mathbf{G}|\,r_a}. \tag{6.25}$$

$$\text{filling fraction:} \quad f = \frac{\pi r_a^2}{a^2}$$

In Fig. 6.4 we show band diagrams for TM-polarized modes in the low-refractive-index-contrast photonic crystal $\varepsilon_a = 2.56$; $\varepsilon_b = 1.96$; $f = 0.5$ computed by resolving the eigenproblem (6.22).

When a small index contrast is introduced, degenerate bands of a periodic lattice with diminishingly small index contrast split, leading to the appearance of small local band gaps. When comparing dispersion relations for TM and TE modes in Fig. 6.5 one notices that, in general, for lattices of high-dielectric-index rods in a low-index background, the formation of band gaps for TM modes is more robust than that for TE modes.

6.2.3 Calculation of the group velocity in the case of 2D photonic crystals

Expressions (6.15), (6.16), and (6.19) for the group velocity of the photonic crystal eigenstate can be further simplified in the case of 2D photonic crystals. Particularly,

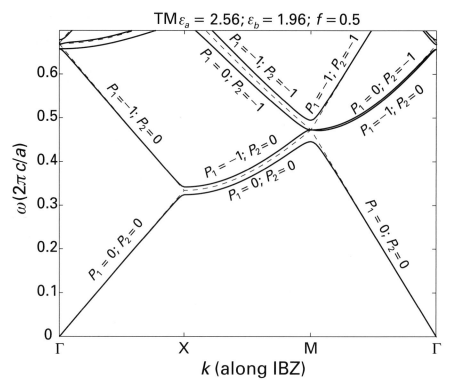

Figure 6.4 When a small index contrast is present, degenerate bands of a 2D periodic lattice split, thus forming local bandgaps. The dotted lines plot the band structure of the same periodic lattice and diminishingly small index contrast.

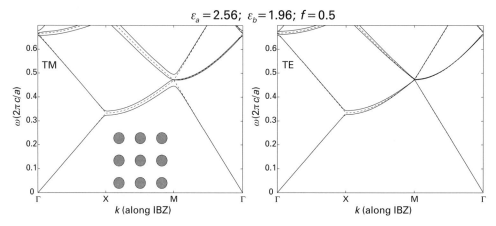

Figure 6.5 Comparing dispersion relations for TE and TM modes. Notice that in the problem of high-index rods in a low-index background, TM modes are more robust in opening sizable bandgaps.

for TE polarization, only the H_z component of the magnetic field is nonzero, therefore, (6.19) is greatly simplified:

$$\text{TE:}\quad \mathbf{v}_g = \frac{\partial \omega_{\mathbf{k}_t}}{\partial \mathbf{k}_t} = \frac{1}{2\omega_{\mathbf{k}_t}} \frac{\sum\limits_{\mathbf{G},\mathbf{G}'} \kappa(\mathbf{G}-\mathbf{G}')(H^*_{z,\mathbf{k}_t}(\mathbf{G})H_{z,\mathbf{k}_t}(\mathbf{G}'))(2\mathbf{k}_t+\mathbf{G}+\mathbf{G}')}{\sum\limits_{\mathbf{G}} |H_{z,\mathbf{k}_t}(\mathbf{G})|^2}. \tag{6.26}$$

The expression for the group velocity of a TM-polarized eigenstate of a photonic crystal is somewhat more complex. First of all, given the solution for the photonic crystal eigenstate in terms of the electric field $E_{z,\mathbf{k}_t}(\mathbf{G})$, the corresponding harmonic magnetic field can be found as:

$$\mathbf{H}_{t,\mathbf{k}_t}(\mathbf{r}_t) = \frac{1}{i\omega_{\mathbf{k}_t}} \nabla \times E_{z,\mathbf{k}_t}(\mathbf{r}_t). \tag{6.27}$$

After substitution of (6.21) into (6.27) we get:

$$\mathbf{H}_{t,\mathbf{k}_t}(\mathbf{r}_t) = \frac{1}{\omega_{\mathbf{k}_t}} \sum_{\mathbf{G}} E_{z,\mathbf{k}_t}(\mathbf{G})((\mathbf{k}_t+\mathbf{G})\times\hat{\mathbf{z}})\exp(i(\mathbf{k}_t+\mathbf{G})\mathbf{r}_t), \tag{6.28}$$

and, therefore:

$$\mathbf{H}_{t,\mathbf{k}_t}(\mathbf{G}) = \frac{1}{\omega_{\mathbf{k}_t}} E_{z,\mathbf{k}_t}(\mathbf{G})((\mathbf{k}_t+\mathbf{G})\times\hat{\mathbf{z}}). \tag{6.29}$$

From (6.29) it follows that the mode normalization is:

$$\langle H_{\mathbf{k}_t} \mid H_{\mathbf{k}_t}\rangle = \sum_{\mathbf{G}} |\mathbf{H}_{\mathbf{k}}(\mathbf{G})|^2 = \frac{1}{\omega_{\mathbf{k}_t}^2}\sum_{\mathbf{G}} |E_{z,\mathbf{k}_t}(\mathbf{G})|^2 |\mathbf{k}_t+\mathbf{G}|^2, \tag{6.30}$$

while the operator derivative average is:

$$\begin{aligned}
\langle H_{\mathbf{k}_t}|&\partial\hat{H}/\partial\mathbf{k}_t|H_{\mathbf{k}_t}\rangle \\
&= \frac{1}{\omega_{\mathbf{k}_t}^2}\sum_{\mathbf{G},\mathbf{G}'}\kappa(\mathbf{G}-\mathbf{G}')(E^*_{z,\mathbf{k}_t}(\mathbf{G})E_{z,\mathbf{k}_t}(\mathbf{G}')) \\
&\quad\times\begin{bmatrix}[((\mathbf{k}_t+\mathbf{G})\times\hat{\mathbf{z}})((\mathbf{k}_t+\mathbf{G}')\times\hat{\mathbf{z}})](2\mathbf{k}_t+\mathbf{G}+\mathbf{G}')\\ -((\mathbf{k}_t+\mathbf{G})\times\hat{\mathbf{z}})((\mathbf{k}_t+\mathbf{G})((\mathbf{k}_t+\mathbf{G}')\times\hat{\mathbf{z}}))\\ -((\mathbf{k}_t+\mathbf{G}')\times\hat{\mathbf{z}})((\mathbf{k}_t+\mathbf{G}')((\mathbf{k}_t+\mathbf{G})\times\hat{\mathbf{z}}))\end{bmatrix} \\
&= \frac{1}{\omega_{\mathbf{k}_t}^2}\sum_{\mathbf{G},\mathbf{G}'}\kappa(\mathbf{G}-\mathbf{G}')(E^*_{z,\mathbf{k}_t}(\mathbf{G})E_{z,\mathbf{k}_t}(\mathbf{G}')) \\
&\quad\times\begin{bmatrix}[(\mathbf{k}_t+\mathbf{G})(\mathbf{k}_t+\mathbf{G}')](2\mathbf{k}_t+\mathbf{G}+\mathbf{G}')\\ +(\hat{\mathbf{z}}\times(\mathbf{G}-\mathbf{G}'))(\hat{\mathbf{z}}((\mathbf{k}_t+\mathbf{G})\times(\mathbf{k}_t+\mathbf{G}')))\end{bmatrix}.
\end{aligned} \tag{6.31}$$

Finally, the group velocity for the TM-polarized photonic crystal state is:

$$\text{TM: } \mathbf{v}_g = \frac{\partial \omega_{\mathbf{k}_t}}{\partial \mathbf{k}_t}$$

$$= \frac{1}{2\omega_{\mathbf{k}_t}} \frac{\sum\limits_{\mathbf{G},\mathbf{G}'} \kappa(\mathbf{G}-\mathbf{G}')(E_{z,\mathbf{k}_t}^*(\mathbf{G})E_{z,\mathbf{k}_t}(\mathbf{G}')) \left[\begin{array}{l} [(\mathbf{k}_t+\mathbf{G})(\mathbf{k}_t+\mathbf{G}')](2\mathbf{k}_t+\mathbf{G}+\mathbf{G}') \\ +(\hat{\mathbf{z}}\times(\mathbf{G}-\mathbf{G}'))(\hat{\mathbf{z}}((\mathbf{k}_t+\mathbf{G})\times(\mathbf{k}_t+\mathbf{G}'))) \end{array} \right]}{\sum\limits_{\mathbf{G}} |E_{z,\mathbf{k}_t}(\mathbf{G})|^2 |\mathbf{k}_t+\mathbf{G}|^2}.$$

$$(6.32)$$

6.2.4   Perturbative formulation for the photonic crystal lattices with small refractive index contrast

We now quantify the appearance of the bandgaps using a perturbative formulation based on (6.22), (6.23). We first consider band splitting at the point M for which $\mathbf{k}_t^M = (\pi/a, \pi/a)$. For a uniform dielectric ε_b, the eigenstate (plane wave) at M will have a corresponding frequency $\omega_M \sqrt{\varepsilon_b} = \sqrt{2}\pi/a$, and according to (6.5) and Fig. 6.3 such a state will be four-fold degenerate with four bands labeled as $(0,0), (-1,0), (0,-1),$ $(-1,-1)$ intersecting at the M point. When the index contrast is zero, this degeneracy in frequency implies that at the M point the general solution can be represented as a linear combination of these four plane waves, namely:

$$E_{z,\mathbf{k}_t^M}(\mathbf{r}_t) = \sum_{\mathbf{G}\in\mathbf{G}_\omega} E_{z,\mathbf{k}_t^M}(\mathbf{G}) \exp(i\left(\mathbf{k}_t^M + \mathbf{G}\right)\mathbf{r}_t), \qquad (6.33)$$

where a set of vectors $\mathbf{G}_\omega$ includes all the reciprocal wave vectors for which the corresponding plane waves in (6.33) have the same frequency ω as computed by (6.5). Thus, for the M point, for example,

$$\mathbf{G}_\omega = [\mathbf{G}_1, \mathbf{G}_2, \mathbf{G}_3, \mathbf{G}_4] = [(0,0), (-2\pi/a, 0), (0, -2\pi/a), (-2\pi/a, -2\pi/a)]. \qquad (6.34)$$

As an example, we consider the case of TM modes. When the index contrast is zero, then in (6.22) the only nonzero coupling element is $\kappa(0)$, thus reducing (6.22) to a system of linear uncoupled equations with a solution:

$$\omega_{\mathbf{k}_t}^2 = \kappa(0) |\mathbf{k}_t + \mathbf{G}|^2$$

$$E_{z,\mathbf{k}_t}(\mathbf{G}) = 1. \qquad (6.35)$$

The expression for the eigenfrequency in (6.35) is the same as in (6.5). As we mentioned before, at the point $\mathbf{k}_t^M$, the eigenstate is four-fold degenerate. In the presence of a small index contrast, the specific linear combinations (6.33) of originally degenerate states will become new eigenstates. Thus, to find new eigenstates, we have to solve (6.22) by retaining only the coefficients $E_{z,\mathbf{k}_t}(\mathbf{G})$ with $\mathbf{G} \in \mathbf{G}_\omega$ in the expansion. Using notation

(6.34), one can thus write (6.22) as:

$$
2\left(\frac{\pi}{a}\right)^2
\begin{pmatrix}
\kappa(0) & \kappa\left(\frac{2\pi}{a}\right) & \kappa\left(\frac{2\pi}{a}\right) & \kappa\left(\sqrt{2}\frac{2\pi}{a}\right) \\
\kappa\left(\frac{2\pi}{a}\right) & \kappa(0) & \kappa\left(\sqrt{2}\frac{2\pi}{a}\right) & \kappa\left(\frac{2\pi}{a}\right) \\
\kappa\left(\frac{2\pi}{a}\right) & \kappa\left(\sqrt{2}\frac{2\pi}{a}\right) & \kappa(0) & \kappa\left(\frac{2\pi}{a}\right) \\
\kappa\left(\sqrt{2}\frac{2\pi}{a}\right) & \kappa\left(\frac{2\pi}{a}\right) & \kappa\left(\frac{2\pi}{a}\right) & \kappa(0)
\end{pmatrix}
\begin{pmatrix}
E_{z,\mathbf{k}_t^M}(\mathbf{G}_1) \\
E_{z,\mathbf{k}_t^M}(\mathbf{G}_2) \\
E_{z,\mathbf{k}_t^M}(\mathbf{G}_3) \\
E_{z,\mathbf{k}_t^M}(\mathbf{G}_4)
\end{pmatrix}
$$

$$
= \omega_{\mathbf{k}_t^M}^2
\begin{pmatrix}
E_{z,\mathbf{k}_t^M}(\mathbf{G}_1) \\
E_{z,\mathbf{k}_t^M}(\mathbf{G}_2) \\
E_{z,\mathbf{k}_t^M}(\mathbf{G}_3) \\
E_{z,\mathbf{k}_t^M}(\mathbf{G}_4)
\end{pmatrix}.
\tag{6.36}
$$

It is straightforward to find the eigenvalues and eigenvectors of (6.36):

$$
\omega_{\mathbf{k}_t^M}^2 = 2\left(\frac{\pi}{a}\right)^2 \left(\kappa(0) - \kappa\left(\sqrt{2}\frac{2\pi}{a}\right)\right)
$$
$$
(E_{z,\mathbf{k}_t^M}(\mathbf{G}_1),
$$
$$
E_{z,\mathbf{k}_t^M}(\mathbf{G}_2),\, E_{z,\mathbf{k}_t^M}(\mathbf{G}_3),\, E_{z,\mathbf{k}_t^M}(\mathbf{G}_4)) = \frac{1}{2}(-1, 0, 0, 1)\,;
$$
$$
\omega_{\mathbf{k}_t^M}^2 = 2\left(\frac{\pi}{a}\right)^2 \left(\kappa(0) - \kappa\left(\sqrt{2}\frac{2\pi}{a}\right)\right)
$$
$$
(E_{z,\mathbf{k}_t^M}(\mathbf{G}_1),
$$
$$
E_{z,\mathbf{k}_t^M}(\mathbf{G}_2),\, E_{z,\mathbf{k}_t^M}(\mathbf{G}_3),\, E_{z,\mathbf{k}_t^M}(\mathbf{G}_4)) = \frac{1}{2}(0, -1, 1, 0)\,;
$$
$$
\omega_{\mathbf{k}_t^M}^2 = 2\left(\frac{\pi}{a}\right)^2 \left(\kappa(0) + \kappa\left(\sqrt{2}\frac{2\pi}{a}\right) - 2\kappa\left(\frac{2\pi}{a}\right)\right)
$$
$$
(E_{z,\mathbf{k}_t^M}(\mathbf{G}_1),\, E_{z,\mathbf{k}_t^M}(\mathbf{G}_2),
$$
$$
E_{z,\mathbf{k}_t^M}(\mathbf{G}_3),\, E_{z,\mathbf{k}_t^M}(\mathbf{G}_4)) = \frac{1}{4}(1, -1, -1, 1)\,;
$$
$$
\omega_{\mathbf{k}_t^M}^2 = 2\left(\frac{\pi}{a}\right)^2 \left(\kappa(0) + \kappa\left(\sqrt{2}\frac{2\pi}{a}\right) + 2\kappa\left(\frac{2\pi}{a}\right)\right)
$$
$$
(E_{z,\mathbf{k}_t^M}(\mathbf{G}_1),\, E_{z,\mathbf{k}_t^M}(\mathbf{G}_2),
$$
$$
E_{z,\mathbf{k}_t^M}(\mathbf{G}_3),\, E_{z,\mathbf{k}_t^M}(\mathbf{G}_4)) = \frac{1}{4}(1, 1, 1, 1).
\tag{6.37}
$$

Note that even the introduction of finite index contrast does not completely lift degeneracy in the band structure. This interesting fact is a consequence of a general group theory consideration applied to a periodic structure exhibiting additional discrete symmetries (like discrete rotational symmetry, reflection planes, etc.). In general, the splitting of degenerate bands requires structural perturbations that reduce the overall number of symmetries in a system (e.g., ellipticity in the shape of all the rods).

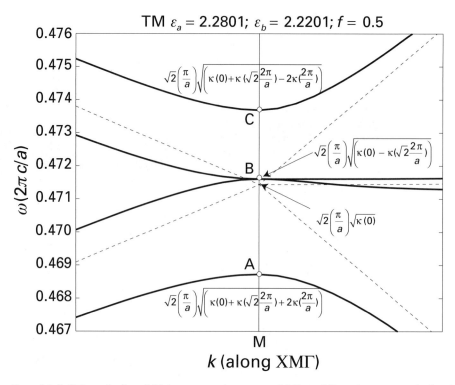

Figure 6.6 Splitting of a four-fold degenerate eigenstate at M. Dotted lines demonstrate the band structure of a uniform dielectric with a dielectric constant $\varepsilon = 1/\kappa(0)$.

In Fig. 6.6 we demonstrate band splitting near the M point for the case of $f = 0.5$; $\varepsilon_a = 2.2801$; $\varepsilon_b = 2.2201$. For these structural parameters, from (6.25) one finds that $\kappa(0) \geq 0$, $\kappa(2\pi/a) \leq 0$, $\kappa(\sqrt{2}(2\pi/a)) \leq 0$, $|\kappa(\sqrt{2}(2\pi/a))| \ll |\kappa(2\pi/a)|$. In Fig. 6.6 solid curves describe the dispersion relation as calculated using full plane-wave method (6.22), dotted curves describe the band structure of a uniform dielectric with a dielectric constant $\varepsilon = 1/\kappa(0)$, and filled circles describe eigenfrequencies at the M point as calculated using perturbation theory (6.37).

Finally, we consider field distributions in the split bands. Using the expansion coefficient as in (6.37) and substituting them in (6.33) one can reconstruct the spatial distribution in the new eigenfields. Thus, one can verify that at the band edges:

$$
\text{A:} \quad \left| E_{z,\mathbf{k}_t^M} \right|_A^2 = \frac{1}{4}\left(1 + \cos\left(\frac{2\pi}{a}x\right) + \cos\left(\frac{2\pi}{a}y\right) + \cos\left(\frac{2\pi}{a}x\right)\cos\left(\frac{2\pi}{a}y\right)\right)
$$

$$
= \cos^2\left(\frac{\pi}{a}x\right)\cos^2\left(\frac{\pi}{a}y\right), \tag{6.38}
$$

$$
\text{B}_1: \quad \left| E_{z,\mathbf{k}_t^M} \right|_B^2 = \sin^2\left(\frac{\pi}{a}(x-y)\right), \tag{6.39}
$$

$$
\text{B}_2: \quad \left| E_{z,\mathbf{k}_t^M} \right|_B^2 = \sin^2\left(\frac{\pi}{a}(x+y)\right), \tag{6.40}
$$

$$\text{TM } \varepsilon_a = 2.2801; \, \varepsilon_b = 2.2201; \, f = 0.5$$

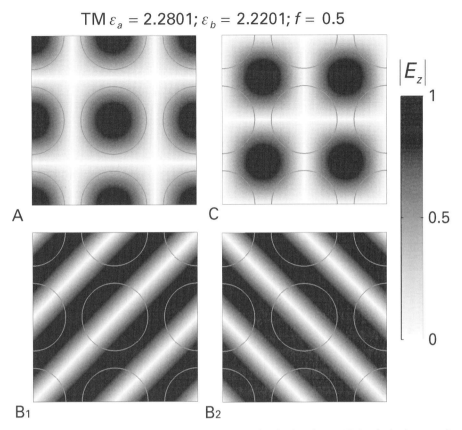

Figure 6.7 Distribution of the electric field amplitude for the four lowest TM-polarized states of a photonic crystal at the M point. White circles give the positions of the rod-cladding interfaces. A low-refractive-index contrast is assumed.

$$C: \quad \left| E_{z,\mathbf{k}_t^M} \right|^2_C = \frac{1}{4} \left(1 - \cos \left(\frac{2\pi}{a} x \right) - \cos \left(\frac{2\pi}{a} y \right) + \cos \left(\frac{2\pi}{a} x \right) \cos \left(\frac{2\pi}{a} y \right) \right)$$

$$= \sin^2 \left(\frac{\pi}{a} x \right) \sin^2 \left(\frac{\pi}{a} y \right). \tag{6.41}$$

In Fig. 6.7 we present electric field distributions at the band edge point M for the eigenmodes of a 2D photonic crystal with a small index contrast. At the lower edge of a bandgap (point A), the electric field tends to concentrate in the high-index rods, therefore states at the lower band edge are called rod states. At the upper band edge (point C), owing to the orthogonality relation with the fundamental state (point A), the electric field is expelled into the cladding region, therefore states at the upper band edge are called cladding states. Degenerate modes at the point B are mixed.

The fact that the modes with electric field distribution outside the high dielectric constant region (Fig. 6.7(c)) exhibit higher frequencies than modes with electric field distribution inside of such a region can be easily understood by recalling the variational principle that states that the fundamental mode must have most of its displacement field concentrated in the high refractive index dielectric.

6.2.5 Photonic crystal lattices with high-refractive-index contrast

When the refractive index contrast is increased, band splitting becomes pronounced, opening the possibility of creating complete bandgaps. Complete bandgaps are defined as frequency regions where there are no extended states propagating inside of the bulk of a crystal. Many applications of photonic crystals rely on the existence of such bandgaps. For example, photonic crystals can be used as omnidirectional mirrors when operated inside a complete bandgap. Indeed, in that case, the incoming radiation has to be reflected completely as there are no bulk states of a photonic crystal to couple to. Generally, opening a complete bandgap is relatively easy for one of the polarizations. For example, in Fig. 6.8(a) we present band diagrams of the TE- and TM-polarized states for a photonic crystal with $f = 0.5; \varepsilon_a = 7.84; \varepsilon_b = 1.0$. In this figure one can notice several complete bandgaps for the TM states, and no complete bandgaps for the TE states. By simple inversion of the material regions $f = 0.5; \varepsilon_a = 1.0; \varepsilon_b = 7.84$ one can open complete bandgaps for TE polarization as demonstrated in Fig. 6.8(b). Finally, we note in passing that designing photonic crystals with complete bandgaps for both TE and TM polarizations is more challenging, however, possible.

6.3 Comparison between various projected band diagrams

As discussed in Section 2.4.4, there are several types of band diagrams that can be used to present the states of a photonic crystal. So far we have only used one type of band diagram, which presented dispersion relations of the photonic crystal states along the ΓXMΓ curve (Fig. 6.9(a)) that traced the edge of an irreducible Brillouin zone in the k-space. Another way of presenting the states of a protonic crystal is by using a so-called projected band diagram. To construct a band diagram projected onto the k_x direction, for example, one fixes the value of k_x and then plots on the same graph the states with all otherwise allowed k_y values. In the projected band diagram (Fig. 6.9(b)) gray regions define allowed photonic crystal states, while empty regions define bandgaps. Note that while both band diagrams of Fig. 6.9(a) and Fig. 6.9(b) give consistent definitions of complete bandgaps, it is, however, much faster to compute the band diagram of Fig. 6.9(a), as it involves only the states on the edge of an irreducible Brillouin zone (1D calculation), while the band diagram of Fig. 6.9(b) requires computation of all the states inside an irreducible Brillouin zone (2D calculation). Nevertheless, when detailed information about state distribution is required, then the band diagram of Fig. 6.9(b) is preferred. For example, from Fig. 6.9(b) it follows that for $\omega = 0.36$, there exists a small bandgap in the vicinity of $k_x = 0.25$; this defines a set of directions along which the propagation in a photonic crystal is effectively suppressed. Such information is difficult to ascertain from the band diagram of Fig. 6.9(a).

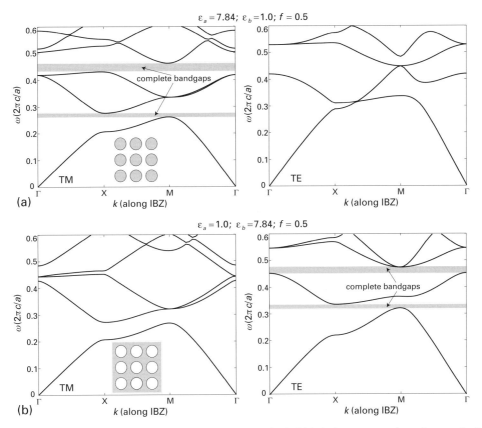

Figure 6.8 Band diagrams for the TE and TM modes in high-index-contrast photonic crystals. (a) For the high refractive index rods in the air structure, complete bandgaps are found for TM polarization. (b) For the air holes in the high-refractive-index background structure, complete bandgaps are found for TE polarization.

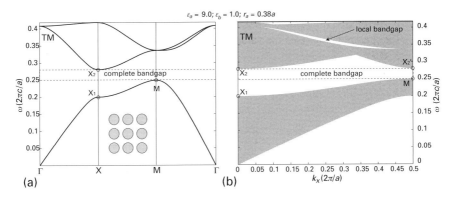

Figure 6.9 (a) Band diagram of the allowed states along the edge of an irreducible Brillouin zone. (b) Projected band diagram of allowed states (along the k_x direction).

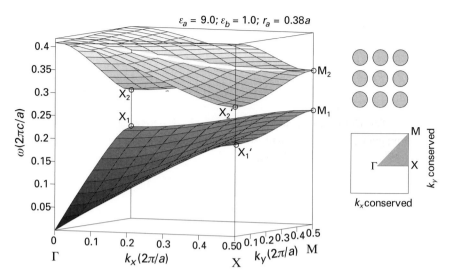

Figure 6.10 Complete 3D band diagram of the allowed TM-polarized states of a photonic crystal of a square lattice of rods in air.

6.4 Dispersion relation at a band edge, density of states and Van Hove singularities

In Fig. 6.10, we present a complete 3D dispersion relation of the allowed states of a 2D photonic crystal with $\varepsilon_a = 9.0$; $\varepsilon_b = 1.0$; $r_a = 0.38a$. Analogous to a 1D case, at the edge of a first Brillouin zone ($X_1 M X_1'$), the group velocity has at least one zero component ($v_{g,y} = 0$ along the $X_1 M$ direction, and $v_{g,x} = 0$ along the $M X_1'$ direction). At the M points all the group velocity components are zero. At the X points, owing to the symmetry of an irreducible Brillouin zone, all the group velocity components are also zero. However, at the X points, the group velocity might not be zero if, instead of rods, the square lattice were made of irregularly shaped objects. The group velocity at the M points is always zero. As seen from Figs. 6.9(a) and 6.10, in the third-lowest band along the $\Gamma X_1'$ direction, $v_{g,y} = 0$, while $v_{g,x}$ stays small. Regions of phase space with small group velocities are known as group-velocity anomalies. Typically, a small group velocity implies a large interaction time of propagation through the photonic crystal state and a photonic crystal material. Such a prolonged interaction (in time or space) enhances coupling of electromagnetic radiation and material properties, leading to an increase in the nonlinear effects, an increase in absorption losses, the enhancement of stimulated emission, etc.

Another important factor responsible for the many unusual optical properties of the photonic crystals is a strongly frequency-dependent distribution of density of states (DOS). The importance of the concept of DOS comes from the fact that given a broad frequency source, the frequencies with larger DOS will, generally speaking, be excited more strongly than the ones with smaller DOS. Therefore, by manipulation of the DOS, one can favour excitation of the particular states with predesigned properties. Inside a complete bandgap of an infinite photonic crystal, there are no guided states and, hence, the DOS is zero. Outside the complete bandgaps, the DOS is nonzero, while exhibiting sharp peaks at some specific frequency values corresponding to the so-called Van Hove singularities. Such singularities are generally found at the frequencies where some of the photonic crystal states exhibit zero group velocity. For comparison, in a uniform dielectric, the DOS is a simple monotonically increasing function of frequency.

The DOS $D(\omega)$ is defined as a frequency derivative of the total number of electromagnetic states $N(\omega)$ with frequencies smaller than ω that can be excited in a volume V^d of a d-dimensional system:

$$N(\omega) = 2\frac{V^d}{(2\pi)^d} \int\limits_{\omega(\mathbf{k})<\omega} d\mathbf{k}; \quad D(\omega) = \frac{\partial N(\omega)}{\partial \omega}. \tag{6.42}$$

For a 2D uniform dielectric of area A, (6.42) gives:

$$2D: N(\omega) = \omega^2 \frac{A\varepsilon}{2\pi^2}; \quad D(\omega) = \omega \frac{A\varepsilon}{\pi^2}. \tag{6.43}$$

To quantify Van Hove singularities, consider the shape of a dispersion relation near the points of zero group velocity at the edge of the first Brillouin zone. Using Taylor expansions near the high symmetry points we can represent the dispersion relations as:

$$\begin{aligned} \text{1st band } X_1': \quad & \omega = \omega_{X_1'} - \alpha\delta k_x^2 + \beta\delta k_y^2 \\ \text{1st band M}: \quad & \omega = \omega_{M_1} - \alpha\delta k_x^2 - \beta\delta k_y^2 \\ \text{2nd band } \Gamma, \text{M}: \quad & \omega = \omega_{M_2} + \alpha\delta k_x^2 + \beta\delta k_y^2 \\ \delta k_x = k_x - \pi/a; \quad & \delta k_y = k_y - \pi/a; \quad \alpha > 0, \quad \beta > 0. \end{aligned} \tag{6.44}$$

From (6.44) and Fig. 6.10 we can also visualize the shapes of surfaces of constant frequency (equivalent to the Fermi surface in solid-state physics) for the optical bands of a 2D square lattice of rods. Several such surfaces are presented in Fig. 6.11 for the fundamental band of Fig. 6.10.

We now compute the density of states at frequencies close to the frequencies of high symmetry points M and X, where the group velocity is zero. From (6.42) we see that the number of states at a particular frequency is proportional to the area of gray regions in Fig. 6.11. Thus, close to the M_1 point, $\omega \sim \omega_{M_1}$, $\omega < \omega_{M_1}$:

$$\delta k_y = \frac{1}{\sqrt{\beta}}\sqrt{\omega_{M_1} - \omega - \alpha\delta k_x^2}, \quad \delta k_x \in \left(0, \sqrt{\frac{\omega_{M_1} - \omega}{\alpha}}\right), \tag{6.45}$$

and

$$N(\omega)_{M_1} = \frac{A}{2\pi^2}\int\limits_{\omega(\mathbf{k})<\omega} d\mathbf{k} \approx \frac{A}{2\pi^2}\left(\left(\frac{2\pi}{a}\right)^2 - \frac{4}{\sqrt{\beta}}\int\limits_0^{\sqrt{(\omega_{M_1}-\omega)/\alpha}} d\delta k_x \sqrt{\omega_{M_1} - \omega - \alpha\delta k_x^2}\right), \tag{6.46}$$

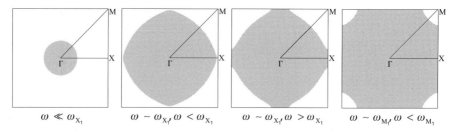

$\omega \ll \omega_{X_1}$ $\omega \sim \omega_{X_1}, \omega < \omega_{X_1}$ $\omega \sim \omega_{X_1}, \omega > \omega_{X_1}$ $\omega \sim \omega_{M_1}, \omega < \omega_{M_1}$

Figure 6.11 Regions of $\omega(\mathbf{k}) < \omega$ (in gray) for the fundamental band of Fig. 6.10, and various values of ω.

which gives:

$$N(\omega)_{M_1} = \frac{A}{2\pi^2} \left(\left(\frac{2\pi}{a}\right)^2 - \pi \frac{\omega_{M_1} - \omega}{\sqrt{\alpha\beta}} \right) \rightarrow D(\omega)_{M_1} = \frac{A}{2\pi\sqrt{\alpha\beta}} = \text{constant.}$$

(6.47)

From (6.47) and from similar consideration for the point M_2 we conclude that the DOS changes discontinuously from zero in the bandgap to a finite value at the band edge.

Now consider point X, inside a fundamental band. Considerations for both $\omega \sim \omega_{X_1}, \omega < \omega_{X_1}$, and $\omega \sim \omega_{X_1}, \omega > \omega_{X_1}$ are essentially the same, therefore we will discuss only the first case. Using (6.44) near the X'_1 point, we write:

$$N(\omega)_{X'_1} = \frac{A}{2\pi^2} \int\limits_{\omega(\mathbf{k})<\omega} d\mathbf{k} \approx \frac{A}{2\pi^2} \left(\frac{4}{\sqrt{\beta}} \int\limits_0^{\pi/a} d\delta k_y \sqrt{\omega_{X'_1} - \omega + \alpha\delta k_y^2} \right), \quad (6.48)$$

which gives:

$$D(\omega)_{X'_1} = \frac{\partial N(\omega)_{X'_1}}{\partial\omega} \approx \frac{A}{2\pi^2} \left(\frac{2}{\sqrt{\beta}} \int\limits_0^{\pi/a} \frac{d\delta k_y}{\sqrt{\omega_{X'_1} - \omega + \alpha\delta k_y^2}} \right) \propto \frac{1}{\sqrt{\omega_{X'_1} - \omega}}, \quad (6.49)$$

and, thus, DOS is weakly divergent near the points of zero group velocity where curvatures along the two principal directions have different signs. Figure 6.12 is a sketch of a typical DOS for TM modes of a square lattice of rods.

One of the applications that rely on singularities in the DOS is lasing. For example, consider a photonic crystal made of a gain media with a broad emission spectrum (laser dyes) centered around one of the singularities in the DOS (Fig. 6.12). When excited with a pump, the broad emission spectrum of a dye will be changed considerably by the highly nonuniform DOS of a photonic crystal. Thus, at the output of a photonic crystal, one expects a much narrower spectrum centered around the frequency of a DOS singularity, with a bandwidth comparable to that of a DOS peak.

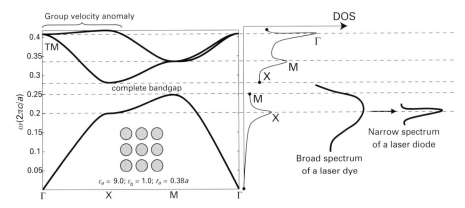

Figure 6.12 Schematic of the DOS of TM modes of a square lattice of rods in air.

6.5 Refraction from photonic crystals

The direction of energy propagation in photonic crystals is determined by the direction of the vector of group velocity:

$$\mathbf{v}_g = \frac{\partial \omega}{\partial \mathbf{k}}. \tag{6.50}$$

From differential calculus it follows that (6.50) defines a gradient in a wave-vector space, which is perpendicular to the corresponding surface of constant frequency. In a uniform dielectric, for example, $\omega = |\mathbf{k}|/\sqrt{\varepsilon}$, and, therefore, $\mathbf{v}_g = (\mathbf{k}/|\mathbf{k}|)/\sqrt{\varepsilon}$, thus coinciding with the direction of a wave vector $\mathbf{k}$. In photonic crystals, the dispersion relation is considerably modified from that of free space, therefore unusual refraction properties can be observed at the interfaces with photonic crystals.

We start by deriving Snell's law of refraction at the interface between two uniform dielectrics with dielectric constants $\sqrt{\varepsilon_1}$ and $\sqrt{\varepsilon_2}$. A general approach to solving the refraction problem is to fix the frequency of operation ω, then to consider the distribution of the allowed wave vectors, and, finally, to impose the constraints due to conservation of the wave vector component along the interface. Thus, for a given frequency of operation ω, on both sides of the interface, the allowed delocalized states are characterized by the wave vectors confined to the semicircles with radii $|\mathbf{k}| = \omega\sqrt{\varepsilon_{1,2}}$ (Fig. 6.13(a)). Owing to continuous translational symmetry along the interface, the component of a wave vector $k_\parallel$ along such an interface is conserved. Therefore, in each of the half spaces, the directions of the transmitted and reflected light can be found by intersecting the constant frequency curves with lines perpendicular to the interface and separated from the center of coordinates by $|k_\parallel|$.

For the case of an interface between the uniform dielectric and a semi-infinite photonic crystal (Fig. 6.13(b)) the projection of a wave vector along the direction of interface is again conserved. This is because the system remains periodic along the interface, therefore, the Bloch theorem can be applied in the direction of the interface, thus defining a corresponding conserved wavenumber $k_\parallel$. Due to the acircular shape of the constant frequency curves in photonic crystals, the direction of the state group velocity can be

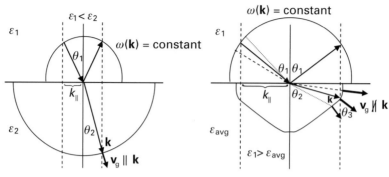

$$k_{\parallel} = \omega\sqrt{\varepsilon_1}\sin(\theta_1) = \omega\sqrt{\varepsilon_2}\sin(\theta_2)$$

$$\sqrt{\varepsilon_1}\sin(\theta_1) = \sqrt{\varepsilon_2}\sin(\theta_2)$$

When probing directions close to the bandgap edge, strong refraction might occur

Figure 6.13 Light transmission and reflection directions at (a) the interface between two uniform dielectrics and (b) the interface between a uniform dielectric and a 2D photonic crystal.

quite different from the direction of the state phase velocity ($\mathbf{v}_p \sim \mathbf{k}$). This effect is especially pronounced in the vicinity of the band edges, as surfaces of constant frequency are distorted more strongly at such frequencies. Therefore, when operated near the band edge, there exist regimes where a considerable change in the angles θ_3 of a transmitted light can be observed when an incident direction (angle θ_1) is only slightly varied (Fig. 6.13(b)). Similar considerations can also be applied to explain the possibility of having strong variation in the propagation direction of refracted light as a function of frequency near the band edge (superprism effect).

6.6   Defects in a 2D photonic crystal lattice

Probably the most important feature of photonic crystals is the ability to support spatially localized modes when a perfectly periodic structure is somewhat perturbed. Localized modes of an imperfect PhC lattice are called defect states. One distinguishes two principal defect types. These are the point defects, also referred to as resonators, and the line defects, commonly known as waveguides. Waveguides and resonators are the building blocks of modern photonics. Much of the current research is thus devoted to the study of the fundamental properties of resonators and waveguides, as well as properties of devices integrating several such basic components.

6.6.1   Line defects

The first defect type that we are going to consider is a line defect, which is also commonly known as a photonic crystal waveguide. Line defects can be created by modifying a row (or several rows) of unit cells in an otherwise periodic lattice. For example, a waveguide can be formed by a row of low refractive index rods in a square lattice of high refractive

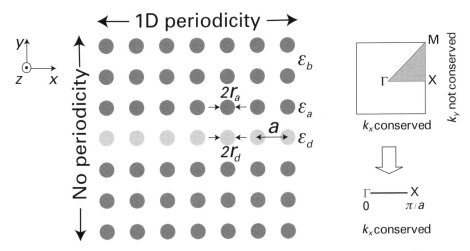

Figure 6.14 Schematic of a 2D photonic crystal waveguide. A line defect is introduced into a periodic lattice by modifying one row of rods along the $\hat{\mathbf{x}}$ direction. As the periodicity in $\hat{\mathbf{x}}$ direction is conserved, the Bloch wavenumber k_x can be used to label the defect modes.

index rods in air (see Fig. 6.14). In this case, periodicity in the $\hat{\mathbf{y}}$ direction is destroyed by the presence of a waveguide. Periodicity in the $\hat{\mathbf{x}}$ direction remains intact, therefore the k_x component of a Bloch wavevector will remain a conserved parameter, labeling eigenmodes of a PhC waveguide.

The band diagram of the bulk states of a PhC with a line defect can be derived by projecting a complete $\omega(k_x, k_y)$ band diagram of a perfectly periodic photonic lattice (Fig. 6.15) along the k_y direction (similar to Fig. 6.9(b)). When a linear defect is introduced, the dispersion relation of modes localized at the line defect (core-guided modes) will appear in the remaining bandgap. Note that a complete bandgap (common bandgap for any direction of a Bloch $\mathbf{k}$ vector) is not necessary to enable guidance in a photonic-crystal waveguide. In fact, for a given k_x, to get a localized guided mode one only needs a bandgap in a k_y direction. The goal of this section is to develop a numerical method to compute such modes, to understand their field distributions and symmetries, and, finally, to investigate how the defect strength influences the position of the guided modes inside a bandgap.

Plane-wave method for a photonic crystal with a line defect

We start by developing a plane-wave method to compute the dispersion relation of a waveguide directed along the $\hat{\mathbf{x}}$ axis. The uniform waveguide cladding comprises a square lattice of rods with dielectric constant ε_a and radius r_a. The lattice period is a. The waveguide core is formed by substitution of a single row of rods with a row of different rods of dielectric constant ε_d and radius r_d, as shown in Fig. 6.14. All the following derivations are made for the TM-polarized mode, while derivations for TE-polarized modes are the subject of Problem 6.2. Based on the symmetry considerations of Chapter 2,

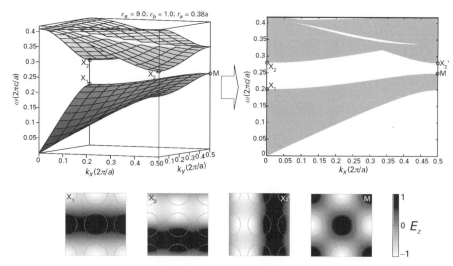

Figure 6.15 For the case of a linear defect, bulk states of a surrounding photonic crystal cladding can be presented using projection of a complete band diagram $\omega(k_x, k_y)$ of a perfectly periodic photonic crystal along the k_y direction. The inset shows the field distributions E_z at several points of high symmetry. Example of $r_a = 0.38a$, $\varepsilon_a = 9.0$ suspended in air, $\varepsilon_b = 1.0$. When a linear defect is introduced, for any k_x, a localized state appears inside a corresponding bandgap.

the general form of the electric field vector of a TM-polarized mode labeled with a Bloch wavenumber k_x will be $\mathbf{E} = (0, 0, E_{z,k_x}(x, y))$:

$$E_{z,k_x}(\mathbf{r}) = \exp(ik_x x)U_{k_x}(\mathbf{r})$$
$$U_{k_x}(\mathbf{r} + a\hat{\mathbf{x}}) = U_{k_x}(\mathbf{r}). \qquad (6.51)$$

Using discrete Fourier transform in the $\hat{\mathbf{x}}$ direction and continuous Fourier transform in the $\hat{\mathbf{y}}$ direction, the electric field (6.51) can be rewritten as:

$$E_{z,k_x}(\mathbf{r}) = \sum_{G_x} \int_{-\infty}^{+\infty} dk_y\, E_z(\mathbf{k} + \mathbf{G}_x) \exp\left(i(\mathbf{k} + \mathbf{G}_x)\mathbf{r}\right)$$

$$\mathbf{G}_x = \bar{b}_x P_x;\ \ \bar{b}_x = \frac{2\pi}{a}\hat{\mathbf{x}};\ \ \ P_x \subset \text{integer} \qquad (6.52)$$

$$\mathbf{k} = k_x \hat{\mathbf{x}} + k_y \hat{\mathbf{y}};\ \ \ k_x \text{ is a constant.}$$

A more convenient form of (6.52) is in terms of the integral over the first Brillouin zone only:

$$E_{z,k_x}(\mathbf{r}) = \sum_{G} \int_{-\pi/a}^{+\pi/a} dk_y\, E_z(\mathbf{k} + \mathbf{G}) \exp\left(i(\mathbf{k} + \mathbf{G})\mathbf{r}\right)$$

$$\mathbf{G} = \bar{b}_x P_x + \bar{b}_y P_y;\ \ \bar{b}_x = \frac{2\pi}{a}\hat{\mathbf{x}};\ \ \bar{b}_y = \frac{2\pi}{a}\hat{\mathbf{y}};\ \ \ (P_x, P_y) \subset \text{integers}$$

$$\mathbf{k} = k_x \hat{\mathbf{x}} + k_y \hat{\mathbf{y}};\ \ \ k_x \text{ is a constant.} \qquad (6.53)$$

The inverse dielectric function of a photonic crystal waveguide can be written as:

$$\frac{1}{\varepsilon(\mathbf{r})} = \frac{1}{\varepsilon_b} + \left(\frac{1}{\varepsilon_a} - \frac{1}{\varepsilon_b}\right) \sum_{\mathbf{R}} S_a(\mathbf{r} - \mathbf{R}) + \left(\frac{1}{\varepsilon_d} - \frac{1}{\varepsilon_a}\right) \sum_{\mathbf{R}_x} S_d(\mathbf{r} - \mathbf{R}_x)$$

$$S_a(\mathbf{r} - \mathbf{R}) = \begin{cases} 1, & |\mathbf{r} - \mathbf{R}| \le r_a \\ 0, & |\mathbf{r} - \mathbf{R}| > r_a \end{cases}; \quad S_d(\mathbf{r} - \mathbf{R}_x) = \begin{cases} 1, & |\mathbf{r} - \mathbf{R}_x| \le r_d \\ 0, & |\mathbf{r} - \mathbf{R}_x| > r_d \end{cases}.$$

$$\mathbf{R} = aP_x\hat{\mathbf{x}} + aP_y\hat{\mathbf{y}}; \quad \mathbf{R}_x = aP_x\hat{\mathbf{x}}; \quad (P_x, P_y) \subset \text{integers} \qquad (6.54)$$

Note that for the case of a waveguide, the dielectric function (6.54) contains three terms. The first term is constant, the second term is periodic in both the $\hat{\mathbf{x}}$ and $\hat{\mathbf{y}}$ directions, and the third term is periodic only in the $\hat{\mathbf{x}}$ direction. Using both discrete and continuous Fourier transforms, the inverse dielectric function can be presented as:

$$\frac{1}{\varepsilon(\mathbf{r})} = \frac{1}{\varepsilon_b} + \left(\frac{1}{\varepsilon_a} - \frac{1}{\varepsilon_b}\right) \sum_{\mathbf{G}} S_a(\mathbf{G}) \exp(i\mathbf{G}\mathbf{r})$$

$$+ \left(\frac{1}{\varepsilon_d} - \frac{1}{\varepsilon_a}\right) \sum_{\mathbf{G}} \frac{a}{2\pi} \int_{-\pi/a}^{+\pi/a} dk_y \, S_d(k_y\hat{\mathbf{y}} + \mathbf{G}) \exp(i(k_y\hat{\mathbf{y}} + \mathbf{G})\mathbf{r})$$

$$S_a(\mathbf{G}) = 2f_a \frac{J_1(|\mathbf{G}|r_a)}{|\mathbf{G}|r_a}; \quad S_d(k_y\hat{\mathbf{y}} + \mathbf{G}) = 2f_d \frac{J_1(|k_y\hat{\mathbf{y}} + \mathbf{G}|r_d)}{|k_y\hat{\mathbf{y}} + \mathbf{G}|r_d}; \quad f_a = \frac{\pi r_a^2}{a^2}; \quad f_d = \frac{\pi r_d^2}{a^2}$$

$$\mathbf{G} = \bar{b}_x P_x + \bar{b}_y P_y; \quad \bar{b}_x = \frac{2\pi}{a}\hat{\mathbf{x}}; \quad \bar{b}_y = \frac{2\pi}{a}\hat{\mathbf{y}}; \quad (P_x, P_y) \subset \text{integers}. \qquad (6.55)$$

Substituting (6.53) and (6.55) into Maxwell's equation (6.8) written in terms of only the electric field, $\omega^2\mathbf{E} = 1/\varepsilon(\mathbf{r}) \cdot \nabla \times (\nabla \times \mathbf{E})$, we get:

$$\omega^2(k_x) \sum_{\mathbf{G}'} \int_{-\pi/a}^{+\pi/a} dk_y' \, E_z(\mathbf{k}' + \mathbf{G}') \exp\left(i(\mathbf{k}' + \mathbf{G}')\mathbf{r}\right)$$

$$= \left[\frac{1}{\varepsilon_b} + \left(\frac{1}{\varepsilon_a} - \frac{1}{\varepsilon_b}\right) \sum_{\mathbf{G}''} S_a(\mathbf{G}'') \exp\left(i\mathbf{G}''\mathbf{r}\right) + \left(\frac{1}{\varepsilon_d} - \frac{1}{\varepsilon_a}\right)\right.$$

$$\times \sum_{\mathbf{G}''} \frac{a}{2\pi} \int_{-\pi/a}^{+\pi/a} dk_y'' S_d(k_y''\hat{\mathbf{y}} + \mathbf{G}'') \exp\left(i(k_y''\hat{\mathbf{y}} + \mathbf{G}'')\mathbf{r}\right)\right]$$

$$\times \left[\sum_{\mathbf{G}'} \int_{-\pi/a}^{+\pi/a} dk_y' \, E_z(\mathbf{k}' + \mathbf{G}') \left|\mathbf{k}' + \mathbf{G}'\right|^2 \exp\left(i(\mathbf{k}' + \mathbf{G}')\mathbf{r}\right)\right].$$

$$(6.56)$$

Multiplying the left- and right-hand sides of (6.56) by $(2\pi)^{-2} \exp(-i(\mathbf{k} + \mathbf{G})\mathbf{r})$, integrating over the 2D vector $\mathbf{r}$, and using the orthogonality of the 2D plane waves in the form:

$$\frac{1}{(2\pi)^2} \int_{\infty} d\mathbf{r} \exp(i\mathbf{k}\mathbf{r}) = \delta(\mathbf{k}), \qquad (6.57)$$

we get:

$$\omega^2(k_x) \sum_{\mathbf{G}'} \int_{-\pi/a}^{+\pi/a} dk'_y E_z(\mathbf{k}' + \mathbf{G}') \delta\left(\mathbf{k}' + \mathbf{G}' - \mathbf{k} - \mathbf{G}\right)$$

$$= \frac{1}{\varepsilon_b} \sum_{\mathbf{G}'} \int_{-\pi/a}^{+\pi/a} dk'_y E_z(\mathbf{k}' + \mathbf{G}') \left|\mathbf{k}' + \mathbf{G}'\right|^2 \delta(\mathbf{k}' + \mathbf{G}' - \mathbf{k} - \mathbf{G})$$

$$+ \left(\frac{1}{\varepsilon_a} - \frac{1}{\varepsilon_b}\right) \sum_{\mathbf{G}''} \sum_{\mathbf{G}'} \int_{-\pi/a}^{+\pi/a} dk'_y E_z(\mathbf{k}' + \mathbf{G}') S_a(\mathbf{G}'') \left|\mathbf{k}' + \mathbf{G}'\right|^2$$

$$\times \delta(\mathbf{k}' + \mathbf{G}' + \mathbf{G}'' - \mathbf{k} - \mathbf{G})$$

$$+ \left(\frac{1}{\varepsilon_d} - \frac{1}{\varepsilon_a}\right) \frac{a}{2\pi} \sum_{\mathbf{G}''} \sum_{\mathbf{G}'} \int_{-\pi/a}^{+\pi/a} dk''_y \int_{-\pi/a}^{+\pi/a} dk'_y E_z(\mathbf{k}' + \mathbf{G}') S_d(k''_y \hat{\mathbf{y}} + \mathbf{G}'') \left|\mathbf{k}' + \mathbf{G}'\right|^2$$

$$\times \delta\left(\mathbf{k}' + \mathbf{G}' + k''_y \hat{\mathbf{y}} + \mathbf{G}'' - \mathbf{k} - \mathbf{G}\right), \tag{6.58}$$

and finally:

$$\omega^2(k_x) E_z(\mathbf{k} + \mathbf{G}) =$$
$$\frac{1}{\varepsilon_b} E_z(\mathbf{k} + \mathbf{G}) |\mathbf{k} + \mathbf{G}|^2 + \left(\frac{1}{\varepsilon_a} - \frac{1}{\varepsilon_b}\right) \sum_{\mathbf{G}'} E_z(\mathbf{k} + \mathbf{G}') S_a(\mathbf{G} - \mathbf{G}') |\mathbf{k} + \mathbf{G}'|^2$$

$$+ \left(\frac{1}{\varepsilon_d} - \frac{1}{\varepsilon_a}\right) \sum_{\mathbf{G}'} \frac{a}{2\pi} \int_{-\pi/a}^{+\pi/a} dk'_y E_z(\mathbf{k}' + \mathbf{G}') S_d(\mathbf{k} + \mathbf{G} - \mathbf{k}' - \mathbf{G}') |\mathbf{k}' + \mathbf{G}'|^2 .$$

$$\mathbf{G} = \bar{b}_x P_x + \bar{b}_y P_y; \quad \mathbf{G}' = \bar{b}_x P'_x + \bar{b}_y P'_y;$$
$$\bar{b}_x = \frac{2\pi}{a} \hat{\mathbf{x}}; \quad \bar{b}_y = \frac{2\pi}{a} \hat{\mathbf{y}}; \quad (P_x, P_y, P'_x, P'_y,) \subset \text{integers}$$
$$\mathbf{k} = k_x \hat{\mathbf{x}} + k_y \hat{\mathbf{y}}; \quad \mathbf{k}' = k_x \hat{\mathbf{x}} + k'_y \hat{\mathbf{y}}; \quad k_x \text{ a constant.}$$
$$\tag{6.59}$$

For a given value of a Bloch wavenumber k_x, (6.59) constitutes an eigenvalue problem with respect to the square of a modal frequency $\omega^2(k_x)$. To solve (6.59) we discretize the integral over k'_y in (6.59). Thus, choosing:

$$\mathbf{G}_{P_x^G, P_x^G} = \bar{b}_x P_x^G + \bar{b}_y P_y^G; \quad \mathbf{G}'_{P'^G, P'^G} = \bar{b}_x P_x'^G + \bar{b}_y P_y'^G$$

$$\bar{b}_x = \frac{2\pi}{a} \hat{\mathbf{x}}; \quad \bar{b}_y = \frac{2\pi}{a} \hat{\mathbf{y}};$$
$$(P_x^G, P_y^G, P_x'^G, P_y'^G,) = [-N_G, N_G]$$

$$\mathbf{k}_{P_y^k} = k_x \hat{\mathbf{x}} + \frac{\pi}{a N_k} P_y^k \hat{\mathbf{y}}; \quad \mathbf{k}'_{P_y'^k} = k_x \hat{\mathbf{x}} + \frac{\pi}{a N_k} P_y'^k \hat{\mathbf{y}}; \quad (P_y^k, P_y'^k) = [-N_k, N_k]; \, k_x \text{ a constant,}$$
$$\tag{6.60}$$

and using a trapezoidal approximation for the 1D integral we finally get a system of $(2N_G + 1)^2(2N_k + 1)$ linear equations:

$$\omega^2(k_x)E_z(\mathbf{k}_{P_y^k} + \mathbf{G}_{P_x^G, P_y^G})$$

$$= \frac{1}{\varepsilon_b}E_z(\mathbf{k}_{P_y^k} + \mathbf{G}_{P_x^G, P_y^G})|\mathbf{k}_{P_y^k} + \mathbf{G}_{P_x^G, P_y^G}|^2$$

$$+ \left(\frac{1}{\varepsilon_a} - \frac{1}{\varepsilon_b}\right) \sum_{\substack{P_x'^G=[-N_g, N_g] \\ P_y'^G=[-N_g, N_g]}} E_z(\mathbf{k}_{P_y^k} + \mathbf{G}'_{P_x'^G, P_y'^G})S_a(\mathbf{G}_{P_x^G, P_y^G} - \mathbf{G}'_{P_x'^G, P_y'^G})|\mathbf{k}_{P_y^k} + \mathbf{G}'_{P_x'^G, P_y'^G}|^2$$

$$+ \left(\frac{1}{\varepsilon_d} - \frac{1}{\varepsilon_a}\right)\frac{1}{2N_k} \sum_{\substack{P_y'^k=[-N_k, N_k] \\ P_x'^G=[-N_g, N_g] \\ P_y'^G=[-N_g, N_g]}} E_z(\mathbf{k}'_{P_y'^k} + \mathbf{G}'_{P_x'^G, P_y'^G})w(\mathbf{k}'_{P_y'^k})S_d(\mathbf{k}_{P_y^k} + \mathbf{G}_{P_x^G, P_y^G} - \mathbf{k}'_{P_y'^k} - \mathbf{G}'_{P_x'^G, P_y'^G})$$

$$\times |\mathbf{k}'_{P_y'^k} + \mathbf{G}'_{P_x'^G, P_y'^G}|^2, \tag{6.61}$$

where the weighting function $w(\mathbf{k}'_{P_y'^k})$ for the trapezoidal integration rule is defined as:

$$w(\mathbf{k}'_{P_y'^k}) = \begin{cases} 1, & |P_y'^k| \neq N_k \\ 1/2, & |P_y'^k| = N_k \end{cases}. \tag{6.62}$$

Example of a low refractive index core waveguide

In what follows, we study dispersion relations of the core-guided modes of a waveguide created in a square lattice of dielectric rods $r_a = 0.38a$, $\varepsilon_a = 9.0$ suspended in air, $\varepsilon_b = 1.0$. We first consider the case of a low-refractive-index-core waveguide formed by reducing the dielectric constant of the core rods $\varepsilon_d < \varepsilon_a$, $r_d = r_a$. In the numerical solution of (6.61) we use 1089 plane waves; $N_g = 5$, $N_k = 4$. In Fig. 6.16, dispersion relations of the core-guided modes are presented in the first photonic bandgap for various values of the waveguide-core dielectric constant ε_d. Note that for a low refractive index defect, waveguide-core modes bifurcate from the lower edge of the bandgap. Core-guided modes appear for the defect of any strength (measured as $|\varepsilon_d^{-1} - \varepsilon_a^{-1}|$), as long as ε_d is different from ε_a.

At the points of high symmetry marked (a) and (b) in Fig. 6.16, we also plot the field distributions of the modal electric field (at these two points, the electric field distribution can be chosen as purely real). Point (a) is located closer to the lower edge of the first bandgap at $k_x = 0$, $\varepsilon_d = 4.84$. The corresponding electric field is localized in the waveguide core, while penetrating considerably into the photonic crystal cladding, owing to the proximity of the mode to the bandgap edge. This defect mode is bifurcated from the lower edge of the first bandgap, which for $k_x = 0$ corresponds to the X_1 symmetry point on the full band diagram in Fig. 6.15. Therefore, the waveguide core mode field distribution at $k_x = 0$ is similar to the field distribution of a mode of a perfectly periodic photonic crystal at the X_1 symmetry point (see Fig. 6.15). This fact will be elaborated in greater detail in the following section on perturbation theory for line defects. Point (b) is located in the center of a bandgap at $k_x = \pi/a$, $\varepsilon_d = 6.76$. The corresponding electric field is well localized in the waveguide core, and exhibits only a small penetration into the cladding. This defect mode is bifurcated from the lower edge of the first bandgap, which

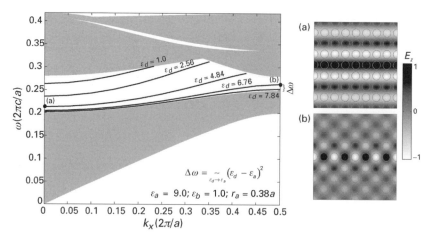

Figure 6.16 Low refractive index core, 2D photonic crystal waveguide. Dispersion relations and field distributions of the core-guided modes of a waveguide created in a square lattice of rods $r_a = 0.38a$, $\varepsilon_a = 9.0$ suspended in air, $\varepsilon_b = 1.0$. The waveguide core is made of the lower dielectric constant rods $\varepsilon_d < \varepsilon_a, r_d = r_a$. Field distributions at points (a) and (b) are similar to the field distributions of the modes of a perfect photonic crystal at the symmetry points X_1 and M (Fig. 6.15), from which the waveguide modes bifurcate when the defect is introduced.

for $k_x = \pi/a$ corresponds to the M symmetry point on the full band diagram in Fig. 6.15. Therefore, the waveguide-core mode field distribution at $k_x = \pi/a$ is similar to the field distribution of a mode of a perfectly periodic photonic crystal at the M symmetry point (see Fig. 6.15).

Example of a high refractive index core waveguide

We now consider the case of a high-refractive-index-core waveguide formed by increasing the dielectric constant of the core rods $\varepsilon_d > \varepsilon_a, r_d = r_a$. Figure 6.17 presents dispersion relations of the core-guided modes in the first photonic bandgap for various values of the waveguide core dielectric constant ε_d. For a high-refractive-index defect, waveguide-core modes bifurcate from the upper edge of the bandgap.

At the points of high symmetry marked (a) and (b) in Fig. 6.17 we also plot field distributions of the modal electric field (at these two points, the electric field distribution can be chosen as purely real). Point (a) is located closer to the lower edge of the first bandgap at $k_x = 0$, $\varepsilon_d = 16$. The corresponding electric field is localized in the waveguide core, while penetrating considerably into the photonic crystal cladding, owing to the proximity of the mode to the bandgap edge. This defect mode is bifurcated from the upper edge of the first bandgap, which for $k_x = 0$ corresponds to the X_2 symmetry point on the full band diagram in Fig. 6.15. Therefore, the waveguide-core mode field distribution at $k_x = 0$ is similar to the field distribution of a mode of a perfectly periodic photonic crystal at the X_2 symmetry point (see Fig. 6.15). Point (b) is located in the center of a bandgap at $k_x = \pi/a$, $\varepsilon_d = 19$. The corresponding electric field is well localized in the waveguide core, exhibiting only a small penetration into the cladding. This defect mode is bifurcated from the upper edge of the first bandgap, which for $k_x = \pi/a$ also corresponds to the X_2' symmetry point on the full band diagram in Fig. 6.15. Therefore,

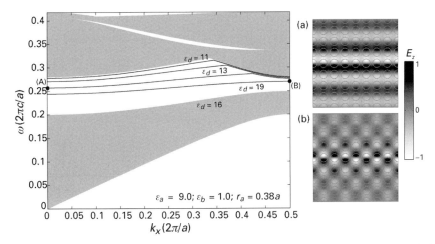

Figure 6.17 Higher-refractive-index-core, 2D photonic crystal waveguide. Dispersion relations and field distributions of the core-guided modes of a waveguide created in a square lattice of rods $r_a = 0.38a$, $\varepsilon_a = 9.0$ suspended in air, $\varepsilon_b = 1.0$. The waveguide core is made of higher-dielectric-constant rods $\varepsilon_d > \varepsilon_a, r_d = r_a$. Field distributions at points (A) and (B) are similar to the field distributions of the modes of a perfect photonic crystal at the symmetry points X_2 and X_2' (Fig. 6.14), from which the waveguide modes bifurcate when a defect is introduced.

the waveguide core mode field distribution at $k_x = \pi/a$ is similar to the field distribution of a mode of a perfectly periodic photonic crystal at the X_2' symmetry point (see Fig. 6.15).

Bifurcation of the core guide mode from the bandgap edge; perturbation theory consideration

In the limit of low-refractive-index contrast between the waveguide core and cladding $\varepsilon_d \simeq \varepsilon_a$, as seen from both Fig. 6.16 and Fig. 6.17, the core-guided mode bifurcates from one of the edges of a bandgap. In this section, we will use perturbation theory to find the functional dependence of the distance of a core-guided mode dispersion relation from the band edge as a function of the refractive index contrast.

We start by identifying what is a low-refractive-index-contrast photonic crystal waveguide. As seen from (6.59), the term describing the waveguide guidance has a $|\varepsilon_d^{-1} - \varepsilon_a^{-1}|$ prefactor, while the term describing the periodic potential of a cladding has a $|\varepsilon_a^{-1} - \varepsilon_b^{-1}|$ prefactor. If the following relation is satisfied:

$$\left|\varepsilon_d^{-1} - \varepsilon_a^{-1}\right| \ll \left|\varepsilon_b^{-1} - \varepsilon_a^{-1}\right|, \tag{6.63}$$

then the effect of a waveguide on the photonic crystal lattice can be considered small, thus constituting a limit of weak defect strength. In the case of a high-refractive-index-contrast photonic crystal $\varepsilon_b \ll \varepsilon_a$, (6.63) can be rewritten as $|\varepsilon_d - \varepsilon_a|/\varepsilon_d \ll \varepsilon_a/\varepsilon_b$. Note that this condition is especially easy to satisfy in the case of a high-index defect $\varepsilon_d > \varepsilon_a$, as $|\varepsilon_d - \varepsilon_a|/\varepsilon_d < 1$ for any value of ε_d.

We now develop the perturbation theory to describe bifurcation of weakly guided modes from the edge of a bandgap of a perfectly periodic photonic crystal. In particular, for a given small refractive index contrast $|\varepsilon_d - \varepsilon_a|/\varepsilon_d \ll 1$, we are interested in finding out how far into the bandgap the guided mode is located. We start with the modes of a perfect photonic crystal satisfying the following Hermitian eigenvalue equation (2.70):

$$\omega_{\mathrm{PC},m}^2(k_x, k_y)\mathbf{H}_{m,k_x,k_y}^{\mathrm{PC}}(\mathbf{r}) = \nabla \times \left(\frac{1}{\varepsilon_{\mathrm{PC}}(\mathbf{r})}\nabla \times \mathbf{H}_{n,k_x,k_y}^{\mathrm{PC}}\right) = \hat{H}_{\mathrm{PC}}\mathbf{H}_{m,k_x,k_y}^{\mathrm{PC}}(\mathbf{r})$$

$$\varepsilon_{\mathrm{PC}}(\mathbf{r} + aP_x\hat{\mathbf{x}} + aP_y\hat{\mathbf{y}}) = \varepsilon_{\mathrm{PC}}(\mathbf{r}); \ (P_x, P_y) \subset \text{integer}$$

$$\mathbf{H}_{m,k_x,k_y}^{\mathrm{PC}} = \exp(\mathrm{i}\mathbf{k}\mathbf{r})\mathbf{U}_{m,k_x,k_y}(\mathbf{r});$$

$$\mathbf{U}_{m,k_x,k_y}(\mathbf{r} + aP_x\hat{\mathbf{x}} + aP_y\hat{\mathbf{y}}) = \mathbf{U}_{m,k_x,k_y}(\mathbf{r}); (P_x, P_y) \subset \text{integer}, \qquad (6.64)$$

where index m is the band number, while k_x, k_y is the Bloch wave vector of a mode. Additionally, such modes satisfy the following orthogonality relation (2.132):

$$\int_{\infty} \mathrm{d}\mathbf{r}\mathbf{H}_{m,k_x,k_y}^{*\mathrm{PC}}(\mathbf{r})\mathbf{H}_{m',k_x',k_y'}^{\mathrm{PC}}(\mathbf{r}) = \delta_{m,m'}\delta(k_x - k_x')\delta(k_y - k_y'), \qquad (6.65)$$

where integration is performed over the whole two-dimensional space.

The dielectric constant of a waveguide can be considered as a perturbation on the dielectric constant of a uniform photonic crystal. In this case, the waveguide Hamiltonian can be written in terms of a Hamiltonian of a uniform photonic crystal plus a perturbation term:

$$\omega_{\mathrm{wg}}^2(k_x)\mathbf{H}_{k_x}^{\mathrm{wg}} = \nabla \times \left(\frac{1}{\varepsilon_{\mathrm{wg}}(\mathbf{r})}\nabla \times \mathbf{H}_{k_x}^{\mathrm{wg}}\right) = \hat{H}_{\mathrm{wg}}\mathbf{H}_{k_x}^{\mathrm{wg}}$$

$$= \left[\nabla \times \left(\left(\frac{1}{\varepsilon_{\mathrm{wg}}(\mathbf{r})} - \frac{1}{\varepsilon_{\mathrm{PC}}(\mathbf{r})}\right)\nabla\times\right) + \nabla \times \left(\frac{1}{\varepsilon_{\mathrm{PC}}(\mathbf{r})}\nabla\times\right)\right]\mathbf{H}_{k_x}^{\mathrm{wg}}$$

$$= \left[\Delta\hat{H} + \hat{H}_{\mathrm{PC}}\right]\mathbf{H}_{k_x}^{\mathrm{wg}} \qquad (6.66)$$

When a waveguide is introduced, the periodicity in the $\hat{\mathbf{x}}$ direction is still preserved, therefore the Bloch wavenumber k_x is conserved. In the $\hat{\mathbf{y}}$ direction, periodicity is destroyed, therefore the mode of a waveguide should be expended into a linear combination of the photonic crystal modes having all possible values of a k_y wavenumber. Moreover, one also has to sum over the contributions of different bands:

$$\mathbf{H}_{k_x}^{\mathrm{wg}} = \sum_{m=1}^{+\infty}\int_{-\pi/a}^{\pi/a}\mathrm{d}k_y\, A_{m,k_y}\mathbf{H}_{m,k_x,k_y}^{\mathrm{PC}}(\mathbf{r}). \qquad (6.67)$$

Substitution of (6.67) into (6.66) leads to the following equation:

$$\omega_{\mathrm{wg}}^2(k_x)\mathbf{H}_{k_x}^{\mathrm{wg}} = \omega_{\mathrm{wg}}^2(k_x)\sum_{m=1}^{+\infty}\int_{-\pi/a}^{\pi/a}\mathrm{d}k_y\, A_{m,k_y}\mathbf{H}_{m,k_x,k_y}^{\mathrm{PC}}(\mathbf{r}) = \left[\Delta\hat{H} + \hat{H}_{\mathrm{pc}}\right]\mathbf{H}_{k_x}^{\mathrm{wg}}$$

$$= \sum_{m=1}^{+\infty}\int_{-\pi/a}^{\pi/a}\mathrm{d}k_y\, A_{m,k_y}\Delta\hat{H}\mathbf{H}_{m,k_x,k_y}^{\mathrm{PC}}(\mathbf{r}) + \sum_{m=1}^{+\infty}\int_{-\pi/a}^{\pi/a}\mathrm{d}k_y\, A_{m,k_y}\omega_{\mathrm{PC},m}^2(k_x, k_y)\mathbf{H}_{m,k_x,k_y}^{\mathrm{PC}}(\mathbf{r}),$$

$$(6.68)$$

which can be rewritten in a more compact form as:

$$\sum_{m=1}^{+\infty} \int_{-\pi/a}^{\pi/a} dk_y \, A_{m,k_y} \left(\omega_{wg}^2(k_x) - \omega_{PC,m}^2(k_x, k_y)\right) \mathbf{H}_{m,k_x,k_y}^{PC}(\mathbf{r})$$

$$= \sum_{m=1}^{+\infty} \int_{-\pi/a}^{\pi/a} dk_y \, A_{m,k_y} \Delta \hat{H} \mathbf{H}_{m,k_x,k_y}^{PC}(\mathbf{r}). \tag{6.69}$$

We now multiply (6.69) on the left and on the right by $\mathbf{H}_{m',k_x',k_y'}^{*PC}(\mathbf{r})$, and use orthogonality relation (6.65) to arrive at the following equation:

$$A_{m',k_y'} \left(\omega_{wg}^2(k_x) - \omega_{PC,m'}^2(k_x, k_y')\right) = \sum_{m=1}^{+\infty} \int_{-\pi/a}^{\pi/a} dk_y \, A_{m,k_y} W_{k_x,k_y,k_y'}^{m,m'}, \tag{6.70}$$

with the following definition of $W_{k_x,k_y,k_y'}^{m,m'}$ ((2.87) is used to simplify the form of a matrix element):

$$\int_{\infty} d\mathbf{r} \mathbf{H}_{m',k_x',k_y'}^{*PC}(\mathbf{r}) \Delta \hat{H} \mathbf{H}_{m,k_x,k_y}^{PC}(\mathbf{r}) = \delta(k_x - k_x') \frac{2\pi}{a} \int_{-a/2}^{+a/2} dx \int_{-\infty}^{+\infty} dy \mathbf{H}_{m',k_x',k_y'}^{*PC}(\mathbf{r}) \Delta \hat{H} \mathbf{H}_{m,k_x,k_y}^{PC}(\mathbf{r})$$

$$= \delta(k_x - k_x') \omega_{m,k_x,k_y} \omega_{m',k_x',k_y'} \frac{2\pi}{a} \int_{-a/2}^{+a/2} dx \int_{-\infty}^{+\infty} dy \left(\frac{1}{\varepsilon_{wg}(\mathbf{r})} - \frac{1}{\varepsilon_{PC}(\mathbf{r})}\right) \times \mathbf{D}_{m',k_x',k_y'}^{*}(\mathbf{r}) \mathbf{D}_{m,k_x,k_y}(\mathbf{r})$$

$$= \delta(k_x - k_x') W_{k_x,k_y,k_y'}^{m,m'}. \tag{6.71}$$

In the limit of small index contrast, from (6.71) it follows that:

$$W_{k_x,k_y,k_y'}^{m,m'} \sim \left(\varepsilon_d^{-1} - \varepsilon_a^{-1}\right) \underset{\varepsilon_d \to \varepsilon_a}{\to} 0. \tag{6.72}$$

To simplify further considerations, we define an auxiliary function ϕ_{m,k_y} as:

$$\phi_{m,k_y} = A_{m,k_y}(\omega_{wg}^2(k_x) - \omega_{PC,m}^2(k_x, k_y)). \tag{6.73}$$

Then, (6.70) can be rewritten as:

$$\phi_{m',k_y'} = \sum_{m=1}^{+\infty} \int_{-\pi/a}^{\pi/a} dk_y \frac{\phi_{m,k_y}}{\omega_{wg}^2(k_x) - \omega_{PC,m}^2(k_x, k_y)} W_{k_x,k_y,k_y'}^{m,m'}. \tag{6.74}$$

Consider, as an example, bifurcation of a guided mode from the lower band edge of the fundamental bandgap in the case of a low dielectric constant defect $\varepsilon_d < \varepsilon_a$ (Fig. 6.16). The dispersion relation of the modes located at the lower edge of a fundamental bandgap ($X_1 - M$ curve in Fig. 6.15) can be described as:

$$\omega_{PC,1}(k_x, k_y)\big|_{k_y \sim \pi/a} \underset{\delta k_y = \tilde{k}_y - \pi/a}{\simeq} \omega_{PC,1}(k_x, \pi/a) - |\gamma(k_x)| \, \delta k_y^2. \tag{6.75}$$

For the guided mode bifurcated from the lower bandgap edge we write:

$$\omega_{wg}(k_x) \underset{\Delta\omega(k_x) \to +0}{=} \omega_{PC,1}(k_x, \pi/a) + \Delta\omega(k_x). \tag{6.76}$$

In the limit when the guided mode is close to the lower edge of the fundamental bandgap, only the term $m = 1$ in (6.74) will contribute significantly to the sum. Therefore, in this limit, (6.74) can be simplified as:

$$\phi_{m',k_y'} \underset{\Delta\omega(k_x) \to 0}{\simeq} \frac{1}{2\omega_{PC,1}(k_x, \pi/a)} \int\limits_{-\pi/a}^{\pi/a} d(\delta k_y) \frac{\phi_{1,\pi/a+\delta k_y} W_{k_x, \pi/a+\delta k_y, k_y'}^{1,m'}}{\Delta\omega(k_x) + |\gamma(k_x)| \, \delta k_y^2}. \tag{6.77}$$

Assuming slow functional dependence of ϕ_{1,k_y} and $W_{k_x,k_y,k_y'}^{1,m'}$ with respect to k_y, for $m' = 1$, $k_y' = \pi/a$ we get:

$$\phi_{1,\pi/a} \underset{\Delta\omega(k_x) \to 0}{\simeq} \frac{\phi_{1,\pi/a} W_{k_x, \pi/a, \pi/a}^{1,1}}{2\omega_{PC,1}(k_x, \pi/a)} \int\limits_{-\pi/a}^{\pi/a} \frac{d(\delta k_y)}{\Delta\omega(k_x) + |\gamma(k_x)| \, \delta k_y^2}$$

$$\underset{\Delta\omega(k_x) \to 0}{=} \frac{\pi}{2\sqrt{\Delta\omega(k_x)}} \frac{\phi_{1,\pi/a} W_{k_x, \pi/a, \pi/a}^{1,1}}{\sqrt{|\gamma(k_x)|}\omega_{PC,1}(k_x, \pi/a)}, \tag{6.78}$$

which finally results in the following perturbative expression for the modal distance from the lower bandgap edge:

$$\Delta\omega(k_x) = \left(\frac{\pi}{2} \frac{W_{k_x, \pi/a, \pi/a}^{1,1}}{\sqrt{|\gamma(k_x)|}\omega_{PC,1}(k_x, \pi/a)} \right)^2 \underset{\varepsilon_d \to \varepsilon_a}{\sim} (\varepsilon_d - \varepsilon_a)^2. \tag{6.77}$$

6.6.2   Point defects

A point defect can be realized by introducing a localized perturbation of a dielectric constant into a perfectly periodic photonic crystal. As an example, a resonator is created when the dielectric constant of a single rod is perturbed from its original value in a photonic crystal lattice (Fig. 6.18). When one of the unit cells in a photonic crystal is different from the rest, periodicity is destroyed in all directions, therefore there is no Bloch wavenumber to label resonator modes, and the only conserved parameter remaining is frequency.

The only band diagram possible in this case is a line along the frequency direction featuring a continuum of delocalized states in a photonic crystal, interrupted by complete bandgaps (Fig. 6.19). When perturbation is introduced into a photonic crystal lattice, the

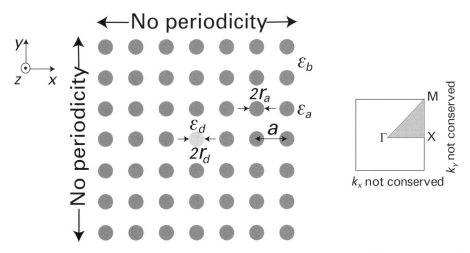

Figure 6.18 Schematic of a resonator in a 2D photonic crystal lattice. A point defect is introduced into a periodic lattice by modifying the dielectric constant of a single rod. As the periodicity in both $\hat{x}$ and $\hat{y}$ directions is destroyed, no Bloch wavenumber can be used to label the modes.

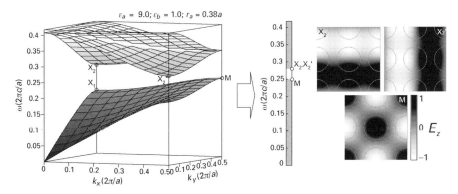

Figure 6.19 For the case of a point defect, bulk states of a surrounding photonic crystal cladding can be presented using projection of a complete band diagram $\omega(k_x, k_y)$ of a perfectly periodic photonic crystal onto a frequency axis. The inset shows the field distributions E_z at the band edges of a complete fundamental bandgap. Example of $r_a = 0.38a$, $\varepsilon_a = 9.0$ suspended in air $\varepsilon_b = 1.0$. When a defect is introduced, a localized state appears inside a complete bandgap.

resonator mode will be pulled into the bandgap from one of the bandgap edges. Unlike the case of a line defect, for a point defect state to exist, the presence of a complete bandgap in the surrounding photonic crystal cladding is essential. In what follows, we start by presenting a plane-wave expansion method to find modes localized at a point defect. We then present a numerical study of the states of a resonator created by changing the dielectric constant of a single rod in the otherwise perfectly periodic lattice of the dielectric rods in air. Finally, for the case of a small refractive-index contrast between the resonator rod and the rest of the dielectric rods in the cladding (weak resonator), we use perturbation theory to find the separation of a defect state frequency from the bandgap edge as a function of the refractive-index contrast.

Plane-wave method for a photonic crystal with a point defect

We start by developing a plane-wave method to compute localized states of a resonator. Uniform cladding comprises a square lattice of rods with dielectric constant ε_a and radius r_a. The lattice period is a. A resonator is formed by replacing a single rod with a different rod of dielectric constant ε_d and radius r_d as demonstrated schematically in Fig. 6.18. All the following derivations are made for the TM-polarized mode, while derivations for TE-polarized modes are the subject of Problem 6.3. Based on the symmetry considerations of Chapter 2, the general form of the electric field vector of a TM-polarized mode will be $\mathbf{E} = (0, 0, E_{z,\omega_{\mathrm{res}}}(x, y))$. Using continuous Fourier transforms in the $\hat{\mathbf{x}}$ and $\hat{\mathbf{y}}$ directions, the electric field can be rewritten as:

$$E_{z,\omega_{\mathrm{res}}}(\mathbf{r}) = \int\limits_{-\infty}^{+\infty} dk_x \int\limits_{-\infty}^{+\infty} dk_y E_{z,\omega_{\mathrm{res}}}(\mathbf{k}) \exp{(i\mathbf{k}\mathbf{r})} = \int\limits_{\infty} d\mathbf{k} E_{z,\omega_{\mathrm{res}}}(\mathbf{k}) \exp{(i\mathbf{k}\mathbf{r})}$$

$$\mathbf{k} = k_x \hat{\mathbf{x}} + k_y \hat{\mathbf{y}}. \tag{6.78}$$

A more convenient form of (6.78) is in terms of the integral over the first Brillouin zone (FBZ) only:

$$E_{z,\omega_{\mathrm{res}}}(\mathbf{r}) = \sum_{\mathbf{G}} \int\limits_{\mathrm{FBZ}} d\mathbf{k} E_z(\mathbf{k} + \mathbf{G}) \exp{(i(\mathbf{k} + \mathbf{G})\mathbf{r})}$$

$$\mathbf{G} = \bar{b}_x P_x + \bar{b}_y P_y; \bar{b}_x = \frac{2\pi}{a}\hat{\mathbf{x}}; \quad \bar{b}_y = \frac{2\pi}{a}\hat{\mathbf{y}}; (P_x, P_y) \subset \text{integers}$$

$$\mathbf{k} = k_x \hat{\mathbf{x}} + k_y \hat{\mathbf{y}}; \quad (k_x, k_y) \subset \left(-\frac{\pi}{a}, \frac{\pi}{a}\right). \tag{6.79}$$

The inverse dielectric function of a resonator in a uniform photonic crystal lattice can be written as:

$$\frac{1}{\varepsilon(\mathbf{r})} = \frac{1}{\varepsilon_b} + \left(\frac{1}{\varepsilon_a} - \frac{1}{\varepsilon_b}\right) \sum_{\mathbf{R}} S_a(\mathbf{r} - \mathbf{R}) + \left(\frac{1}{\varepsilon_d} - \frac{1}{\varepsilon_a}\right) S_d(\mathbf{r})$$

$$S_a(\mathbf{r} - \mathbf{R}) = \begin{cases} 1, |\mathbf{r} - \mathbf{R}| \le r_a \\ 0, |\mathbf{r} - \mathbf{R}| > r_a \end{cases}; S_d(\mathbf{r}) = \begin{cases} 1, |\mathbf{r}| \le r_d \\ 0, |\mathbf{r}| > r_d \end{cases}$$

$$\mathbf{R} = a P_x \hat{\mathbf{x}} + a P_y \hat{\mathbf{y}}; \quad (P_x, P_y) \subset \text{integers}. \tag{6.80}$$

Note that for the case of a resonator, the dielectric function (6.80) contains three terms. The first term is constant, the second term is periodic in both the $\hat{\mathbf{x}}$ and $\hat{\mathbf{y}}$ directions, and the third term is a localized function. Using both discrete and continuous Fourier transforms, the inverse dielectric function (6.80) can be presented as:

$$\frac{1}{\varepsilon(\mathbf{r})} = \frac{1}{\varepsilon_b} + \left(\frac{1}{\varepsilon_a} - \frac{1}{\varepsilon_b}\right) \sum_{\mathbf{G}} S_a(\mathbf{G}) \exp(i\mathbf{G}\mathbf{r})$$

$$+ \left(\frac{1}{\varepsilon_d} - \frac{1}{\varepsilon_a}\right) \sum_{\mathbf{G}} \left(\frac{a}{2\pi}\right)^2 \int_{\text{FBZ}} d\mathbf{k} S_d(\mathbf{k} + \mathbf{G}) \exp(i(\mathbf{k} + \mathbf{G})\mathbf{r})$$

$$S_a(\mathbf{G}) = 2 f_a \frac{J_1(|\mathbf{G}|r_a)}{|\mathbf{G}|r_a}; \; S_d(\mathbf{k} + \mathbf{G}) = 2 f_d \frac{J_1(|\mathbf{k} + \mathbf{G}|r_d)}{|\mathbf{k} + \mathbf{G}|r_d}; \; f_a = \frac{\pi r_a^2}{a^2}; \; f_d = \frac{\pi r_d^2}{a^2}$$

$$\mathbf{G} = \bar{b}_x P_x + \bar{b}_y P_y; \bar{b}_x = \frac{2\pi}{a}\hat{\mathbf{x}}; \bar{b}_y = \frac{2\pi}{a}\hat{\mathbf{y}}; (P_x, P_y) \subset \text{integers}$$

$$\mathbf{k} = k_x\hat{\mathbf{x}} + k_y\hat{\mathbf{y}}; (k_x, k_y) \subset \left(-\frac{\pi}{a}, \frac{\pi}{a}\right). \tag{6.81}$$

Substituting (6.79) and (6.81) into Maxwell's equations (6.8) written in terms of only the electric field $\omega^2 \mathbf{E} = 1/\varepsilon(\mathbf{r}) \cdot \nabla \times (\nabla \times \mathbf{E})$ we get:

$$\omega_{\text{res}}^2 \sum_{\mathbf{G}'} \int_{\text{FBZ}} d\mathbf{k}' E_z(\mathbf{k}' + \mathbf{G}') \exp(i(\mathbf{k}' + \mathbf{G}')\mathbf{r})$$

$$= \left[\frac{1}{\varepsilon_b} + \left(\frac{1}{\varepsilon_a} - \frac{1}{\varepsilon_b}\right) \sum_{\mathbf{G}''} S_a(\mathbf{G}'') \exp(i\mathbf{G}''\mathbf{r})\right.$$

$$\left. + \left(\frac{1}{\varepsilon_d} - \frac{1}{\varepsilon_a}\right) \sum_{\mathbf{G}''} \left(\frac{a}{2\pi}\right)^2 \int_{\text{FBZ}} d\mathbf{k}'' S_d(\mathbf{k}'' + \mathbf{G}'') \exp(i(\mathbf{k}'' + \mathbf{G}'')\mathbf{r})\right]$$

$$\times \left[\sum_{\mathbf{G}'} \int_{\text{FBZ}} d\mathbf{k}' E_z(\mathbf{k}' + \mathbf{G}') |\mathbf{k}' + \mathbf{G}'|^2 \exp(i(\mathbf{k}' + \mathbf{G}')\mathbf{r})\right]. \tag{6.82}$$

Multiplying the left- and right-hand sides of (6.82) by $\frac{1}{(2\pi)^2} \exp(-i(\mathbf{k} + \mathbf{G})\mathbf{r})$, integrating over the 2D vector $\mathbf{r}$, and using the orthogonality of the 2D plane waves in the form:

$$\frac{1}{(2\pi)^2} \int_{\infty} d\mathbf{r} \exp(i\mathbf{k}\mathbf{r}) = \delta(\mathbf{k}), \tag{6.83}$$

we get:

$$\omega_{\text{res}}^2 \sum_{\mathbf{G}'} \int_{\text{FBZ}} d\mathbf{k}' E_z(\mathbf{k}' + \mathbf{G}')\delta(\mathbf{k}' + \mathbf{G}' - \mathbf{k} - \mathbf{G})$$

$$= \frac{1}{\varepsilon_b} \sum_{\mathbf{G}'} \int_{\text{FBZ}} d\mathbf{k}' E_z(\mathbf{k}' + \mathbf{G}') |\mathbf{k}' + \mathbf{G}'|^2 \delta(\mathbf{k}' + \mathbf{G}' - \mathbf{k} - \mathbf{G})$$

$$+ \left(\frac{1}{\varepsilon_a} - \frac{1}{\varepsilon_b}\right) \sum_{\mathbf{G}'} \sum_{\mathbf{G}''} \int_{\text{FBZ}} d\mathbf{k}' E_z(\mathbf{k}' + \mathbf{G}')S_a(\mathbf{G}'') |\mathbf{k}' + \mathbf{G}'|^2 \delta(\mathbf{k}' + \mathbf{G}' + \mathbf{G}'' - \mathbf{k} - \mathbf{G})$$

$$+\left(\frac{1}{\varepsilon_d}-\frac{1}{\varepsilon_a}\right)\left(\frac{a}{2\pi}\right)^2\sum_{\mathbf{G}'}\sum_{\mathbf{G}''}\int_{\text{FBZ}}d\mathbf{k}'\int_{\text{FBZ}}d\mathbf{k}''E_z(\mathbf{k}'+\mathbf{G}')S_d(\mathbf{k}''+\mathbf{G}'')\left|\mathbf{k}'+\mathbf{G}'\right|^2$$

$$\times\delta\left(\mathbf{k}'+\mathbf{G}'+\mathbf{k}''+\mathbf{G}''-\mathbf{k}-\mathbf{G}\right),\tag{6.84}$$

and finally:

$$\omega_{\text{res}}^2E_z(\mathbf{k}+\mathbf{G})=\frac{1}{\varepsilon_b}E_z(\mathbf{k}+\mathbf{G})\,|\mathbf{k}+\mathbf{G}|^2+\left(\frac{1}{\varepsilon_a}-\frac{1}{\varepsilon_b}\right)\sum_{\mathbf{G}'}E_z(\mathbf{k}+\mathbf{G}')S_a(\mathbf{G}-\mathbf{G}')\left|\mathbf{k}+\mathbf{G}'\right|^2$$

$$+\left(\frac{1}{\varepsilon_d}-\frac{1}{\varepsilon_a}\right)\left(\frac{a}{2\pi}\right)^2\sum_{\mathbf{G}'}\int_{\text{FBZ}}d\mathbf{k}'E_z(\mathbf{k}'+\mathbf{G}')S_d(\mathbf{k}-\mathbf{k}'+\mathbf{G}-\mathbf{G}')\left|\mathbf{k}'+\mathbf{G}'\right|^2$$

$$\mathbf{G}=\bar{b}_xP_x+\bar{b}_yP_y;\;\mathbf{G}'=\bar{b}_xP_x'+\bar{b}_yP_y';\;\bar{b}_x=\frac{2\pi}{a}\hat{\mathbf{x}};\;\bar{b}_y=\frac{2\pi}{a}\hat{\mathbf{y}};$$

$$\left(P_x,P_y,P_x',P_y',\right)\subset\text{integers}$$

$$\mathbf{k}=k_x\hat{\mathbf{x}}+k_y\hat{\mathbf{y}};\quad\mathbf{k}'=k_x'\hat{\mathbf{x}}+k_y'\hat{\mathbf{y}};\quad\left(k_x,k_y,k_x',k_y'\right)\subset\left(-\frac{\pi}{a},\frac{\pi}{a}\right).\tag{6.85}$$

Equation (6.85) constitutes an eigenvalue problem with respect to the square of a modal frequency ω_{res}^2. To solve (6.85) we discretize the 2D integral over $\mathbf{k}'$ in (6.85). Thus, choosing:

$$\mathbf{G}_{P_x^G,P_x^G}=\bar{b}_xP_x^G+\bar{b}_yP_y^G;\mathbf{G}'_{P_x^{\prime G},P_y^{\prime G}}=\bar{b}_xP_x^{\prime G}+\bar{b}_yP_y^{\prime G}\;\;\bar{b}_x=\frac{2\pi}{a}\hat{\mathbf{x}};\;\bar{b}_y=\frac{2\pi}{a}\hat{\mathbf{y}};$$

$$\left(P_x^G,P_y^G,P_x^{\prime G},P_y^{\prime G},\right)=[-N_G,N_G]$$

$$\mathbf{k}_{P_x^k,P_y^k}=\frac{\pi}{aN_k}P_x^k\hat{\mathbf{x}}+\frac{\pi}{aN_k}P_y^k\hat{\mathbf{y}};\mathbf{k}'_{P_x^{\prime k},P_y^{\prime k}}=\frac{\pi}{aN_k}P_x^{\prime k}\hat{\mathbf{x}}+\frac{\pi}{aN_k}P_y^{\prime k}\hat{\mathbf{y}}$$

$$\left(P_x^k,P_y^k,P_x^{\prime k},P_y^{\prime k}\right)=[-N_k,N_k],$$

$$\tag{6.86}$$

and using a trapezoidal approximation for the 2D integral we finally get a system of $(2N_G+1)^2(2N_k+1)^2$ linear equations:

$$\omega_{\text{res}}^2E_z(\mathbf{k}_{P_x^k,P_y^k}+\mathbf{G}_{P_x^G,P_y^G})=\frac{1}{\varepsilon_b}E_z(\mathbf{k}_{P_x^k,P_y^k}+\mathbf{G}_{P_x^G,P_y^G})\left|\mathbf{k}_{P_x^k,P_y^k}+\mathbf{G}_{P_x^G,P_y^G}\right|^2$$

$$+\left(\frac{1}{\varepsilon_a}-\frac{1}{\varepsilon_b}\right)\sum_{\substack{P_x^{\prime G}=[-N_g,N_g]\\P_y^{\prime G}=[-N_g,N_g]}}E_z(\mathbf{k}_{P_x^k,P_y^k}+\mathbf{G}'_{P_x^{\prime G},P_y^{\prime G}})$$

$$\times S_a(\mathbf{G}_{P_x^G,P_y^G}-\mathbf{G}'_{P_x^{\prime G},P_y^{\prime G}})\left|\mathbf{k}_{P_x^k,P_y^k}+\mathbf{G}'_{P_x^{\prime G},P_y^{\prime G}}\right|^2$$

$$+\left(\frac{1}{\varepsilon_d}-\frac{1}{\varepsilon_a}\right)\left(\frac{1}{2N_k}\right)^2\sum_{\substack{P_x^{\prime k}=[-N_k,N_k]\\P_y^{\prime k}=[-N_k,N_k]\\P_x^{\prime G}=[-N_g,N_g]\\P_y^{\prime G}=[-N_g,N_g]}}E_z(\mathbf{k}'_{P_x^{\prime k},P_y^{\prime k}}+\mathbf{G}'_{P_x^{\prime G},P_y^{\prime G}})w(\mathbf{k}'_{P_x^{\prime k},P_y^{\prime k}})$$

$$\times S_d(\mathbf{k}_{P_x^k,P_y^k}+\mathbf{G}_{P_x^G,P_y^G}-\mathbf{k}'_{P_x^{\prime k},P_y^{\prime k}}-\mathbf{G}'_{P_x^{\prime G},P_y^{\prime G}})\left|\mathbf{k}'_{P_x^{\prime k},P_y^{\prime k}}+\mathbf{G}'_{P_x^{\prime G},P_y^{\prime G}}\right|^2,\tag{6.87}$$

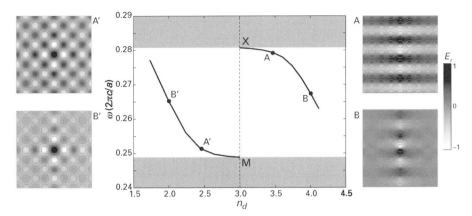

Figure 6.20 Localized states of a point defect in a 2D photonic crystal. Frequencies and field distributions of the resonator modes are presented as a function of a resonator rod of refractive index n_d. The resonator is created in a square lattice of rods $r_a = 0.38a$, $\varepsilon_a = 9.0$ suspended in air, $\varepsilon_b = 1.0$, by changing the dielectric constant of a single rod to ε_d. For the lower-index defect $\varepsilon_d < \varepsilon_a$, the resonator state is a singlet bifurcating from the lower bandgap edge. For the higher index defect $\varepsilon_d > \varepsilon_a$, resonator state is a doubly degenerate mode bifurcating from the upper bandgap edge.

where the weighting function $w(\mathbf{k}'_{P_x^{ik}, P_y^{ik}})$ for the trapezoidal integration rule is defined as:

$$w(\mathbf{k}'_{P_x^{ik}, P_y^{ik}}) = \begin{cases} 1, & (|P_x'^k| \neq N_k) \cap (|P_y'^k| \neq N_k) \\ 1/2, & ((|P_x'^k| = N_k) \cap (|P_y'^k| \neq N_k)) \cup ((|P_x'^k| \neq N_k) \cap (|P_y'^k| = N_k)). \\ 1/4, & (|P_x'^k| = N_k) \cap (|P_y'^k| = N_k) \end{cases}$$

$$(6.88)$$

Example of a localized resonator state

In what follows we study localized states of a resonator created in a square lattice of dielectric rods $r_a = 0.38a$, $\varepsilon_a = 9.0$ suspended in air, $\varepsilon_b = 1.0$. The resonator is formed by replacing one of the dielectric rods with a rod of the same radius but different dielectric constant $\varepsilon_d \neq \varepsilon_a$. In our numerical solution of (6.87) we use 9801 plane waves $N_g = 5$, $N_k = 4$. In Fig. 6.20, the frequency of a resonator state is presented in the first photonic bandgap for various values of the resonator rod refractive index n_d. Note that for a low-refractive-index defect, the resonator state bifurcates from the lower bandgap edge, while for the high-refractive-index defect, the resonator state bifurcates from the upper bandgap edge. Core-guided modes appear for the defect of any strength (measured as $|\varepsilon_d^{-1} - \varepsilon_a^{-1}|$), as long as ε_d is different from ε_a.

For the low index defect $\varepsilon_d < \varepsilon_a$, the resonator state is a singlet bifurcating from the lower bandgap edge (the left part of Fig. 6.20). The field distributions at points (A′) and (B′) are similar to the field distribution of the mode of a perfect photonic crystal at the M symmetry point (see Fig. 6.15). The field distribution at point (B′) is localized more strongly at a defect site than the field distribution at point (A′). This is easy to rationalize as point B′ is situated deeper inside a bandgap than point A′. For the high-index defect $\varepsilon_d > \varepsilon_a$, the resonator state is a doubly degenerate mode bifurcating from the upper bandgap edge (the right part of Fig. 6.20). The field distributions at points (A) and (B) are similar to the field distribution of the mode of a perfect photonic crystal at the symmetry point X_2. The field distributions for the other degenerate modes are related to the ones shown in Fig. 6.20 (right) by a 90° rotation, and are similar to the field distribution of the mode of a perfect photonic crystal at the symmetry point X_2'.

In the above-mentioned example, a defect is introduced into the infinitely periodic photonic crystal lattice. In this case, the photonic bandgap of a surrounding photonic crystal completely suppresses radiation loss from the resonator site. In the absence of material losses, the lifetime of a resonator state is infinite as it is a true eigenstate of a Maxwell Hamiltonian. In the density-of-states diagram, the resonator state is defined by a delta function located inside the photonic bandgap. Experimentally, photonic crystal cladding is finite and material loss is not negligible, thus leading to the finite lifetime of a resonator state due to irradiation and absorption. In the density-of-states diagram, such a state is no longer characterized by a delta function, but rather by a peak of finite width, which is inversely proportional to the lifetime of a defect state. Of special interest are the high-quality resonators that support localized states with lifetimes that are considerably longer than the period of a corresponding oscillation frequency. Such resonators are typically used to filter a narrow frequency band from the wavelength multiplexed signal, or to enhance a light–matter interaction (due to the considerable lifetime of a trapped photon) for nonlinear optics and lasing applications. In the other extreme, several low quality resonators can be coupled to each other in a chain, resulting in filters with complex frequency response, optical delay lines, and slow light devices.

Bifurcation of a resonator state from the bandgap edge; perturbation theory consideration

In the limit of low-refractive-index contrast between the resonator and photonic crystal cladding $\varepsilon_d \simeq \varepsilon_a$, as seen from Fig. 6.20, the resonator mode bifurcates from one of the edges of a bandgap. In this section we will use perturbation theory to find the functional dependence of the distance of a resonator mode frequency from a corresponding bandgap edge as a function of the refractive-index contrast. As before, we consider the case of low-refractive-index contrast $|\varepsilon_d - \varepsilon_a|/\varepsilon_d \ll \varepsilon_a/\varepsilon_b$.

We now develop the perturbation theory to describe bifurcation of weakly localized modes from the bandgap edge of a perfectly periodic photonic crystal. As in the case of a photonic crystal waveguide, we start with the orthogonal set of modes of a perfect photonic crystal cladding satisfying the Hermitian eigenvalue equation (6.10), and characterized by the band number m, and a Bloch wave vector k_x, k_y.

The dielectric constant of a resonator can be considered as a perturbation of the dielectric constant of a uniform photonic crystal. In this case, the waveguide Hamiltonian can be written in terms of a Hamiltonian of a uniform photonic crystal plus a perturbation term:

$$\omega_{\text{res}}^2 \mathbf{H}_{\omega_{\text{res}}}^{\text{res}} = \nabla \times \left(\frac{1}{\varepsilon_{\text{res}}(\mathbf{r})} \nabla \times \mathbf{H}_{\omega_{\text{res}}}^{\text{res}} \right) = \hat{H}_{\text{res}} \mathbf{H}_{\omega_{\text{res}}}^{\text{res}}$$

$$= \left[\nabla \times \left(\left(\frac{1}{\varepsilon_{\text{res}}(\mathbf{r})} - \frac{1}{\varepsilon_{\text{PC}}(\mathbf{r})} \right) \nabla \times \right) + \nabla \times \left(\frac{1}{\varepsilon_{\text{PC}}(\mathbf{r})} \nabla \times \right) \right] \mathbf{H}_{\omega_{\text{res}}}^{\text{res}}$$

$$= [\Delta \hat{H} + \hat{H}_{\text{PC}}] \mathbf{H}_{\omega_{\text{res}}}^{\text{res}}. \tag{6.89}$$

When a point defect is introduced, the periodicity in both the $\hat{x}$ and the $\hat{y}$ directions is destroyed. Therefore, the resonator state should be expended into a linear combination of the photonic crystal modes having all possible values of a $\mathbf{k}$ wave vector. Moreover, one also has to sum over the contributions of different bands:

$$\mathbf{H}_{\omega_{\text{res}}}^{\text{res}} = \sum_{m=1}^{+\infty} \int_{-\pi/a}^{\pi/a} dk_x \int_{-\pi/a}^{\pi/a} dk_y A_{m,k_x,k_y} \mathbf{H}_{m,k_x,k_y}^{\text{PC}}(\mathbf{r}). \tag{6.90}$$

Substitution of (6.90) into (6.89) leads to the following equation:

$$\omega_{\text{res}}^2 \mathbf{H}_{\omega_{\text{res}}}^{\text{res}} = \omega_{\text{res}}^2 \sum_{m=1}^{+\infty} \int_{-\pi/a}^{\pi/a} dk_x \int_{-\pi/a}^{\pi/a} dk_y A_{m,k_x,k_y} \mathbf{H}_{m,k_x,k_y}^{\text{PC}}(\mathbf{r}) = [\Delta \hat{H} + \hat{H}_{\text{pc}}] \mathbf{H}_{\omega_{\text{res}}}^{\text{res}}$$

$$= \sum_{m=1}^{+\infty} \int_{-\pi/a}^{\pi/a} dk_x \int_{-\pi/a}^{\pi/a} dk_y A_{m,k_x,k_y} \Delta \hat{H} \mathbf{H}_{m,k_x,k_y}^{\text{PC}}(\mathbf{r})$$

$$+ \sum_{m=1}^{+\infty} \int_{-\pi/a}^{\pi/a} dk_x \int_{-\pi/a}^{\pi/a} dk_y A_{m,k_x,k_y} \omega_{\text{PC},m}^2(k_x, k_y) \mathbf{H}_{m,k_x,k_y}^{\text{PC}}(\mathbf{r}), \tag{6.91}$$

which can be rewritten in a more compact form as:

$$\sum_{m=1}^{+\infty} \int_{-\pi/a}^{\pi/a} dk_x \int_{-\pi/a}^{\pi/a} dk_y A_{m,k_x,k_y} \left(\omega_{\text{res}}^2 - \omega_{\text{PC},m}^2(k_x, k_y) \right) \mathbf{H}_{m,k_x,k_y}^{\text{PC}}(\mathbf{r})$$

$$= \sum_{m=1}^{+\infty} \int_{-\pi/a}^{\pi/a} dk_x \int_{-\pi/a}^{\pi/a} dk_y A_{m,k_x,k_y} \Delta \hat{H} \mathbf{H}_{m,k_x,k_y}^{\text{PC}}(\mathbf{r}). \tag{6.92}$$

We now multiply (6.92) on the left and right by $\mathbf{H}_{m',k_x',k_y'}^{*\text{PC}}(\mathbf{r})$, and use the orthogonality relation (6.65) to arrive at the following equation:

$$A_{m,k_x',k_y'} \left(\omega_{\text{res}}^2 - \omega_{\text{PC},m}^2(k_x', k_y') \right) = \sum_{m=1}^{+\infty} \int_{-\pi/a}^{\pi/a} dk_x \int_{-\pi/a}^{\pi/a} dk_y A_{m,k_x,k_y} W_{k_x,k_y,k_x',k_y'}^{m,m'}, \tag{6.93}$$

with the following definition of $W_{k_x,k_y,k_x',k_y'}^{m,m'}$ ((2.87) is used to simplify the form of a matrix element):

$$
\int_{\infty} d\mathbf{r} H_{m',k_x',k_y'}^{*\mathrm{PC}}(\mathbf{r}) \Delta \hat{H} H_{m,k_x,k_y}^{\mathrm{PC}}(\mathbf{r}) = \int_{-\infty}^{+\infty} dx \int_{-\infty}^{+\infty} dy H_{m',k_x',k_y'}^{*\mathrm{PC}}(\mathbf{r}) \Delta \hat{H} H_{m,k_x,k_y}^{\mathrm{PC}}(\mathbf{r})
$$

$$
= \omega_{m,k_x,k_y} \omega_{m',k_x,k_y} \int_{-\infty}^{+\infty} dx \int_{-\infty}^{+\infty} dy \left(\frac{1}{\varepsilon_{\mathrm{res}}(\mathbf{r})} - \frac{1}{\varepsilon_{\mathrm{PC}}(\mathbf{r})} \right) \mathbf{D}_{m',k_x',k_y'}^{*}(\mathbf{r}) \mathbf{D}_{m,k_x,k_y}(\mathbf{r})
$$

$$
= W_{k_x,k_y,k_x',k_y'}^{m,m'}. \tag{6.94}
$$

In the limit of small index contrast, from (6.94) it follows that:

$$
W_{k_x,k_y,k_x',k_y'}^{m,m'} \sim \left(\varepsilon_d^{-1} - \varepsilon_a^{-1} \right) \underset{\varepsilon_d \to \varepsilon_a}{\to} 0. \tag{6.95}
$$

To simplify further considerations, we define an auxiliary function ϕ_{m,k_x,k_y} as:

$$
\phi_{m,k_x,k_y} = A_{m,k_x,k_y} \left(\omega_{\mathrm{res}}^2 - \omega_{\mathrm{PC},m}^2(k_x,k_y) \right). \tag{6.96}
$$

Then, (6.93) can be rewritten as:

$$
\phi_{m',k_x',k_y'} = \sum_{m'=1}^{+\infty} \int_{-\pi/a}^{\pi/a} dk_x \int_{-\pi/a}^{\pi/a} dk_y \frac{\phi_{m,k_x,k_y}}{\omega_{\mathrm{res}}^2 - \omega_{\mathrm{PC},m}^2(k_x,k_y)} W_{k_x,k_y,k_x',k_y'}^{m,m'}. \tag{6.97}
$$

Consider, as an example, the bifurcation of a resonator state from the lower edge of the fundamental bandgap in the case of a low dielectric constant defect $\varepsilon_d < \varepsilon_a$ (the left part in Fig. 6.20). The dispersion relation of the modes located at the lower edge of a fundamental bandgap (vicinity of the M point in Fig. 6.14) can be described as:

$$
\omega_{\mathrm{PC},1}(k_x,k_y)\big|_{\substack{k_x \sim \pi/a, k_y \sim \pi/a \\ \delta k_x = k_x - \pi/a \\ \delta k_y = k_y - \pi/a}} \cong \omega_{\mathrm{PC},1}(\pi/a,\pi/a) - |\gamma_x| \, \delta k_x^2 - |\gamma_y| \, \delta k_y^2.
$$

$$\tag{6.98}$$

For the resonator state bifurcated from the lower bandgap edge we write:

$$
\omega_{\mathrm{res}} \underset{\Delta\omega \to +0}{=} \omega_{\mathrm{PC},1}(\pi/a,\pi/a) + \Delta\omega. \tag{6.99}
$$

In the limit when resonator mode is close to the lower edge of the fundamental bandgap, only the term $m = 1$ in (6.97) will contribute significantly to the sum. Therefore, in this limit, (6.97) can be simplified as:

$$
\phi_{m',k_x',k_y'} \underset{\Delta\omega \to +0}{\cong} \frac{1}{2\omega_{\mathrm{PC},1}(\pi/a,\pi/a)}
$$

$$
\times \int_{-\pi/a}^{\pi/a} d(\delta k_x) \int_{-\pi/a}^{\pi/a} d(\delta k_y) \frac{\phi_{1,k_x,k_y}}{\Delta\omega + |\gamma_x| \, \delta k_x^2 + |\gamma_y| \, \delta k_y^2} W_{\pi/a,\pi/a,k_x',k_y'}^{1,m'}. \tag{6.100}
$$

Assuming slow functional dependence of ϕ_{1,k_x,k_y} and $W^{1,m'}_{k_x,k_y,k'_x,k'_y}$ with respect to k_x and k_y, for $k'_x = \pi/a$, $k'_y = \pi/a$ we get:

$$\phi_{1,\pi/a,\pi/a} \underset{\Delta\omega \to +0}{\simeq} \frac{\phi_{1,\pi/a,\pi/a} W^{1,1}_{\pi/a,\pi/a,\pi/a,\pi/a}}{2\omega_{PC,1}(\pi/a,\pi/a)} \int_{-\pi/a}^{\pi/a} \int_{-\pi/a}^{\pi/a} \frac{d(\delta k_x) d(\delta k_y)}{\Delta\omega + |\gamma_x| \delta k_x^2 + |\gamma_y| \delta k_y^2}$$

$$\underset{\Delta\omega \to +0}{\simeq} -\log(\Delta\omega) \frac{\pi}{\sqrt{|\gamma_x||\gamma_y|}} \frac{\phi_{1,\pi/a,\pi/a} W^{1,1}_{\pi/a,\pi/a,\pi/a,\pi/a}}{2\omega_{PC,1}(\pi/a,\pi/a)}, \tag{6.101}$$

which finally results in the following perturbative expression for the modal distance from the lower bandgap edge:

$$\Delta\omega = \exp\left(-\frac{2\sqrt{|\gamma_x||\gamma_y|}\omega_{PC,1}(\pi/a,\pi/a)}{\pi W^{1,1}_{\pi/a,\pi/a,\pi/a,\pi/a}}\right), \tag{6.102}$$

or alternatively:

$$\log(\omega_{res} - \omega_{PC,1}(\pi/a,\pi/a)) \underset{\substack{\varepsilon_d \to \varepsilon_a \\ \varepsilon_d < \varepsilon_a}}{\sim} 1/(\varepsilon_d - \varepsilon_a). \tag{6.103}$$

Note that for the same defect strength, the separation of a resonator mode from the bandgap edge (6.103) is much smaller than a corresponding separation of a waveguide-guided mode (6.77). Therefore, it is generally easier to create a guided state that is well localized in one dimension than it is to create the resonator state localized in both dimensions.

6.7 Problems

6.1 Electromagnetic eigenstates of the low-refractive-index-contrast photonic crystal lattice

In Section 6.2.4 we have introduced a perturbative formulation to describe formation of the TM bandgaps at the M symmetry point of a square lattice of high-refractive-index rods placed in a low refractive index cladding (see Fig. 6.2). In this problem we study further bandgap formation and modal structure of the TE- and TM-polarized states along the X–M direction in the reciprocal space.

The Bloch wave vector along the line joining the X and M symmetry points is described as $\mathbf{k}_t^{X-M} = (\pi/a, k_y)$; $k_y \in [0, \pi/a]$. For a negligible index contrast $\varepsilon_a = \varepsilon_b$, the fundamental band along the X–M direction is doubly degenerate and consists of the two bands labeled in Fig. 6.3 as $p_1 = -1$, $p_2 = 0$ and $p_1 = 0$, $p_2 = 0$. The dispersion relations of the two degenerate bands along the X–M direction are described as:

$$\omega^2_{\mathbf{k}_t^{X-M}} = \kappa(0) \left|\mathbf{k}_t^{X-M} + \mathbf{G}\right|^2; \quad \kappa(0) = 1/\varepsilon_b,$$
$$\mathbf{G} \in \mathbf{G}_{X-M} = [(0,0), (-2\pi/a, 0)] \tag{P6.1.1}$$

while the electromagnetic field of a degenerate eigenstate is described as a linear super-position of all the degenerate (in terms of frequency) plane waves:

$$F_{z,\mathbf{k}_t^{X-M}}(\mathbf{r}_t) = \sum_{\mathbf{G} \in \mathbf{G}_{X-M}} F_{z,\mathbf{k}_t^{X-M}}(\mathbf{G}) \exp(\mathrm{i} \left(\mathbf{k}_t^{X-M} + \mathbf{G}\right) \mathbf{r}_t), \tag{P6.1.2}$$

where $F_z = E_z$ for the case of TM polarization, and $F_z = H_z$ for the case of TE polarization. For a negligible refractive index contrast, the expansion coefficients $F_{z,\mathbf{k}_t^{X-M}}(\mathbf{G})$ can take any values.

Now consider a small but sizable index contrast $\varepsilon_a = \varepsilon_b + \delta\varepsilon$, $\delta\varepsilon \to 0$ for which coupling between the degenerate plane waves in (P6.1.2) can no longer be neglected. Substituting expansion (P6.1.2) into Maxwell's equations in the plane-wave formulation (6.22) for TM polarization or (6.23) for TE polarization, find the eigenfields and the eigenfrequencies of the two newly formed states in terms of the coupling coefficients $\kappa(|\mathbf{G}|)$, $\mathbf{G} \in \mathbf{G}_{X-M}$ (see (6.25)) for all the allowed values of parameter k_y.

Finally, plot the dispersion relations of the two perturbation split modes for all the allowed values of the parameter k_y. Additionally, plot the field intensities of the two modal fields. To generate the required plots assume the following values of the structural parameters: $r = a/\sqrt{2\pi}$; $f = 0.5$; $\varepsilon_a = 2.56$; $\varepsilon_b = 1.96$.

6.2   Plane-wave method for the TE waveguide modes

Following the derivations of Section 6.6.1 (see the subsection on the plane-wave method for a photonic crystal with a line defect), develop equations of a plane-wave method to compute TE-polarized modes of a photonic-crystal waveguide directed along the $\hat{\mathbf{x}}$ axis. As before, a uniform waveguide cladding comprises a square lattice of rods with dielectric constant ε_a and radius r_a. The lattice period is a. The waveguide core is formed by substitution of a single row of rods with a row of different rods of dielectric constant ε_d and radius r_d, as shown in Fig. 6.14. Note that the general form of the magnetic field vector of a TE-polarized guided mode labeled with a Bloch wavenumber k_x is $\mathbf{H} = (0, 0, H_{z,k_x}(x, y))$:

$$H_{z,k_x}(\mathbf{r}) = \exp(\mathrm{i}k_x x)U_{k_x}(\mathbf{r})$$
$$U_{k_x}(\mathbf{r} + a\hat{\mathbf{x}}) = U_{k_x}(\mathbf{r})$$

and it satisfies the corresponding Maxwell's equation $\omega^2 \mathbf{H} = \nabla \times (1/\varepsilon(\mathbf{r}) \cdot \nabla \times \mathbf{H})$.

6.3   Plane-wave method for the TE-resonator modes

Following the derivations of Section 6.6.2 (see the subsection on the plane-wave method for a photonic crystal with a point defect), develop equations of a plane-wave method to compute the TE-polarized modes of a resonator embedded in a photonic crystal lattice. As before, the photonic crystal cladding comprises a square lattice of rods with dielectric constant ε_a and radius r_a. The lattice period is a. The resonator is formed by substitution of a single dielectric rod with a different rod of dielectric constant ε_d and radius r_d, as shown in Fig. 6.18. Note that the general form of the magnetic field vector of a TE-polarized resonator state is $\mathbf{H} = (0, 0, H_{z,\omega_{res}}(x, y))$, and it satisfies the corresponding Maxwell's equation $\omega^2 \mathbf{H} = \nabla \times (1/\varepsilon(\mathbf{r}) \cdot \nabla \times \mathbf{H})$.

6.4   Guided-mode field distribution for the case of a weak linear defect

In this problem, we will use the results of the perturbation theory developed in Section 6.6.1 (see the subsection on bifurcation of the core-guide mode from the bandgap edge) to characterize the spatial localization of a photonic-crystal-waveguide-guided mode as a function of its location inside a photonic bandgap. In particular, we will consider the magnetic field distribution of a guided mode bifurcating from the lower edge of a fundamental bandgap (see Fig. 6.16). As before, the linear defect is created by lowering the dielectric constant of the rods along a single row in an otherwise perfectly periodic square lattice of dielectric rods in air.

Using (6.67) for the magnetic field of a defect state:

$$\mathbf{H}_{k_x}^{\text{wg}} = \sum_{m=1}^{+\infty} \int_{-\pi/a}^{\pi/a} dk_y \, A_{m,k_y} \mathbf{H}_{m,k_x,k_y}^{\text{PC}}(\mathbf{r}) = \sum_{m=1}^{+\infty} \int_{-\pi/a}^{\pi/a} dk_y \, A_{m,k_y} \exp(\mathrm{i} \mathbf{k} \mathbf{r}) \mathbf{U}_{m,k_x,k_y}^{\text{PC}}(\mathbf{r})$$

$$\mathbf{U}_{m,k_x,k_y}^{\text{PC}}(\mathbf{r} + \hat{x} a P_x + \hat{y} a P_y) = \mathbf{U}_{m,k_x,k_y}^{\text{PC}}(\mathbf{r}); \ (P_x, P_y) \subset \text{integer},$$

and (6.73) for the expansion coefficients A_{m,k_y} in terms of the auxiliary functions ϕ_{m,k_y}, as well as the approximation to the dispersion relation of the bulk states of a uniform photonic crystal for any value k_x along the $X_1 - M$ direction:

$$\omega_{\text{PC},1}(k_x, k_y) = \omega_{\text{PC},1}(k_x, \pi/a) - |\gamma(k_x)| \, \delta k_y^2; \, \delta k_y = k_y - \frac{\pi}{a},$$

find the functional form of a guided-mode magnetic field for a frequency:

$$\omega_{k_x}^{\text{wg}} \underset{\Delta\omega \to 0}{=} \omega_{\text{PC},1}(k_x, \pi/a) + \Delta\omega(k_x),$$

by retaining the terms of the highest order only, and assuming that functional dependence of the auxiliary functions and periodic parts of Bloch modes are slow:

$$\phi_{1,k_y} \underset{k_y \sim \pi/a}{\approx} \phi_{1,\pi/a};$$

$$\mathbf{U}_{1,k_x,k_y}^{\text{PC}}(\mathbf{r}) \underset{k_y \sim \pi/a}{\approx} \mathbf{U}_{1,k_x,\pi/a}^{\text{PC}}(\mathbf{r}).$$

Demonstrate that the magnetic field of a resonator state has the functional form:

$$\mathbf{H}_{k_x}^{\text{wg}} \underset{\mathbf{k}_{\text{wg}}=(k_x,\pi/a)}{\approx} \exp(\mathrm{i} \mathbf{k}_{\text{wg}} \mathbf{r}) \mathbf{U}_{1,k_x,\pi/a}^{\text{PC}}(\mathbf{r}) \phi(\Delta\omega(k_x), |y|, |\gamma(k_x)|),$$

and find the explicit form of an envelope function $\phi(\Delta\omega(k_x), |y|, |\gamma(k_x)|)$ when $|\mathbf{r}| \gg a$. What is the characteristic length scale associated with a localized field distribution of a resonator state, and what is the scaling of this length scale with the distance from the bandgap edge $\Delta\omega$?

6.5   Resonator-state field distribution for the case of a weak point defect

In this problem we will use the results of the perturbation theory developed in Section 6.6.2 (see the subsection on bifurcation of a resonator state from the bandgap edge) to characterize the spatial localization of a resonator state as a function of its location inside a photonic bandgap. In particular, we will consider the magnetic field distribution of a resonator state bifurcating from the lower band edge of a fundamental bandgap (see Fig. 6.20). As before, the resonator is created by lowering the dielectric constant of one of the rods in an otherwise perfectly periodic square lattice of dielectric rods in air.

Using (6.90) for the magnetic field of a defect state:

$$\mathbf{H}^{\text{res}}_{\omega_{\text{res}}} = \sum_{m=1}^{+\infty} \int_{-\pi/a}^{\pi/a} dk_x \int_{-\pi/a}^{\pi/a} dk_y \, A_{m,k_x,k_y} \exp(i\mathbf{k}\mathbf{r}) \mathbf{U}^{\text{PC}}_{m,k_x,k_y}(\mathbf{r})$$

$$\mathbf{U}^{\text{PC}}_{m,k_x,k_y}(\mathbf{r} + \hat{\mathbf{x}}aP_x + \hat{\mathbf{y}}aP_y) = \mathbf{U}^{\text{PC}}_{m,k_x,k_y}(\mathbf{r}); (P_x, P_y) \subset \text{integer},$$

and (6.96) for the expansion coefficients A_{m,k_x,k_y} in terms of the auxiliary functions ϕ_{m,k_x,k_y}, as well as the approximation to the dispersion relation of the bulk states of a uniform photonic crystal near the M point:

$$\omega_{\text{PC},1}(k_x, k_y) = \omega_{\text{PC},1}(\pi/a, \pi/a) - |\gamma|\left(\delta k_x^2 + \delta k_y^2\right)$$

$$\delta k_x = k_x - \tfrac{\pi}{a}; \quad \delta k_y = k_y - \tfrac{\pi}{a},$$

find the functional form of a resonator state magnetic field of frequency:

$$\omega_{\text{res}} \underset{\Delta\omega\to 0}{=} \omega_{\text{PC},1}(\pi/a, \pi/a) + \Delta\omega,$$

by retaining the terms of the highest order only, and assuming that the functional dependences of the auxiliary functions and periodic parts of Bloch modes are slow:

$$\phi_{1,k_x,k_y} \underset{\substack{k_x \sim \pi/a \\ k_y \sim \pi/a}}{\approx} \phi_{1,\pi/a,\pi/a}; \mathbf{U}^{\text{PC}}_{1,k_x,k_y}(\mathbf{r}) \underset{\substack{k_x \sim \pi/a \\ k_y \sim \pi/a}}{\approx} \mathbf{U}^{\text{PC}}_{1,\pi/a,\pi/a}(\mathbf{r}).$$

Demonstrate that the magnetic field of a resonator state has the function form:

$$\mathbf{H}^{\text{res}}_{\omega_{\text{res}}} \approx \exp(i\mathbf{k}_M\mathbf{r})\mathbf{U}^{\text{PC}}_{1,\pi/a,\pi/a}(\mathbf{r})\phi(\Delta\omega, |\mathbf{r}|, |\gamma|),$$

and find the explicit form of an envelope function $\phi(\Delta\omega, |\mathbf{r}|, |\gamma|)$ when $|\mathbf{r}| \gg a$. What is the characteristic length scale associated with a localized field distribution of a resonator state, and what is the scaling of this length scale with the distance from the bandgap edge $\Delta\omega$?

References

[1]   A MATLAB implementation of a plane-wave expansion method that computes Bloch modes of the 2D uniform lattices and photonic crystal fibers, as well as eigenmodes of the line and point defects in a 2D periodic lattice can be found at: www.photonics.phys.polymtl.ca/fund_phot_cryst_guid/planewave2D.m

[2]   A description of a general implementation of a planewave method can be found at S. G. Johnson and J. D. Joannopoulos. Block-iterative frequency-domain methods for Maxwell's equations in a planewave basis, *Opt. Express* **8** (2001), 173–190.

7   Quasi-2D photonic crystals

This chapter is dedicated to periodic structures that are geometrically more complex than 2D photonic crystals, but not as complex as full 3D photonic crystals. In particular, we consider the optical properties of photonic crystal fibers, optically induced photonic lattices, and photonic crystal slabs.

7.1   Photonic crystal fibers

First, we consider electromagnetic modes that propagate along the direction of continuous translational symmetry $\hat{z}$ of a 2D photonic crystal (Fig. 7.1(a)). In this case, modes carry electromagnetic energy along the $\hat{z}$ direction, and, therefore, can be considered as modes of an optical fiber extended in the $\hat{z}$ direction and having a periodic dielectric profile in its cross-section. Fibers of this type are called photonic crystal fibers. From Section 2.4.5, it follows that fiber modes can be labeled with a conserved wave vector of the form $\mathbf{k} = \mathbf{k}_t + \hat{z}k_z$, where $\mathbf{k}_t$ is a transverse Bloch wave vector, and $k_z \neq 0$. To analyze the modes of a photonic crystal fiber with periodic cross-section, we will employ the general form of a plane-wave expansion method presented in Section 6.2.

Furthermore, if a defect that is continuous along the $\hat{z}$ direction is introduced into a 2D photonic crystal lattice (Fig. 7.1(b)), such a defect could support a localized state (see Section 6.6), thus, effectively, becoming the core of a photonic crystal fiber. To illustrate this phenomenon, we will present an example of guidance in optically induced photonic lattices as such systems allow a semi-analytical treatment, while retaining all the key elements of general photonic crystal fiber guidance.

7.1.1   Plane-wave expansion method

We start by detailing the plane-wave expansion method for calculation of the modes of photonic crystal fibers with strictly periodic transverse profiles. As an example, we consider the case of a square lattice of holes extended infinitely along the $\hat{z}$ direction, and having dielectric constant ε_a and radii r_a (Fig. 7.1(a)). The dielectric constant of a background material (cladding) is defined as ε_b. From Section 2.4.5 (2D discrete

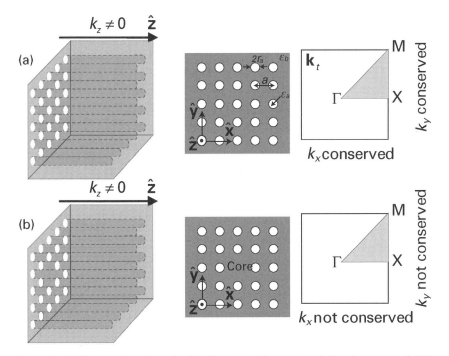

Figure 7.1 (a) Propagation along the direction of continuous translational symmetry in 2D photonic crystal lattices. Modes can be labeled with a wave vector $\mathbf{k} = \mathbf{k}_t + \hat{\mathbf{z}} k_z$, where $k_z \neq 0$, and $\mathbf{k}_t$ is a transverse Bloch wave vector confined to a fiber cross-section. (b) A defect in a 2D photonic crystal lattice can localize light, while still guiding the light along the $\hat{\mathbf{z}}$ direction. This is a case of a photonic crystal fiber, where the defect is a fiber core.

translational symmetry plus 1D continuous symmetry), it follows that the general form of a modal field in such a fiber is:

$$\mathbf{H_k}(\mathbf{r}) = \exp(i k_z z + i \mathbf{k}_t \mathbf{r}_t) \mathbf{U_k}(\mathbf{r}_t) \\ \mathbf{U_k}(\mathbf{r}_t + a\hat{\mathbf{x}} N_1 + a\hat{\mathbf{y}} N_2) = \mathbf{U_k}(\mathbf{r}_t)$$

(7.1)

where N_1, N_2 are any integers. Similarly to (6.6), the periodic part of the vector field can then be expanded into plane waves using a discrete Fourier transform. Therefore, the vector fields of a photonic crystal fiber mode characterized by a conserved wave vector $\mathbf{k} = \mathbf{k}_t + \hat{\mathbf{z}} k_z$ can be written as:

$$\mathbf{H_k}(\mathbf{r}) = \sum_{\mathbf{G}} \mathbf{H}(\mathbf{k} + \mathbf{G}) \exp(i(\mathbf{k} + \mathbf{G})\mathbf{r})$$

$$\mathbf{G} = \overline{b}_x P_x + \overline{b}_y P_y; \ \overline{b}_x = \tfrac{2\pi}{a} \hat{\mathbf{x}}; \ \overline{b}_y = \tfrac{2\pi}{a} \hat{\mathbf{y}}; \ (P_x, P_y) \subset \text{integers}$$

$$\mathbf{k} = \mathbf{k}_t + k_z \hat{\mathbf{z}}; \mathbf{k}_t \text{ a constant vector}; k_z \text{ a constant scalar} \neq 0$$

$$\mathbf{k}_t = k_x \hat{\mathbf{x}} + k_y \hat{\mathbf{y}}; \ k_x \subset \left(-\tfrac{\pi}{a}, \tfrac{\pi}{a}\right); \ k_y \subset \left(-\tfrac{\pi}{a}, \tfrac{\pi}{a}\right).$$

(7.2)

Note that substitution of the magnetic field expansion (7.2) into the Maxwell Hamiltonian (6.10) leads to exactly the same form of the plane-wave expansion method as before (6.13). The only difference is that in the problem of calculation of modes of photonic crystal fibers the projection of a wave vector along the direction of continuous translational symmetry is nonzero $k_z \neq 0$. In this case, as can be seen from the form of (6.13), we can no longer define two simple polarizations analogous to the TE and TM polarizations of the modes of a 2D photonic crystal where $k_z = 0$.

Furthermore, to satisfy the divergence-free condition for the magnetic field (6.11) one has to ensure that for any wave vector $\mathbf{k} + \mathbf{G}$, the corresponding Fourier component of a magnetic field $\mathbf{H}(\mathbf{k} + \mathbf{G})$ is orthogonal to such a wave vector:

$$(\mathbf{k} + \mathbf{G}) \perp \mathbf{H}(\mathbf{k} + \mathbf{G}). \tag{7.3}$$

Therefore, the first challenge is to find a convenient expansion basis for the magnetic field components $\mathbf{H}(\mathbf{k} + \mathbf{G})$, so that the orthogonality relation (7.3) is satisfied. In what follows, consider the following basis vectors:

$$\mathbf{V}_{\mathbf{k}+\mathbf{G}}^{1} = \frac{(\mathbf{k}_t + \mathbf{G}) \times \hat{\mathbf{z}}}{|\mathbf{k}_t + \mathbf{G}|};$$

$$\mathbf{V}_{\mathbf{k}+\mathbf{G}}^{2} = \frac{(\mathbf{k}_t + \mathbf{G})k_z}{|\mathbf{k}_t + \mathbf{G}| \, |\mathbf{k} + \mathbf{G}|} - \hat{\mathbf{z}} \frac{|\mathbf{k}_t + \mathbf{G}|}{|\mathbf{k} + \mathbf{G}|}. \tag{7.4}$$

One can verify that the vectors defined in (7.4) are orthonormal and that $\left(\mathbf{V}_{\mathbf{k}+\mathbf{G}}^{1}, \mathbf{V}_{\mathbf{k}+\mathbf{G}}^{2}, \mathbf{k} + \mathbf{G}\right)$ form a triad:

$$\left|\mathbf{V}_{\mathbf{k}+\mathbf{G}}^{1}\right| = \left|\mathbf{V}_{\mathbf{k}+\mathbf{G}}^{2}\right| = 1;$$

$$\left(\mathbf{V}_{\mathbf{k}+\mathbf{G}}^{1} \cdot \mathbf{V}_{\mathbf{k}+\mathbf{G}}^{2}\right) = \left(\mathbf{V}_{\mathbf{k}+\mathbf{G}}^{1} \cdot (\mathbf{k} + \mathbf{G})\right) = \left(\mathbf{V}_{\mathbf{k}+\mathbf{G}}^{2} \cdot (\mathbf{k} + \mathbf{G})\right) = 0;$$

$$(\mathbf{k} + \mathbf{G}) \times \mathbf{V}_{\mathbf{k}+\mathbf{G}}^{1} = |\mathbf{k} + \mathbf{G}| \, \mathbf{V}_{\mathbf{k}+\mathbf{G}}^{2};$$

$$(\mathbf{k} + \mathbf{G}) \times \mathbf{V}_{\mathbf{k}+\mathbf{G}}^{2} = - |\mathbf{k} + \mathbf{G}| \, \mathbf{V}_{\mathbf{k}+\mathbf{G}}^{1}. \tag{7.5}$$

Using (7.4), we can now expand the magnetic field vector component $\mathbf{H}(\mathbf{k} + \mathbf{G})$ with the resultant expansion automatically satisfying the orthogonality relation (7.3):

$$\mathbf{H}(\mathbf{k} + \mathbf{G}) = h_{\mathbf{k}+\mathbf{G}}^{1} \mathbf{V}_{\mathbf{k}+\mathbf{G}}^{1} + h_{\mathbf{k}+\mathbf{G}}^{2} \mathbf{V}_{\mathbf{k}+\mathbf{G}}^{2}. \tag{7.6}$$

According to (6.13), Maxwell's equations for the magnetic field vector (7.2) can be written in terms of its Fourier components as:

$$-\sum_{\mathbf{G}'} \kappa(\mathbf{G} - \mathbf{G}') (\mathbf{k} + \mathbf{G}) \times \left[(\mathbf{k} + \mathbf{G}') \times \mathbf{H}(\mathbf{k} + \mathbf{G}')\right] = \omega_{\mathbf{k}}^{2} \mathbf{H}_{\mathbf{k}}(\mathbf{G}). \tag{7.7}$$

Substituting (7.6) into (7.7) leads to the following:

$$-\sum_{\mathbf{G}'} \kappa(\mathbf{G} - \mathbf{G}') (\mathbf{k} + \mathbf{G}) \times \left[(\mathbf{k} + \mathbf{G}') \times \left(h_{\mathbf{k}+\mathbf{G}'}^{1} \mathbf{V}_{\mathbf{k}+\mathbf{G}'}^{1} + h_{\mathbf{k}+\mathbf{G}'}^{2} \mathbf{V}_{\mathbf{k}+\mathbf{G}'}^{2}\right)\right] = \omega_{\mathbf{k}}^{2} \mathbf{H}(\mathbf{k} + \mathbf{G}). \tag{7.8}$$

Using standard vector identities and properties of the expansion basis (7.5) one can further simplify (7.8) to become:

$$-\sum_{\mathbf{G}'} \kappa(\mathbf{G} - \mathbf{G}')|\mathbf{k} + \mathbf{G}'| (\mathbf{k} + \mathbf{G}) \times \left(h_{\mathbf{k}+\mathbf{G}'}^{1} \mathbf{V}_{\mathbf{k}+\mathbf{G}'}^{2} - h_{\mathbf{k}+\mathbf{G}'}^{2} \mathbf{V}_{\mathbf{k}+\mathbf{G}'}^{1}\right)$$

$$= \omega_{\mathbf{k}}^{2} \left(h_{\mathbf{k}+\mathbf{G}}^{1} \mathbf{V}_{\mathbf{k}+\mathbf{G}}^{1} + h_{\mathbf{k}+\mathbf{G}}^{2} \mathbf{V}_{\mathbf{k}+\mathbf{G}}^{2}\right). \tag{7.9}$$

We can now get two sets of equations for the components h_{k+G}^1, h_{k+G}^2 by multiplying (7.9) from the left by either $\mathbf{V}_{k+G}^1$ or $\mathbf{V}_{k+G}^2$. In particular, multiplying (7.9) by $\mathbf{V}_{k+G}^1$ from the left and using orthogonality relations (7.5), we get:

$$-\sum_{\mathbf{G}'} \kappa(\mathbf{G}-\mathbf{G}')|\mathbf{k}+\mathbf{G}'|\mathbf{V}_{k+G}^1 \cdot \left[(\mathbf{k}+\mathbf{G}) \times \left(h_{k+G'}^1 \mathbf{V}_{k+G'}^2 - h_{k+G'}^2 \mathbf{V}_{k+G'}^1\right)\right]$$
$$= \omega_k^2 h_{k+G}^1, \tag{7.10}$$

which after several straightforward manipulations can be transformed into:

$$\sum_{\mathbf{G}'} \kappa(\mathbf{G}-\mathbf{G}')|\mathbf{k}+\mathbf{G}'|\,|\mathbf{k}+\mathbf{G}|\left(h_{k+G'}^1 \left(\mathbf{V}_{k+G}^2 \cdot \mathbf{V}_{k+G'}^2\right) - h_{k+G'}^2 \left(\mathbf{V}_{k+G}^2 \cdot \mathbf{V}_{k+G'}^1\right)\right)$$
$$= \omega_k^2 h_{k+G}^1. \tag{7.11}$$

Similarly, multiplying (7.9) by $\mathbf{V}_{k+G}^2$ from the left, and after several straightforward manipulations one gets:

$$\sum_{\mathbf{G}'} \kappa(\mathbf{G}-\mathbf{G}')|\mathbf{k}+\mathbf{G}'|\,|\mathbf{k}+\mathbf{G}|\left(-h_{k+G'}^1 \left(\mathbf{V}_{k+G}^1 \cdot \mathbf{V}_{k+G'}^2\right) + h_{k+G'}^2 \left(\mathbf{V}_{k+G}^1 \cdot \mathbf{V}_{k+G'}^1\right)\right)$$
$$= \omega_k^2 h_{k+G}^2. \tag{7.12}$$

The matrix elements in (7.11) and (7.12) can be computed by using definitions (7.4) of the expansion vectors to result in:

$$\begin{aligned}
M_{(k+G',k+G)}^{1,1} &= \left|\mathbf{k}+\mathbf{G}'\right|\,|\mathbf{k}+\mathbf{G}|\left(\mathbf{V}_{k+G}^2 \cdot \mathbf{V}_{k+G'}^2\right)\\
&= \frac{(\mathbf{k}_t+\mathbf{G})(\mathbf{k}_t+\mathbf{G}')}{|\mathbf{k}_t+\mathbf{G}|\,|\mathbf{k}_t+\mathbf{G}'|}k_z^2 + |\mathbf{k}_t+\mathbf{G}|\,\left|\mathbf{k}_t+\mathbf{G}'\right|\\
M_{(k+G',k+G)}^{1,2} &= -\left|\mathbf{k}+\mathbf{G}'\right|\,|\mathbf{k}+\mathbf{G}|\left(\mathbf{V}_{k+G}^2 \cdot \mathbf{V}_{k+G'}^1\right)\\
&= \frac{\hat{\mathbf{z}} \cdot \left[(\mathbf{k}_t+\mathbf{G}') \times (\mathbf{k}_t+\mathbf{G})\right]}{|\mathbf{k}_t+\mathbf{G}|\,|\mathbf{k}_t+\mathbf{G}'|}\left|\mathbf{k}+\mathbf{G}'\right| k_z\\
M_{(k+G',k+G)}^{2,1} &= -\left|\mathbf{k}+\mathbf{G}'\right|\,|\mathbf{k}+\mathbf{G}|\left(\mathbf{V}_{k+G}^1 \cdot \mathbf{V}_{k+G'}^2\right)\\
&= \frac{\hat{\mathbf{z}} \cdot \left[(\mathbf{k}_t+\mathbf{G}) \times (\mathbf{k}_t+\mathbf{G}')\right]}{|\mathbf{k}_t+\mathbf{G}|\,|\mathbf{k}_t+\mathbf{G}'|}\left|\mathbf{k}+\mathbf{G}\right| k_z\\
M_{(k+G',k+G)}^{2,2} &= \left|\mathbf{k}+\mathbf{G}'\right|\,|\mathbf{k}+\mathbf{G}|\left(\mathbf{V}_{k+G}^1 \cdot \mathbf{V}_{k+G'}^1\right)\\
&= \frac{(\mathbf{k}_t+\mathbf{G}')(\mathbf{k}_t+\mathbf{G})}{|\mathbf{k}_t+\mathbf{G}|\,|\mathbf{k}_t+\mathbf{G}'|}\,|\mathbf{k}+\mathbf{G}|\,\left|\mathbf{k}+\mathbf{G}'\right|. \tag{7.13}
\end{aligned}$$

Finally, in a matrix form, (7.11) and (7.12) can be written as:

$$\sum_{\mathbf{G}'} \kappa(\mathbf{G}-\mathbf{G}') \begin{pmatrix} M_{(k+G',k+G)}^{1,1} & M_{(k+G',k+G)}^{1,2} \\ M_{(k+G',k+G)}^{2,1} & M_{(k+G',k+G)}^{2,2} \end{pmatrix} \begin{pmatrix} h_{k+G'}^1 \\ h_{k+G'}^2 \end{pmatrix} = \omega_k^2 \begin{pmatrix} h_{k+G}^1 \\ h_{k+G}^2 \end{pmatrix}. \tag{7.14}$$

In calculating (7.13), one has to take special care of the case $|\mathbf{k}_t| = 0$. There, the problem arises when one or both of the reciprocal vectors $\mathbf{G}$, $\mathbf{G}'$ are zero. Assume, for example, that $\mathbf{G} = 0$. Then, expansion vectors can be taken as $\mathbf{V}_{k+G}^1 = \hat{\mathbf{x}}$; $\mathbf{V}_{k+G}^2 = \hat{\mathbf{y}}$. It

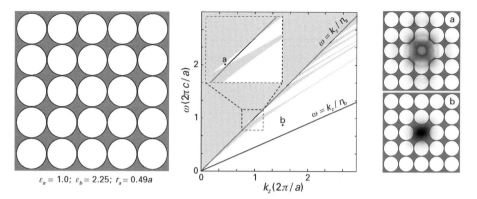

Figure 7.2 Schematic of a high-air-filling-fraction photonic crystal lattice and a corresponding projected band diagram. States corresponding to the air core defect (a) can be localized in the air and are found inside the photonic bandgaps above the light line of air. States corresponding to the solid core defect (b) are guided in the material of the core and are typically found below the region of the photonic bandgaps and closer to the light line of a core material.

can be further demonstrated that in this case matrix elements can still be computed as in (7.13) by using a symbolic substitution:

$$\frac{\mathbf{k_t} + \mathbf{G}}{|\mathbf{k_t} + \mathbf{G}|} \rightarrow \hat{\mathbf{x}}; \ |\mathbf{k} + \mathbf{G}| \rightarrow k_z; \ |\mathbf{k_t} + \mathbf{G}| \rightarrow 0.$$

7.1.2   Band diagram of modes of a photonic crystal fiber

We now use the plane-wave expansion method developed in the previous section to characterize the modes of a photonic crystal fiber with a periodic transverse profile. In particular, we consider the fiber with a cross-section made of a square array of large air holes $\varepsilon_a = 1.0, r_a = 0.49a$ placed in the cladding material with $\varepsilon_b = 2.25$ (Fig. 7.2).

The most frequently used band diagram to represent the modes of a fiber plots the frequencies of all the allowed states of a fiber as a function of the propagation constant k_z. To construct such a band diagram using the plane-wave expansion method, one first fixes the value of a propagation constant k_z, then for all the transverse wave vectors $\mathbf{k_t}$ in the first Brillouin zone one solves (7.14) for the modal frequencies, and finally one plots all the found frequencies along a single line. The same procedure is then repeated for the various values of k_z.

Figure 7.2 presents an example of a band diagram of the modes of a photonic crystal fiber with a periodic cross-section. In the plot, the continuum of the allowed states is shown in gray. A remarkable feature of this band diagram is the existence of the "finger-like" bandgap regions where no states exist. Moreover, when overlaying the material light lines one notices that some of the bandgap regions are partially located above the light line of the lower refractive index material (air) (inset of Fig. 7.2). As demonstrated in Section 6.6, the introduction of a defect into the photonic crystal lattice can lead to the appearance of localized-at-a-defect states with the frequencies inside the bandgaps of an underlying photonic crystal lattice. Similarly, for the case of photonic crystal fibers one finds that a structural defect in a fiber cross-section can localize light. Frequencies of these new defect-guided states appear within the bandgaps of an original band diagram. Consider the high-air-filling-fraction structure in Fig. 7.2. When an air defect in the form of a large air hole is introduced (Fig. 7.2(a)), then the mode localized in such a hollow core can appear with a frequency located above the light line of the air and in the bandgap of an original projected band diagram. On the other hand, when a high-refractive-index defect is introduced in the form of a filled hole (Fig. 7.2(b)), then the mode localized in such a solid core typically appears with a frequency located between the light line of the air and the light line of a cladding material.

Finally, a modified plane-wave expansion method can be developed in a manner analogous to the one presented in Section 6.6 to calculate dispersion relations of the defect states of a general photonic crystal fiber. Instead of repeating these derivations we rather focus on the case of a low-refractive-index-contrast photonic crystal fiber, where fiber modes and their dispersion relations can be computed semi-analytically. Such fibers have been made and studied. [1] Similar low-index-contrast structures are also realized in optically induced photonic lattices, [2,3] laser-written waveguides, [4] and planar-etched waveguide arrays on AlGaAs substrates (see [5,6] for instance). Theoretical treatments of light guiding in different types of low-index-contrast structures are similar. Next, we consider optically induced photonic lattices to develop the theory.

7.2   Optically induced photonic lattices

In recent experiments in biased photorefractive crystals (such as SBN:60), an ordinarily polarized photonic lattice is first imprinted onto the crystal by the beam interference method [2] or the amplitude mask method [3]. This photonic lattice can be made uniform along its propagation direction inside the crystal. This lattice light distribution creates a weak transverse refractive-index variation (on the order of 10^{-3} to 10^{-4}), thus forming a low-index-contrast photonic crystal fiber. When an extraordinarily polarized probe beam is launched into this photonic lattice, it will "feel" this index variation and significantly change its propagation behavior.

A schematic of the experimental set-up creating uniform photonic lattices by the beam interference method is shown in Fig. 7.3(a). [2] There, four plane waves of the form

$$E_1 = A \exp^{ik_x x + ik_z z}$$
$$E_2 = A \exp^{-ik_x x + ik_z z}$$
$$E_3 = A \exp^{ik_y y + ik_z z}$$
$$E_4 = A \exp^{-ik_y y + ik_z z} \tag{7.15}$$

interfere on a photorefractive crystal, with a resultant light intensity distribution:

$$I = \left| \sum_{j=1,\dots,4} E_j \right|^2$$
$$= 16A^2 \cos^2 \frac{k_x x + k_y y}{2} \cos^2 \frac{k_x x - k_y y}{2}. \tag{7.16}$$

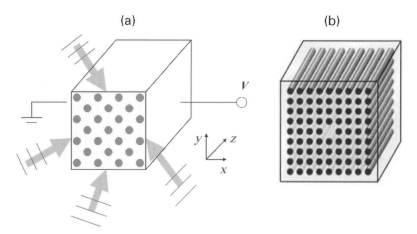

Figure 7.3 (a) A uniform photonic lattice created by the beam interference method in a photorefractive crystal. [2] (b) A photonic lattice with a single-site defect created by the amplitude mask method. [3]

Such light intensity distribution, and hence refractive index variation, can be kept uniform along the crystal $\hat{z}$ axis, thus being analogous to a photonic crystal fiber. To create a local defect inside a uniform photonic lattice that can persist along the $\hat{z}$ axis, one can use another experimental technique of amplitude masks together with frequency filtering. [3] A defective photonic lattice created that way is shown in Fig. 7.3(b).

7.2.1   Light propagation in low-index-contrast periodic photonic lattices

In optically induced photonic lattices, the resulting refractive index variation is weak. In this regime a so-called scalar approximation of Maxwell's equations is valid, which simplifies considerably the task of finding the eigenmodes of a fiber. In this section, we derive a scalar approximation and apply it to the case of optically induced photonic lattices.

In an isotropic charge-free medium, Maxwell's equations take the form:

$$\nabla \times \mathbf{E} = -\frac{\partial \mathbf{B}}{\partial t}, \tag{7.17}$$

$$\nabla \times \mathbf{H} = \frac{\partial \mathbf{D}}{\partial t}, \tag{7.18}$$

$$\nabla \cdot \mathbf{D} = 0, \tag{7.19}$$

$$\nabla \cdot \mathbf{B} = 0, \tag{7.20}$$

and the corresponding medium equations are

$$\mathbf{D} = \varepsilon \mathbf{E},$$
$$\mathbf{B} = \mu_0 \mathbf{H}, \tag{7.21}$$

where $\mathbf{E}$ and $\mathbf{H}$ are the electric and magnetic field vectors; $\mathbf{D}$ and $\mathbf{B}$ are the electric displacement and magnetic induction field vectors, and ε and μ_0 are the dielectric constant and the medium's permeability, respectively.

Assuming that the fields are monochromatic and harmonic, i.e., $\mathbf{E}(x, y, z, t) = \mathbf{E}(x, y, z)e^{i\omega t}$ and $\mathbf{B}(x, y, z, t) = \mathbf{B}(x, y, z)e^{i\omega t}$, taking the curl of both sides of (7.17) and substituting (7.18) and (7.21) into the resulting equation, we get:

$$\nabla \times (\nabla \times \mathbf{E}) = \nabla \times (-i\omega\mu_0\mathbf{H}) = \omega^2\mu_0\varepsilon\mathbf{E}. \tag{7.22}$$

The left-hand side of (7.22) can be further expanded by using the following vector identity:

$$\nabla \times (\nabla \times \mathbf{E}) \equiv \nabla(\nabla \cdot \mathbf{E}) - \nabla^2\mathbf{E}. \tag{7.23}$$

Moreover, using (7.19) we can write:

$$\nabla \cdot \mathbf{D} = \nabla \cdot (\varepsilon\mathbf{E}) = \mathbf{E} \cdot \nabla\varepsilon + \varepsilon\nabla \cdot \mathbf{E} = 0. \tag{7.24}$$

Replacing the left-hand side of (7.22) by (7.23) and using (7.24) we finally find:

$$\nabla^2\mathbf{E} + \omega^2\mu_0\varepsilon\mathbf{E} = -\nabla\left(\mathbf{E} \cdot \frac{\nabla\varepsilon}{\varepsilon}\right). \tag{7.25}$$

If the dielectric constant ε changes little over a single optical wavelength, the right side of (7.25) can be neglected. [7] This is known as a scalar approximation resulting in the Helmholtz equation:

$$\nabla^2\mathbf{E} + k_0^2 n^2\mathbf{E} = 0, \tag{7.26}$$

where $k_0 = \omega/c$, $n = \sqrt{\mu_0\varepsilon}\, c$ is the refractive index of the medium, and c is the speed of light in vacuum. Note that if the electric field $\mathbf{E}$ is linearly polarized along one transverse axis, then the Helmholtz equation (7.26) becomes scalar (hence the name of the approximation). Therefore, in what follows we use a scalar notation for the electric field E in place of a vector one.

The scalar approximation is usually satisfied if the refractive index n does not vary strongly in the medium; this is the case in low-refractive-index-contrast fibers, photorefractive crystals imprinted with photonic lattices, and many nonlinear optics problems. In this case, the refractive index n can be considered as a constant background index n_b plus a weak variation Δn:

$$n = n_b + \Delta n, \quad \Delta n \ll n_b. \tag{7.27}$$

The variation Δn generally depends on the spatial coordinates of the medium and the light intensity $|E|^2$.

Consider an electric field E in the form of a slowly varying wave envelope along the propagation direction $\hat{z}$:

$$E = U(x, y, z)e^{ikz}, \tag{7.28}$$

where $k = k_0 n_b$, and $U(x, y, z)$ changes slowly with z (hence $\left|\frac{\partial^2 U}{\partial z^2}\right| \ll \left|\frac{\partial U}{\partial z}\right|$). Substituting (7.28) into (7.26) and recalling assumption (7.27), the Helmholtz equation (7.26) becomes:

$$i\frac{\partial U}{\partial z} + \frac{1}{2k}\nabla_\perp^2 U + \frac{k\Delta n}{n_b}U = 0. \tag{7.29}$$

Here the second order of Δn and $\partial^2 U/\partial z^2$ have been omitted, owing to their smallness, and $\nabla_\perp^2 = \partial^2/\partial x^2 + \partial^2/\partial y^2$ is a transverse Laplace operator.

Equation (7.29) is known as a Schrödinger equation. When describing the light propagation, the first term in the left-hand side of (7.29) represents the envelope propagation, the second term represents the transverse diffraction of the light, and the last term represents the effect of the refractive index change on the light propagation.

In PCF fibers made of linear materials, Δn is typically a piece-wise constant function varying across the fiber cross-section. If fiber materials are nonlinear, Δn also contains a term proportional to the light intensity $|E|^2$. In photorefractive crystals, on the other hand, the expression for Δn is somewhat different (see below).

In a photorefractive crystal, the crystal line, as well as the direction of an external DC electric field (with voltage differential V across the crystal), are both chosen along the $\hat{x}$ axis. In this configuration, the refractive index along $\hat{x}$ direction is n_e, and the refractive index along $\hat{y}$ direction is n_o, with the corresponding electro-optic coefficients r_{33} and r_{13}, respectively. When an extraordinarily polarized (along the $\hat{x}$ direction), or ordinarily polarized (along the $\hat{y}$ direction) light is launched into the crystal, the changes in the refractive indices can be described according to [2,8] as:

$$\Delta\left(\frac{1}{n_e^2}\right) = r_{33}E_{sc}$$
$$\Delta\left(\frac{1}{n_o^2}\right) = r_{13}E_{sc}, \tag{7.30}$$

where $E_{sc} = E_0/(1 + I)$ is a steady-state space-charge field, I is the total beam intensity normalized to the background illumination in the crystal, and E_0 is the amplitude of an external DC electric field.

In photorefractive crystals, the lattice beam is ordinarily polarized and is made uniform along the $\hat{z}$ direction. [2,3] For the other beam (the probe beam), which is extraordinarily polarized, in view of (7.27), we have:

$$\Delta\left(\frac{1}{n_e^2}\right) = -2\frac{1}{n_e^3}\Delta n_e, \tag{7.31}$$

and

$$\Delta n_e = -\frac{n_e^3}{2}r_{33}E_{sc}. \tag{7.32}$$

After substitution of this equation into (7.29), the guiding equation for the probe beam becomes:

$$i\frac{\partial U}{\partial z} + \frac{1}{2k}\nabla_\perp^2 U - \frac{kn_e^2 r_{33}}{2}\frac{E_0}{1+I}U = 0. \tag{7.33}$$

By introducing the following variable transformations

$$z \rightarrow z \cdot 2kD^2/\pi^2,$$

$$(x, y) \rightarrow (x \cdot D/\pi, y \cdot D/\pi),$$

$$E_0 \rightarrow E_0 \cdot \pi^2/(k^2 n_e^2 D^2 r_{33}),$$

where D is the lattice spacing, (7.33) can be made dimensionless as:

$$i\frac{\partial U}{\partial z} + \nabla_{\perp}^2 U - \frac{E_0}{1+I}U = 0, \tag{7.34}$$

where $I = I_L + |U|^2$ and I_L is the intensity distribution of the periodic lattice. In typical experiments on photorefractive crystals, $D \sim 20\,\mu m$, $\lambda_0 \sim 0.5\,\mu m$, $n_e \sim 2.3$, $r_{33} \sim 280\,pm/V$. Thus, one normalized (x, y) unit corresponds to 6.4 μm, one normalized z unit corresponds to 2.3 mm, and one normalized E_0 unit corresponds to 20 V/mm in physical units. If the probe beam is weak, then (7.32) becomes linear (i.e., $|U|^2$ can be neglected).

7.2.2 Defect modes in 2D photonic lattices with localized defects

Consider an ordinarily polarized 2D square-lattice beam with a localized defect and one extraordinarily polarized probe beam with very low intensity launched into a photorefractive crystal. In our further considerations we assume that the defective lattice beam is uniform along the propagation direction. In this case we rewrite the guiding equation (7.34) for the probe beam as:

$$iU_z + U_{xx} + U_{yy} - \frac{E_0}{1+I_L(x,y)}U = 0. \tag{7.35}$$

The 2D optically induced defective lattice I_L is taken as:

$$I_L = I_0 \cos^2(x)\cos^2(y)[1 + \varepsilon F_D(x,y)], \tag{7.36}$$

where I_0 is the peak intensity of the otherwise uniform photonic lattice, and $F_D(x, y)$ describes the shape of a lattice defect of strength ε. Note that ε in the following text denotes the defect strength, and not the dielectric constant. In addition, the coordinate system that we will use in the following is rotated 45° from that in Fig 7.3(a) and (7.16) for convenience.

The shape of a localized single-site defect in (7.36) is chosen as:

$$F_D(x, y) = \exp[-(x^2 + y^2)^4/128]. \tag{7.37}$$

Defective lattices (7.36) with (7.37) have been realized experimentally. [3,9] When $\varepsilon > 0$, the lattice intensity I_L at the defect site is higher than that of the surrounding regions, and a defect like this is called an attractive defect; otherwise, the defect is called a repulsive defect. Figures 7.4(a) and (b) show the intensity profiles I_L of the repulsive and attractive defective lattices with $\varepsilon = -0.9$ and $\varepsilon = 0.9$, respectively.

Solutions for the localized defect modes of (7.35) are sought in the form of

$$U(x, y, z) = u(x, y)\exp(-i\mu z), \tag{7.38}$$

(a)   (b)

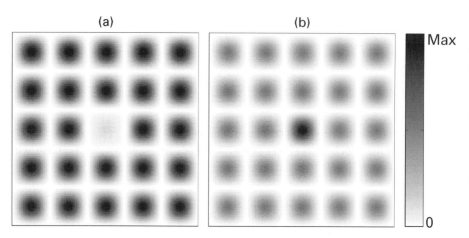

Figure 7.4 Defective lattice intensity profiles I_L as calculated using (7.36). (a) Repulsive defect with $\varepsilon = -0.9$. (b) Attractive defect with $\varepsilon = 0.9$.

where μ is a propagation constant, and $u(x, y)$ is a localized function in a waveguide transverse direction. After substituting (7.38) into (7.35), a linear eigenvalue problem for $u(x, y)$ is obtained:

$$u_{xx} + u_{yy} + \left[\mu - \frac{E_0}{1 + I_0(x, y)} \right] u = 0. \tag{7.39}$$

Note that in (7.39) the modal frequency enters indirectly through the definition of the dimensionless modal propagation constant μ. In what follows we consider the modal frequency to be fixed, while (7.39) will be cast into an eigenvalue problem with respect to the modal propagation constant. All the band diagrams studied in the following sections will present allowed values of the modal propagation constant μ as a function of other conserved parameters. Such band diagrams are known as diffraction relations.

7.2.3   Bandgap structure and diffraction relation for the modes of a uniform lattice

As we have seen in the previous chapters, the spectrum of the allowed electromagnetic states of a uniform periodic lattice constitutes a collection of Bloch bands separated by bandgaps. Eigenmodes forming the Bloch bands are periodic or quasi-periodic Bloch waves, while inside the bandgaps no eigenmodes exist. When a defect is introduced into the lattice, localized eigenmodes (defect modes) may appear inside the bandgaps of an original uniform lattice. These defect modes are confined at a defect site as there are no delocalized states in the surrounding lattice to couple to.

Before proceeding with the analysis of defect modes, we first study the bandgap structure of optically induced lattice (7.39) with a perfectly periodic intensity distribution $I_L = I_0 \cos^2(x) \cos^2(y)$. In what follows, we use the following definition:

$$V(x, y) = -\frac{E_0}{1 + I_0 \cos^2(x) \cos^2(y)}. \tag{7.40}$$

According to the Bloch theorem, eigenfunctions of (7.39) are of the form:

$$u(x, y) = e^{ik_1x+ik_2y} G(x, y; k_1, k_2)$$
$$\mu = \mu(k_1, k_2), \tag{7.41}$$

where $\mu = \mu(k_1, k_2)$ is the diffraction relation, wavenumbers k_1, k_2 are in the first Brillouin zone, i.e., $-1 \le k_1, k_2 \le 1$, and $G(x, y; k_1, k_2)$ is a periodic function in x and y with the same period π as the uniform lattice of (7.36) with $\varepsilon = 0$. Substitution of the Bloch form (7.41) into (7.39) leads to the eigenvalue problem with respect to $\mu(k_1, k_2)$:

$$[(\partial_x + ik_1)^2 + (\partial_y + ik_2)^2 + V(x, y)]G(x, y) = -\mu G(x, y). \tag{7.42}$$

To solve (7.42), we first expand the periodic function $V(x, y)$ and $G(x, y)$ into the summation of plane waves:

$$V(x, y) = \sum_{m,n} V_{mn} e^{iK_m x + iK_n y}$$
$$K_m = 2m, \, K_n = 2n, \tag{7.43}$$
$$G(x, y) = \sum_{p,q} G_{pq} e^{iK_p x + iK_q y}$$
$$K_p = 2p, \, K_q = 2q. \tag{7.44}$$

Substituting $V(x, y)$ and $G(x, y)$ into (7.42), we get:

$$\sum_{p,q} [-(k_1 + K_p)^2 - (k_2 + K_q)^2]G_{p,q} e^{iK_p x + iK_q y}$$
$$+ \sum_{m,n} \sum_{p,q} V_{m,n} G_{p,q} e^{i(K_m + K_p)x + i(K_n + K_q)y} = -\mu \sum_{p,q} G_{p,q} e^{iK_p x + iK_q y}. \tag{7.45}$$

Equating the corresponding Fourier coefficients on both sides, we reduce (7.39) to a matrix eigenvalue problem:

$$[(k_1 + K_i)^2 + (k_2 + K_j)^2]G_{i,j} - \sum_{m,n} V_{m,n} G_{i-m, j-n} = \mu G_{i,j}, \, -\infty < i, j < \infty. \tag{7.46}$$

This eigenvalue problem can then be solved numerically by truncating the number of Fourier coefficients to $-N \le i, j \le N$. For the typical parameter values $E_0 = 15$ and $I_0 = 6$, Fig. 7.5 shows the diffraction relation and bandgaps of (7.39) along the band edge of the irreducible Brillouin zone ($\Gamma \to X \to M \to \Gamma$). Empty areas in Fig. 7.5(c) correspond to complete gaps which are named the semi-infinite $\mu < 5.54$, first $5.76 < \mu < 9.21$, and second $10.12 < \mu < 12.68$ bandgaps, respectively. Finally, the modal intensity distributions at several high symmetry points marked in Fig. 7.5(c) are demonstrated in Fig. 7.6.

From the results of Section 6.6, it can be expected that when a localized perturbation is introduced into the photonic lattice, defect modes will bifurcate out from either the lower or the upper edges of every Bloch band. Moreover, from the intensity profiles of the modes at the edges of the Bloch bands, we can also infer, qualitatively, the field

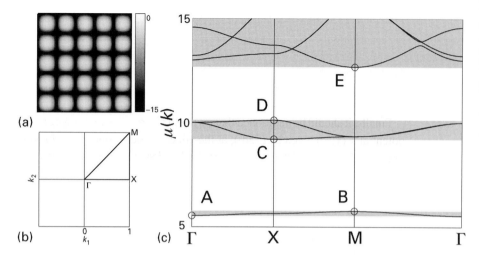

Figure 7.5 (a) Distribution of the effective refractive index change $V(x, y)$ of (7.40). (b) The first Brillouin zone of a 2D square lattice in reciprocal space. (c) Diffraction relation for the uniform lattice potential (7.40) with $E_0 = 15$ and $I_0 = 6$ plotted along the edge $\Gamma \to X \to M \to \Gamma$ of the irreducible Brillouin zone. Shaded regions: first three Bloch bands. Empty regions: photonic bandgaps.

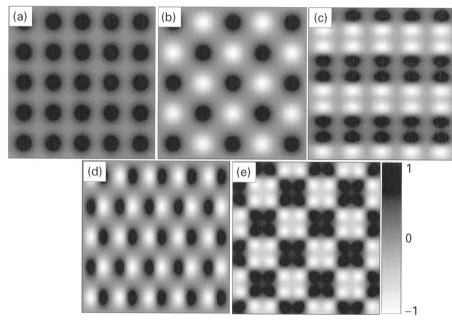

Figure 7.6 Intensity of the Bloch modes (7.41) at various edge points of the Bloch bands corresponding to the diffraction relation in Fig. 7.5(c).

distributions in the defect modes bifurcating from these edges. Thus, defect modes bifurcating from the edges of the first band (points A, B in Fig. 7.5(c)) are expected to exhibit a monopole-like intensity distribution at the defect site. Similarly, defect modes bifurcating from the edges of the second band (points C, D in Fig. 7.5(c)) will have a dipole-like intensity distribution, while defect modes bifurcating from the edge of the third band (point E in Fig. 7.5(c)) will have a quadrupole-like distribution.

Finally, it should be mentioned that owing to discrete rotational symmetry of the lattice potential $V(x, y)$ given by (7.40), at the edges of the second Bloch band (points C and D in Fig. 7.5(c)) the Bloch modes are doubly degenerate. In particular, if we rotate by $90°$ the modal field distributions $u(x, y)$ at points C or D in Fig. 7.6, the resulting field distribution $u(y, x)$ is also a solution of (7.39).

7.2.4 Bifurcations of the defect modes from Bloch band edges for localized weak defects

When the defect is weak ($|\varepsilon| \ll 1$), (7.39) can be further simplified:

$$u_{xx} + u_{yy} + [\mu + V(x, y)]u = \varepsilon f(x, y)u + O(\varepsilon^2), \qquad (7.47)$$

where $V(x, y)$ is given by (7.40), and:

$$f(x, y) = -\frac{E_0 I_0 \cos^2(x) \cos^2(y) F_D(x, y)}{[1 + I_0 \cos^2(x) \cos^2(y)]^2}. \qquad (7.48)$$

Equation (7.47) is a so-called two-dimensional perturbed Hill's equation, in which $f(x, y)$ is a 2D localized perturbation (i.e., $f(x, y) \to 0$ as $(x, y) \to \infty$) to a periodic potential $V(x, y)$. In this section, we study analytically the defect modes described by the general perturbed Hill's equation (7.47) for arbitrary periodic potentials $V(x, y)$ and 2D localized perturbations $f(x, y)$. For convenience, we assume that the potential $v(x, y)$ is Π-periodic along both x and y directions. Application of this analysis to the special case of photonic lattices (7.39) is given in the next section.

When $\varepsilon = 0$, as discussed in the previous section, (7.47) admits Bloch solutions in the form:

$$u(x, y) = B_n(x, y; k_1, k_2)$$
$$\equiv e^{ik_1 x + ik_2 y} G_n(x, y; k_1, k_2), \quad \mu = \mu_n(k_1, k_2), \qquad (7.49)$$

where $\mu = \mu_n(k_1, k_2)$ is the diffraction relation of the nth Bloch band (also called the diffraction surface), vector (k_1, k_2) is confined to the first Brillouin zone ($-1 \leq k_1 \leq 1$, $-1 \leq k_2 \leq 1$), and $G_n(x, y; k_1, k_2)$ is a periodic function in both x and y with a period π. A collection of all the Bloch modes $B_n(x, y; k_1, k_2)$, $n = 1, 2, \ldots$, $(k_1, k_2) \in$ first Brillouin zone, forms a complete basis set. In addition, the orthogonality relation between the Bloch modes is:

$$\int_{-\infty}^{\infty} \int_{-\infty}^{\infty} B_m^*(x, y; k_1, k_2) B_n(x, y; \widehat{k}_1, \widehat{k}_2) dx\, dy$$
$$= (2\pi)^2 \delta(k_1 - \widehat{k}_1)\delta(k_2 - \widehat{k}_2)\delta_{m,n}. \qquad (7.50)$$

Here the Bloch modes have been normalized by

$$\frac{1}{\pi^2} \int_0^\pi dx \int_0^\pi dy \, |G_n(x, y; k_1, k_2)|^2 = 1, \tag{7.51}$$

where $\delta()$ is the delta function, $\delta_{m,n}$ is a Kronecker delta, and the superscript * represents complex conjugation.

In what follows we demonstrate that when $\varepsilon \neq 0$, defect modes bifurcate out from the edges of the Bloch bands and into the bandgaps. In particular, we consider bifurcation of a defect mode from the edge point $\mu_n^e = \mu_n(k_1^e, k_2^e)$ of the nth diffraction surface at the (k_1^e, k_2^e) point in the first Brillouin zone. After introduction of a defect, the propagation constant μ of a defect mode will appear inside the corresponding bandgap in the vicinity of the Bloch band in question. Thus, when $\varepsilon \neq 0$, defect modes can be expanded into Bloch waves as:

$$u(x, y) = \sum_{n=1}^\infty \int_{-1}^1 dk_1 \int_{-1}^1 dk_2 \alpha_n(k_1, k_2) B_n(x, y; k_1, k_2), \tag{7.52}$$

where $\alpha_n(k_1, k_2)$ are the expansion coefficients. In the remainder of this section, unless otherwise indicated, integrals for dk_1 and dk_2 are always over the first Brillouin zone, so the lower and upper limits in the integration formula will be omitted.

When (7.52) is substituted into the left-hand side of (7.47), we get:

$$\sum_{n=1}^\infty \iint \phi_n(k_1, k_2) B_n(x, y; k_1, k_2) dk_1 dk_2 = \varepsilon f(x, y) u(x, y), \tag{7.53}$$

where $\phi_n(k_1, k_2)$ is defined as:

$$\phi_n(k_1, k_2) \equiv \alpha_n(k_1, k_2)[\mu - \mu_n(k_1, k_2)]. \tag{7.54}$$

Substituting (7.52) into the right-hand side of (7.53) and using the orthogonality relation (7.50), we find that $\phi_n(k_1, k_2)$ satisfies the following integral equation:

$$\phi_n(k_1, k_2) = \frac{\varepsilon}{(2\pi)^2} \sum_{m=1}^\infty \iint \frac{\phi_m(\hat{k}_1, \hat{k}_2)}{\mu - \mu_m(\hat{k}_1, \hat{k}_2)} W_{m,n}(k_1, k_2; \hat{k}_1, \hat{k}_2) d\hat{k}_1 d\hat{k}_2, \tag{7.55}$$

where the kernel function $W_{m,n}$ is defined as:

$$W_{m,n}(k_1, k_2; \hat{k}_1, \hat{k}_2) = \int_{-\infty}^\infty \int_{-\infty}^\infty f(x, y) B_n^*(x, y; k_1, k_2) B_m(x, y; \hat{k}_1, \hat{k}_2) dx dy. \tag{7.56}$$

Since $f(x, y)$ is a 2D localized function, $W_{m,n}$ is uniformly bounded for all the (k_1, k_2) and $(\hat{k}_1, \hat{k}_2)$ points in the first Brillouin zone. At the edge point $\mu = \mu_n^e$, $\partial \mu_n/\partial k_1 = \partial \mu_n/\partial k_2 = 0$. For simplicity, we also assume that $\partial^2 \mu_n/\partial k_1 \partial k_2 = 0$ at this edge point – an assumption that is always satisfied for (7.35)–(7.37) owing to symmetries of the defect potential. Under these assumptions, the local diffraction function near the (k_1^e, k_2^e) edge point can then be expanded as:

$$\mu_n(k_1, k_2) = \mu_n^e + \gamma_1 \delta k_1^2 + \gamma_2 \delta k_2^2 + O(\delta k_1^2, \delta k_2^2, \delta k_1 \delta k_2), \tag{7.57}$$

where:

$$\gamma_1 = \frac{1}{2} \frac{\partial^2 \mu_n}{\partial k_1^2}\Big|_{(k_1^e, k_2^e)} ; \; \gamma_2 = \frac{1}{2} \frac{\partial^2 \mu_n}{\partial k_2^2}\Big|_{(k_1^e, k_2^e)}$$

$$\delta k_1 = k_1 - k_1^e; \delta k_2 = k_2 - k_2^e. \qquad (7.58)$$

By definition, the edge point is a local maximum or minimum of the nth diffraction surface, therefore γ_1 and γ_2 must be of the same sign. The defect mode eigenvalue can then be written as:

$$\mu = \mu_n^e + \sigma h^2, \qquad (7.59)$$

where $\sigma = \pm 1$, and $0 < h(\varepsilon) \ll 1$ when $\varepsilon \ll 1$. Substituting (7.57) and (7.59) into (7.55), we find that only a single term in the summation with index $m = n$ makes an $O(\phi_n)$ contribution. In this term, the denominator $\mu - \mu_n(\widehat{k}_1, \widehat{k}_2)$ is $O(h^2)$, is small near the band edge bifurcation point $(\widehat{k}_1, \widehat{k}_2) = (k_1^e, k_2^e)$, and results in an $O(\phi_n)$ contribution in the summation. For the rest of the bands, the terms in the summation give $O(\varepsilon\phi_m)$ contributions, as the denominator $\mu - \mu_m(\widehat{k}_1, \widehat{k}_2)$ does not vanish for $\varepsilon \to 0$. As a result, the summation (7.55) over the bands $m \neq n$ can be omitted:

$$\phi_n(k_1, k_2) \simeq \frac{\varepsilon}{(2\pi)^2} \iint \frac{\phi_n(\widehat{k}_1, \widehat{k}_2)}{\mu - \mu_n(\widehat{k}_1, \widehat{k}_2)} W_{n,n}(k_1, k_2; \widehat{k}_1, \widehat{k}_2) d\widehat{k}_1 d\widehat{k}_2 + O(\varepsilon\phi_n).$$

$$(7.60)$$

For the denominator in the integral of (7.60) not to vanish one has to choose:

$$\sigma = -\text{sgn}(\gamma_1) = -\text{sgn}(\gamma_2), \qquad (7.61)$$

which simply means that the defect mode is positioned inside the bandgap.

Substituting (7.57) and (7.59) into (7.60) we thus find:

$$\phi_n(k_1, k_2) =$$

$$\frac{\varepsilon\sigma}{(2\pi)^2} \iint \frac{\phi_n(\delta\widehat{k}_1, \delta\widehat{k}_2)}{h^2 + |\gamma_1| \delta\widehat{k}_1^2 + |\gamma_2| \delta\widehat{k}_2^2} W_{n,n}(k_1, k_2; \delta\widehat{k}_1, \delta\widehat{k}_2) d(\delta\widehat{k}_1) d(\delta\widehat{k}_2) + O(\varepsilon\phi_n).$$

$$(7.62)$$

This equation can be further simplified, up to an error $O(\varepsilon\phi_n)$, as:

$$\phi_n(k_1, k_2) = \frac{\varepsilon\sigma}{(2\pi)^2} \phi_n(k_1^e, k_2^e) W_{n,n}(k_1, k_2; k_1^e, k_2^e)$$

$$\iint \frac{1}{h^2 + |\gamma_1| \delta\widehat{k}_1^2 + |\gamma_2| \delta\widehat{k}_2^2} d(\delta\widehat{k}_1) d(\delta\widehat{k}_2) + O(\varepsilon\phi_n). \qquad (7.63)$$

To calculate the integral in the above equation, we first rescale the integration variables $\delta\widetilde{k}_1 = \sqrt{|\gamma_1|}\delta\widehat{k}_1$, $\delta\widetilde{k}_2 = \sqrt{|\gamma_2|}\delta\widehat{k}_2$ and then perform the integration over the scaled Brillouin zone with $|\delta\widetilde{k}_1| \leq \sqrt{|\gamma_1|}$ and $|\delta\widetilde{k}_2| \leq \sqrt{|\gamma_2|}$. With the $O(\varepsilon\phi_n)$ error, the rectangular integration region can be replaced by a circle of radius $\sqrt{1-h^2}$ in the $(\delta\widetilde{k}_1, \delta\widetilde{k}_2)$

plane. With this assumption, integral (7.63) can be evaluated analytically in polar coordinates to result in:

$$\phi_n(k_1, k_2) = \frac{\varepsilon\sigma \ln h}{(2\pi)^2\sqrt{\gamma_1\gamma_2}}\phi_n(k_1^e, k_2^e)W_{n,n}(k_1, k_2; k_1^e, k_2^e) + O(\varepsilon\phi_n).$$ (7.64)

Finally, for the values $k_1 = k_1^e; k_2 = k_2^e$ for (7.64) it follows that:

$$\ln h = -\frac{2\pi\sigma\sqrt{\gamma_1\gamma_2}\ln h}{\varepsilon W_{n,n}(k_1, k_2; k_1^e, k_2^e)} + O(1).$$ (7.65)

By substituting (7.65) into (7.59), we finally find:

$$\mu = \mu_n^e + \sigma C e^{-\beta/\varepsilon}; \quad \beta = \frac{4\pi\sigma\sqrt{\gamma_1\gamma_2}}{W_{n,n}(k_1, k_2; k_1^e, k_2^e)},$$ (7.66)

and C is some positive constant. Note that β and ε must have the same sign, thus ε and $\sigma W_{n,n}(k_1, k_2; k_1^e, k_2^e)$ must have the same sign. Since we have shown that σ and γ_1, γ_2 have opposite signs, we conclude that the condition for defect mode bifurcations from a Γ symmetry edge point is that:

$$\text{sgn}[\varepsilon W_{n,n}(k_1, k_2; k_1^e, k_2^e)] = -\text{sgn}(\gamma_1) = -\text{sgn}(\gamma_2).$$ (7.67)

Under this condition, the defect mode eigenvalue μ bifurcated from the edge point μ_n^e is given by (7.66). Its distance from the edge point, i.e., $|\mu - \mu_n^e|$, is exponentially small with the defect strength $|\varepsilon|$. This contrasts with the 1D case where such dependence is quadratic. [10]

Finally, substituting $f(x, y)$ of (7.48) into (7.56), we find that $W(k_1, k_2; k_1^e, k_2^e)$ is always negative, thus the defect mode bifurcation condition (7.67) becomes:

$$\text{sgn}(\varepsilon) = \text{sgn}(\gamma_1) = \text{sgn}(\gamma_2).$$ (7.68)

Thus, in the case of an attractive defect ($\varepsilon > 0$), defect modes bifurcate out from the lower band edges of Fig. 7.5(c); in the case of a repulsive defect ($\varepsilon < 0$), defect modes bifurcate out from the upper edges of Fig. 7.5(c).

7.2.5   Dependence of the defect modes on the strength of localized defects

The analytical result (7.66) holds under the weak defect strength approximation $|\varepsilon| \ll 1$. If the defect is strong ($|\varepsilon| \sim 1$), this result becomes invalid. For strong defects, defect modes of (7.39) must be determined numerically. The numerical method we use is a squared-operator iteration method. [11] Consecutive approximations for the defect mode field according to this method are given by:

$$u_{n+1} = u_n - M^{-1}(L + \mu_n)M^{-1}(L + \mu_n)u_n\,\Delta t,$$ (7.69)

$$L = \partial_{xx} + \partial_{yy} - \frac{E_0}{1 + I_L},$$ (7.70)

$$\mu_n = -\frac{\langle M^{-1}Lu_n, u_n\rangle}{\langle M^{-1}u_n, u_n\rangle},$$ (7.71)

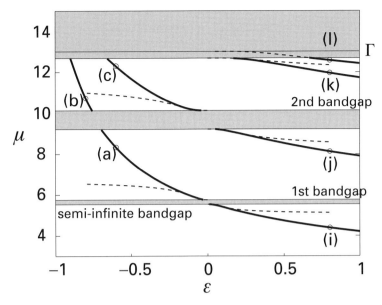

Figure 7.7 Defect mode propagation constants as a function of the defect strength. A point defect in a periodic photoinduced lattice is described by (7.37), and $E_0 = 15$, $I_0 = 6$. Solid lines: numerical results using numerical scheme (7.69)–(7.71). Dashed lines: analytical results from (7.66). The shaded regions are the Bloch bands of a uniform periodic lattice (after [12], © 2007 APS).

where $M = C - \partial_{xx} - \partial_{yy}$ is an accelerator operator for speeding up the convergence of (7.69), the inner product (7.69) is defined as $\langle f, g \rangle = \int_{-\infty}^{\infty} f^* g \, dx$, and C and Δt are positive constants chosen by the user. When implementing this method, it is desirable to use a discrete Fourier transform method to calculate the spatial derivatives as well as inverting the operator M for high accuracy.

Figure 7.7 presents the defect state propagation constants μ as functions of the defect strength ε, found using the numerical scheme (7.69)–(7.71). In this simulation we assume $E_0 = 15$, $I_0 = 6$, and the defect strength values $\varepsilon \subset [-1, 1]$. As predicted in the previous section, from every edge of the Bloch bands, there is one defect mode that bifurcates out. The bifurcation diagram is shown in Fig. 7.7. For comparison, the analytical results from (7.66) are also displayed in Fig. 7.7 (dashed lines).

As can be seen from Fig. 7.7, for the weak attractive defects ($|\varepsilon| \ll 1, \varepsilon > 0$) the bifurcation occurs at the lower edges of Bloch bands, while for the repulsive defect ($|\varepsilon| \ll 1, \varepsilon < 0$), the bifurcation occurs at the upper edges of the Bloch bands. We can further make quantitative comparisons between the numerical values of μ and the theoretical formula (7.66). In particular, in the limit $|\varepsilon| \ll 1$, numerical data for μ are fitted into the form (7.66) with two fitting parameters β_{num} and C_{num}. The value of the fitting parameter β_{num} can then be compared with its analytical estimate (7.66). For example, for the defect mode branch in the semi-infinite bandgap in Fig. 7.7, numerical data fitting for $0 < \varepsilon \ll 1$ give $\beta_{num} = 0.1289$, $C_{num} = 0.4870$. The theoretical value $\beta_{th} = 0.1297$ obtained from (7.66) agrees very well with this numerical value. Another

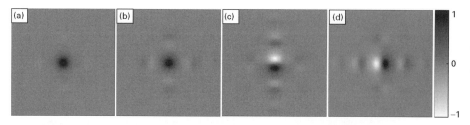

Figure 7.8 Field distributions of the repulsive defect modes marked as (a), (b), and (c) in Fig. 7.7. (d) Field distribution of the second mode degenerate with mode (c).

method of quantitative comparison between numerical results and semi-analytical estimates is to plot the values of μ given by (7.66) alongside the numerical curves (as presented in Fig. 7.7). To do so, we first calculate the theoretical value β from (7.66) at each band edge. As an analytical expression for the constant C in (7.66) is not available we fit this single parameter to the numerical values. Finally, the resultant semi-analytical curves are presented in the same plot (Fig. 7.7) as numerical data, and good quantitative agreement is achieved for weak defects $|\varepsilon| \ll 1$.

As the strength $|\varepsilon|$ of a defect increases, the defect mode branches move away from the band edges (downwards when $\varepsilon > 0$, and upwards when $\varepsilon < 0$). We now examine the intensity distributions in various defect modes. For this purpose, we select one representative point from each branch (marked by circles and labeled by letters in Fig. 7.7). The modal field distributions at these points are displayed in Fig. 7.8 and Fig. 7.10. First, we discuss the modes of the repulsive defects $\varepsilon < 0$. Field distributions for the defect modes marked as (a) and (b) in Fig. 7.7 are symmetric with respect to the reflection in both $\hat{x}$ and $\hat{y}$ axes $u(x, y) = u(y, x)$, and the maximum of the field intensity located in the coordinate origin; in what follows we call such defect modes fundamental. These modes have been experimentally observed. [9] This modal behavior is expected from the discussion of Section 7.3.3. Indeed, defect states (a) and (b) both bifurcate from the upper edge of a first band, marked as (b) in Fig. 7.5, therefore one expects that the field distribution in these states will be similar to the field distribution in the corresponding mode of a uniform periodic lattice (inset (b) in Fig. 7.6).

Similarly, the defect mode branching out from the upper edge of a second band (point (c) in Fig. 7.7) is a dipole-like mode. This behavior is in accordance with the modal distribution at the symmetry point (d) of a uniform lattice shown in Fig 7.4. Moreover, the defect mode that corresponds to point (c) in Fig. 7.7 is doubly degenerate. Assuming that $u(x, y)$ describes the field distribution in one of the two degenerate modes, the field distributions in the other mode can be described as $u(y, x)$. Since the defect mode branch containing point (c) in Fig. 7.7 admits two linearly independent modes, any linear superposition of such modes is again a defect mode. Such linear superpositions enable a variety of interesting field distributions. For example, if the two defect modes in Fig. 7.8(c) and (d) are superimposed with a $\pi/2$ phase shift, the resultant field distribution $u(x, y) + iu(y, x)$ describes a vortex (Fig. 7.9(e)). If the two modes are superimposed with 0 or π phase shift, the resultant field distributions $u(x, y) + u(y, x)$ (Fig. 7.9(f)) or $u(x, y) - u(y, x)$ define the rotated dipole modes.

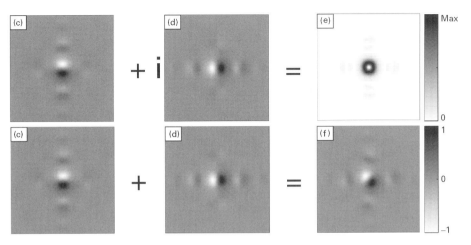

Figure 7.9 Superpositions of the degenerate modes (c) and (d) of Fig. 7.8 with various phase shifts: $\pi/2$ (top row), 0 (bottom row). Note: inset (e) shows the intensity profile of the vortex mode, while the rest of the insets show field distributions.

We now examine the modes of attractive defects (Fig. 7.10). At point (i) in Fig. 7.7, inside the semi-infinite gap, the defect mode is bell-shaped and is strongly confined to the defect site. This field distribution is expected as the defect mode bifurcates from the lower edge of the first Bloch band (point (a) in Fig. 7.5) with a corresponding field distribution in the mode of a uniform lattice shown in Fig. 7.6(a). The guiding mechanism for this mode is total internal reflection, which is different from the bandgap-guiding mechanism for the defect modes in the higher bandgaps.

The doubly degenerate branch containing point (j) in Fig. 7.7 bifurcates from the lower edge of the second Bloch band. Two linearly independent dipole-like defect modes of this branch have field distributions of the form $u(x, y)$ and $u(y, x)$. One such mode is shown in Fig. 7.10(j). As with the case of repulsive defects, a linear superposition of these two modes can generate vortex-like and dipole-like modes. We now consider the branch containing point (k) in Fig. 7.7 and bifurcating from the lower edge of the third semi-infinite Bloch band. This is a singlet branch with a quadrupole-like modal field distribution in accordance with the field distribution at the bifurcation point (e) of Fig. 7.5 and Fig. 7.6.

Finally, the modal field distribution in the branch containing point (l) in Fig. 7.7 is a degenerate doublet having a tripolar-like field distribution. As before, field distributions in the linearly independent degenerate modes can be chosen in the form $u(x, y)$ and $u(y, x)$. The origin of this branch is somewhat nontrivial as the branch bifurcates not from the Bloch band edge (point (e) in Fig. 7.5), but rather from within the third semi-infinite Bloch band. Unlike the defect modes at the points (c) and (j) this defect mode, when superimposed with a $\pi/2$ phase shift with its degenerate pair, does not generate vortex modes. However, modal superposition with 0 and π phase shifts leads to two structurally different defect modes, shown in Fig. 7.10(m) and (n). The defect mode in Fig. 7.10(m) has an intensity maximum in the coordinate origin, surrounded by

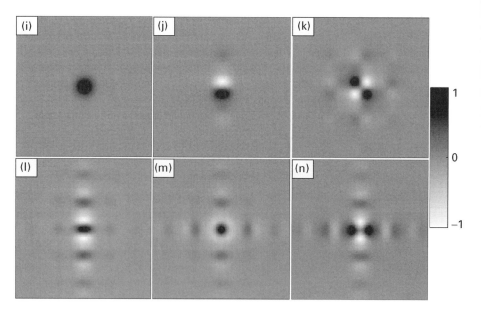

Figure 7.10 (i)-(l) Field profiles of the defect modes in attractive defects corresponding to the points (i), (j), (k), (l) in Fig. 3.3.1. (m)-(n) Degenerate defect modes obtained by superimposition of the two doubly degenerate states at the point (l) with 0 and π phase shifts $u(x, y) + u(y, x)$ and $u(x, y) - u(y, x)$, respectively.

a zero-intensity ring. The defect mode in Fig. 7.10(n) is quadrupole-like, however it is oriented differently from the quadrupole defect mode of point (k) in Fig. 7.10.

Most of the defect mode branches in Fig. 7.7 bifurcate from edges of the Bloch bands. Even the branch of point (b) in the second gap of Fig. 7.7 can be traced to the defect mode bifurcation from the upper edge of the first Bloch band. However, the defect mode branch of point (l) does not bifurcate directly from any Bloch-band edge state. Careful examination shows that the modal field distribution in this branch resembles Bloch modes at the lowest Γ symmetry point in the third semi-infinite Bloch band (see Fig. 7.5(c)). However, from Fig. 7.5(c), we can see that this lowest Γ symmetry point, while being the local minimum of a diffraction surface, is not an edge point of the third Bloch band, and, actually, is located inside the third Bloch band. Because of this, such a special bifurcation point is known as a "quasi-edge point" of a Bloch band. [12]

7.2.6   Defect modes in 2D photonic lattices with nonlocalized defects

In this section, we briefly discuss defect modes in nonlocalized defects. There are two significant differences between the defect modes of the localized and nonlocalized defects. One is that for nonlocalized defects their corresponding eigenvalues can bifurcate out from the edges of the continuous spectrum algebraically, and not exponentially, with the defect strength ε. The other difference is that for nonlocalized defects, defect modes can be embedded inside the continuous spectrum.

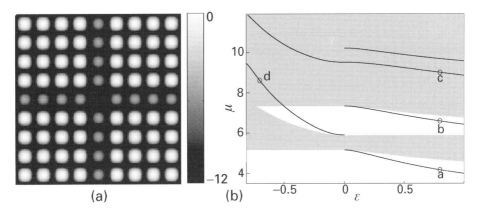

Figure 7.11 (a) Profile of a 2D separable potential $V_D(x) + V_D(y)$ of (7.73) describing a nonlocalized defect ($I_0 = 3$, $E_0 = 6$, $\varepsilon = -0.7$). (b) Defect mode branches supported by the nonlocalized defect in the (μ, ε) parameter space. Shaded: the continuous part of a spectrum of (7.72). Field distributions in defect modes are shown for various points (circled and labeled) in Fig. 7.12 (after [12], © 2007 APS).

For simplicity, we consider the linear Schrödinger equation with nonlocalized defects described by a separable potential:

$$u_{xx} + u_{yy} + \{V_D(x) + V_D(y)\} u = -\mu u, \qquad (7.72)$$

where V_D is a one-dimensional function of the form:

$$V_D(x) = -\frac{E_0}{1 + I_0 \cos^2(x)[1 + \varepsilon F_D(x)]}, \qquad (7.73)$$

and $F_D(x) = \exp(-x^8/128)$ is a defect function as described in the previous section (ε is the defect strength). We chose this particular form of the potential as all the eigenvalues of (7.72) can then be found analytically. The nonlocalized nature of this defect is seen in Fig. 7.11(a), where $V_D(x) + V_D(y)$ for $I_0 = 3$, $E_0 = 6$, and $\varepsilon = -0.7$ is displayed. We see that along the $\hat{\mathbf{x}}$ and $\hat{\mathbf{y}}$ axes in Fig. 7.11(a), the defect extends to infinity, thus the name nonlocalized defect. The I_0 and E_0 parameters above are chosen to be the same as in the 1D defect-mode analysis of [10]. These 1D eigenmodes will be used in the remainder of this section to construct the eigenmodes of the 2D equation (7.72).

Since the potential in (7.72) is separable, the eigenmodes of this equation can be written in the following form:

$$u(x, y) = u_a(x)u_b(y)$$
$$\mu = \mu_a + \mu_b, \qquad (7.74)$$

where u_a, u_b, μ_a, μ_b satisfy the following one-dimensional eigenvalue equations:

$$u_{a,xx} + V_D(x)u_a = -\mu_a u_a, \qquad (7.75)$$
$$u_{b,yy} + V_D(y)u_b = -\mu_b u_b. \qquad (7.76)$$

These 1D equations for u_a and u_b were studied extensively in [10], where their eigen-values and eigenmodes were characterized. Using (7.74) and the 1D results of [10], we can now construct the entire spectrum of (7.72).

Before we construct the eigenvalue spectrum of (7.72) we need to clarify the definitions of discrete and continuous eigenvalues of the 2D equation (7.72). Here, an eigenvalue is called discrete if its eigenfunction is square-integrable (i.e., localized along all directions in the (x, y) plane), otherwise the eigenvalue is called continuous. Note that an eigenfunction that is localized along one direction (say $\hat{x}$ axis) and nonlocalized along the other direction ($\hat{y}$ axis) corresponds to a continuous and not a discrete eigenvalue.

Now we construct the spectra of (7.72) for a specific example with $I_0 = 3$, $E_0 = 6$, and $\varepsilon = 0.8$. At these parameter values, the discrete eigenvalues and continuous-spectrum intervals (1D Bloch bands) of the 1D eigenvalue problem (7.75) are (see [10]):

$$\{\lambda_1, \lambda_2, \lambda_3, \ldots\} = \{2.0847, 4.5002, 7.5951, \ldots\}, \tag{7.77}$$

$$\{[I_1, I_2], [I_3, I_4], [I_5, I_6], \ldots\} = \{[2.5781, 2.9493], [4.7553, 6.6010],$$
$$[7.6250, 11.8775], \ldots\}. \tag{7.78}$$

Using the relation (4.3), we find that the discrete eigenvalues and continuous-spectrum intervals of the 2D eigenvalue problem (7.72) are:

$$\{\mu_1, \mu_2, \mu_3, \mu_4, \ldots\} = \{2\lambda_1, \lambda_1 + \lambda_2, 2\lambda_2, \lambda_1 + \lambda_3, \ldots\}$$
$$= \{4.1694, 6.5849, 9.0004, 9.6798, \ldots\}, \tag{7.79}$$

$$\{\mu_{\text{continuum}}\} = \{[\lambda_1 + I_1, 2I_2], [\lambda_1 + I_3, \infty]\} = \{[4.6628, 5.8986], [6.8400, \infty]\}. \tag{7.80}$$

Note that at the lower edges of the two continuous-spectrum bands ($\mu = 4.6628$ and $\mu = 6.8400$), the eigenfunctions are nonlocalized along one direction, but localized along its orthogonal direction, thus they are not the usual 2D Bloch modes, which would have been nonlocalized along all directions. Repeating the same calculations with other ε values, one can construct the whole spectra of (7.72) in the (μ, ε) plane for $I_0 = 3$ and $E_0 = 6$. The results are displayed in Fig. 7.11(b). Here, solid curves show branches of the defect modes with discrete eigenvalues, while shaded regions define the continuous spectrum of the defect modes. Notably, unlike in the case of localized defects described in the previous section, several defect mode branches (such as the (c) and (d) branches) are either partially or completely embedded inside the continuous part of a spectrum. Another interesting feature of Fig. 7.11(b) is a quadratic scaling $\mu = \mu_n^e + C\varepsilon^2$ of the distance between the bifurcated mode and the corresponding band edge for the defects of small strength $|\varepsilon| \ll 1$. It is not surprising to find this relation as it is the one particular to the branched-out modes in 1D lattices with defects [10]. This scaling should be compared with the exponential scaling of (7.66) for the case of localized defects in 2D lattices.

Even though the defect in (7.72) is nonlocalized, the corresponding defect modes with discrete eigenvalues can be quite similar to those for localized defects. To demonstrate this point, we picked four representative points on various branches of Fig. 7.11(b). These

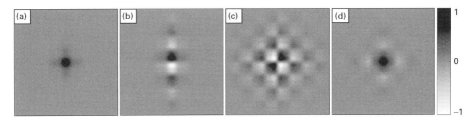

Figure 7.12 Field distributions of the defect modes of a 2D lattice with nonlocalized defects of separable form (7.72), (7.73) at the points (a), (b), (c), (d) of Fig. 7.11(b).

points are marked by circles and labeled (a), (b), (c), and (d). Field distribution profiles of the defect modes at these four points are displayed in Fig. 7.12. Note that the field distributions of the modes in (a), (b), (c), and (d) branches in Fig. 7.11(b) resemble the field distributions of the defect modes in the (i), (j), (k), and (a) branches in Fig. 7.7, respectively. We would also like to point out that, even though the defect modes at the points (c) and (d) are embedded in the continuous part of a spectrum, their corresponding eigenfunctions are truly 2D-localized and square-integrable, as seen in Fig. 7.12(c),(d).

7.3   Photonic-crystal slabs

In this section we consider optical properties of photonic-crystal slabs and photonic-crystal slab waveguides, which is another example of quasi-2D photonic crystals. These structures play a particularly important role in the practical implementations of photonic-crystal devices. The main goal of this section is to explain the principal features of a photonic-crystal slab band diagram. In particular, we demonstrate that such band diagrams can be rationalized in terms of the eigenstates of a corresponding two-dimensional photonic crystal of infinite thickness, however, allowing for the nonzero discrete components of a wave vector along the direction of photonic-crystal uniformity. We apply our analysis to the case of a photonic-crystal slab featuring an underlying hexagonal lattice of air holes (Fig. 7.13). In this section, however, we do not aim at classifying the optical properties of various photonic-crystal slab geometries, which can be found elsewhere. [13]

7.3.1   Geometry of a photonic-crystal slab

By definition, photonic-crystal slabs exhibit two-dimensional discrete translational symmetry in the plane of a crystal, while in the third dimension, translational symmetry is broken. A particular example of a photonic-crystal slab is a thin dielectric layer of thickness h and dielectric constant ε_b perforated with a hexagonal array of holes of radius r_a and dielectric constant ε_a, separated by lattice constant a. We assume that the dielectric index of the holes is the same as that of a cladding surrounding the photonic-crystal slab.

In Fig. 7.13(a), several unit cells in the $\hat{\mathbf{x}}\hat{\mathbf{y}}$ plane of a photonic-crystal slab are shown. The photonic-crystal vertical cross-section (z plane) along the longer diagonal of a unit

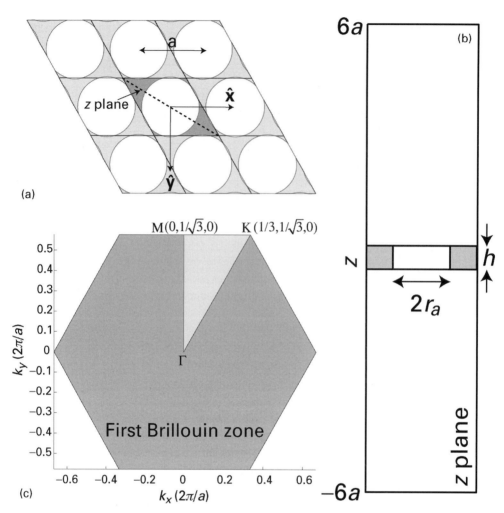

Figure 7.13 Example of a photonic-crystal slab. (a) The photonic-crystal slab possesses discrete translation symmetry in the $\hat{x}\hat{y}$ plane, and is formed by a hexagonal array of holes perforating the dielectric layer of a finite thickness. (b) In the vertical direction (z plane), the photonic-crystal slab has a finite thickness. The dielectric constant of the cladding is assumed to be the same as that of the holes.

cell (the dotted line in Fig. 7.13(a)) is shown in Fig. 7.13(b). Basis vectors defining hexagonal lattice can be chosen as:

$$\begin{aligned} \bar{a}_1 &= a \cdot (1, 0, 0) \\ \bar{a}_2 &= a \cdot (1/2, \sqrt{3}/2, 0) \end{aligned}, \tag{7.81}$$

thus defining the corresponding reciprocal lattice vectors

$$\begin{aligned} \bar{b}_1 &= 2\pi/a \cdot (1, -1/\sqrt{3}, 0) \\ \bar{b}_2 &= 2\pi/a \cdot (0, 2/\sqrt{3}, 0) \end{aligned}, \tag{7.82}$$

according to (2.129). The first Brillouin zone of the hexagonal lattice is presented in Fig. 7.13(c), with the irreducible Brillouin zone marked as ΓKM.

7.3.2 Eigenmodes of a photonic-crystal slab

A standard way of computing eigenmodes of a photonic-crystal slab is by using the plane-wave expansion method (6.13) derived in Section 6.2. This method, however, has to be modified to be able to compute eigenmodes of a system in which translational symmetry is broken along a certain direction ($\hat{z}$ direction in Fig. 7.13(a)). Note that in prior chapters we have used the plane-wave method to compute eigenmodes of the two-dimensional photonic crystals, as well as photonic-crystal fibers. Moreover, from the form of (6.13) it is clear that the method also naturally handles three-dimensional periodic structures. In particular, if we assume that the photonic-crystal slab of Fig. 7.13 is repeated periodically along the $\hat{z}$ direction with a period $a N_z$ ($N_z = 12$ in Fig. 7.13(b)), then the plane-wave method can be applied directly to compute the eigenmodes of such a structure after introduction of a third basis vector and its corresponding reciprocal basis vector as:

$$\begin{aligned}
\bar{a}_3 &= a \cdot (0, 0, N_z) \\
\bar{b}_3 &= 2\pi \left/ a \cdot (0, 0, 1/N_z,) \right.
\end{aligned} \tag{7.83}$$

If the modes of a photonic-crystal slab are strongly localized along the $\hat{z}$ direction (typically in the vicinity of a dielectric layer), and if the modal fields of such localized states do not overlap significantly between the neighbouring unit cells in the $\hat{z}$ direction, then the eigenmodes of an artificially periodic three-dimensional structure approximate well the eigenmodes of a photonic-crystal slab with infinite cladding. This observation forms the core of a supercell method, which is a method of choice for the calculation of band structure of photonic-crystal slabs. From Section 2.4.5 (the subsection on 3D discrete translational symmetry) it follows that eigenmodes of a three-dimensional system exhibiting discrete translational symmetries in all directions have the form (2.128) and can be labeled by a conserved wave vector $\mathbf{k}$ confined to a three-dimensional first Brillouin zone. In particular, the k_z component of a wave vector is confined to the interval $-\pi/(a N_z) < k_z < \pi/(a N_z)$, and, therefore, tends to zero when the supercell size $a N_z$ increases. Thus, in the practical supercell simulations of a photonic-crystal slab shown in Fig. 7.13, the modal wave vector is typically chosen in the form $\mathbf{k} = (k_x, k_y, 0)$, where $\mathbf{k}$ is then confined to a two-dimensional first Brillouin zone of an underlying two-dimensional photonic crystal (Fig. 7.13(c)).

For 2D photonic crystals we were able to classify the eigensolutions further as TE- or TM-polarized. As discussed in Section 2.4.7 (the subsection on mirror symmetry), this was a direct consequence of the presence of the mirror symmetry of a 2D photonic-crystal dielectric profile with respect to reflection in the plane perpendicular to the direction of continuous translational symmetry. In general, if the dielectric profile is truly three-dimensional, the eigensolutions are hybrid modes. However, if a three-dimensional, structure possesses a plane of reflection symmetry, a classification analogous to the

TE- and TM- polarizations is again possible. Consider, for example, the photonic-crystal slab of Fig. 7.13, which has a mirror symmetry with respect to reflection in the $z = 0$ plane. In this case, following the same arguments as in Section 2.4.7 (the subsection on inversion symmetry), one can demonstrate that the eigenfields of a photonic-crystal slab transform according to either one of two possible ways:

even modes:
$$\mathbf{E} = (E_x(x, y, z), E_y(x, y, z), E_z(x, y, z))$$
$$= (E_x(x, y, -z), E_y(x, y, -z), -E_z(x, y, -z)),$$
$$\mathbf{H} = (H_x(x, y, z), H_y(x, y, z), H_z(x, y, z))$$
$$= (-H_x(x, y, -z), -H_y(x, y, -z), H_z(x, y, -z)),$$
(7.84)

odd modes:
$$\mathbf{E} = (E_x(x, y, z), E_y(x, y, z), E_z(x, y, z)),$$
$$= (-E_x(x, y, -z), -E_y(x, y, -z), E_z(x, y, -z)),$$
$$\mathbf{H} = (H_x(x, y, z), H_y(x, y, z), H_z(x, y, z))$$
$$= (H_x(x, y, -z), H_y(x, y, -z), -H_z(x, y, -z)).$$
(7.85)

Interestingly, in the plane $z = 0$, classification of the 3D slab modes into even or odd becomes identical to the TM and TE classification of the 2D modes of planar photonic crystals. Namely, from (7.84) and (7.85) it follows that at $z = 0$ the electromagnetic fields of the eigenmodes of 3D structures with a $z = 0$ mirror symmetry plane have the following nonzero components:

even modes at $z = 0$:
$$\mathbf{E} = (E_x(x, y, 0), E_y(x, y, 0), 0)$$
$$\mathbf{H} = (0, 0, H_z(x, y, 0)),$$
(7.86)

odd modes at $z = 0$:
$$\mathbf{E} = (0, 0, E_z(x, y, 0)),$$
$$\mathbf{H} = (H_x(x, y, 0), H_y(x, y, 0), 0).$$
(7.87)

Finally, from the definition (2.146) of the TM and TE modes it follows that the even modes of a photonic-crystal slab having mirror symmetry plane $z = 0$ correspond to the TM modes of a 2D photonic crystal, while the odd modes of a photonic-crystal slab correspond to the TE modes of a 2D photonic crystal. We now consider a particular example of a photonic-crystal slab and detail the structure of its eigenmodes.

In Fig. 7.14, we show a band diagram of the even modes of a photonic-crystal slab with the following parameters: $h = 2.0a$, $\varepsilon_b = \sqrt{12}$, $r_a = 0.45a$, $\varepsilon_a = 1$. The dispersion relations of the modes are presented along the ΓKM edge of an irreducible Brillouin zone of a corresponding 2D photonic crystal. Modes were computed using MPB implementation of the plane-wave method. [14] In Fig. 7.14(a), in gray, we present a cladding radiation continuum also known as a light cone of a cladding. The boundary of a light cone of a cladding is described by the equation $\varepsilon_a \omega^2 = |\mathbf{k}|^2$. From Section 2.1 it follows that modes located inside the light cone are delocalized in the cladding, and, therefore, are not confined to the photonic-crystal slab. In what follows, we concentrate on the modes that are located below the light cone of a cladding and, therefore, truly guided

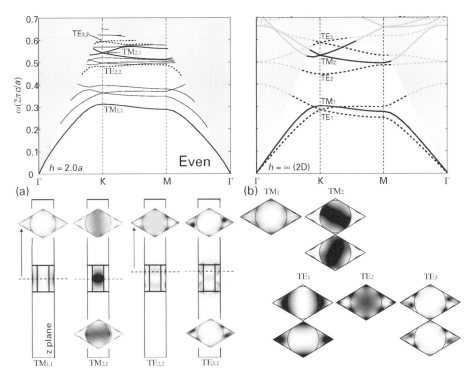

Figure 7.14 (a) Dispersion relations and electric field intensity profiles (shown in a single unit cell) of even modes of a photonic crystal slab of Fig. 7.13; slab thickness is $h = 2.0a$. (b) Dispersion relations and electric field intensity profiles of TM- and TE-polarized modes of a corresponding 2D photonic crystal of infinite thickness.

by the photonic crystal. For comparison, in Fig. 7.14(b) we show band diagrams of the TM- and TE-polarized modes of a corresponding 2D photonic crystal with the same structural parameters.

As we have just established, the even modes of a photonic crystal slab are analogous to the TM-polarized modes of a 2D photonic crystal. Therefore, when comparing the corresponding dispersion relations shown in Fig. 7.14(a) and Fig. 7.14(b) we find clear similarities between the two. In particular, the dispersion relations of the TM modes of a 2D photonic crystal shown in Fig. 7.14(b) as thick solid curves and labeled as TM_1, TM_2 resemble closely dispersion relations of the even modes of a photonic-crystal slab labeled as $TM_{1,1}$, $TM_{2,1}$ and shown in Fig. 7.14(a) as thick solid curves. This similarity becomes particularly clear after inspection of the corresponding modal electric field intensity profiles presented in the lower part of Fig. 7.14 for the K symmetry point. Thus, like the fundamental TM_1 mode of a 2D photonic crystal, the fundamental $TM_{1,1}$ mode of a photonic-crystal slab has most of its electric field intensity concentrated in the dielectric veins, while the electric field intensity in the air hole is small. By further inspection we find that an analog of a doubly degenerate TM_2 mode of a 2D photonic crystal is a doubly degenerate $TM_{2,1}$ mode of a photonic crystal slab with most of its electric field intensity concentrated in the air hole region.

Inspection of the field distributions of slab modes in the vertical cross-section (z plane) shows that modal fields are mostly confined to the slab region with little penetration into the air cladding. The fraction of the modal electric field intensity in the air cladding increases for higher-order modes (compare $TM_{1,1}$ and $TE_{3,2}$, for example). Overall, dispersion relations of the $TM_{1,1}$, $TM_{2,1}$ slab modes have somewhat higher frequencies than dispersion relations of the TM_1, TM_2 modes of 2D photonic crystals. Equivalently, slab modes have lower effective refractive indices than the corresponding modes of a 2D photonic crystal owing to partial penetration of the slab mode fields into the low-refractive-index cladding surrounding the slab.

When comparing band diagrams of a photonic crystal slab with those of a 2D photonic crystal, one also finds that some of the even slab modes (dashed lines in Fig. 7.14(a)) are similar to the TE-polarized modes of a 2D photonic crystal (dashed lines in Fig. 7.14(b)). For example, the dispersion relations and field distributions of the even $TE_{2,2}$, $TE_{3,2}$ slab modes are similar to those of the TE_2, TE_3 2D photonic-crystal modes. This seemingly contradictory observation is rationalized in the next section by tracing the origin of such modes to the 2D photonic-crystal hybrid modes having the nonzero out-of-the-plane component $k_z \neq 0$ of a wave vector. The origin of the other modes (thin lines in Fig. 7.14(a)) is then elucidated in a similar fashion.

7.3.3   Analogy between the modes of a photonic-crystal slab and the modes of a corresponding 2D photonic crystal

An important difference between a photonic-crystal slab and a 2D photonic crystal is the presence of a third dimension. The modes of a photonic-crystal slab of finite thickness h can be thought of as properly symmetrized superpositions of the modes of a 2D photonic crystal having nonzero out-of-plane wave-vector components $k_z \neq 0$. In particular, using eigenfields of a 2D photonic crystal $\mathbf{E}^{2D}_{\mathbf{k}_t,k_z}(x, y)\exp(ik_z z)$ with $k_z \neq 0$, the electric field inside a slab $0 < z < h$ can be presented as:

$$\mathbf{E}^{slab}_{\mathbf{k}_t}(x, y, z) = \int_{-\infty}^{+\infty} dk_z\, A(k_z) \mathbf{E}^{2D}_{\mathbf{k}_t,k_z}(x, y)\exp(ik_z z), \qquad (7.88)$$

where $A(k_z)$ are the expansion coefficients to be determined. Note that by construction the electric field (7.88) automatically satisfies Maxwell's equations inside the slab region. Across the upper and lower slab boundaries (in $\hat{\mathbf{z}}$ direction), the in-plane electric field vector component of slab-guided modes is continuous and matches the field of outgoing plane waves in the cladding regions. Imposing these boundary conditions allows us, in principle, to find an expansion of the photonic-crystal slab modes into the modes of a 2D photonic crystal.

Owing to the final thickness h of a photonic crystal slab, the absolute values of the expansion coefficients $|A(k_z)|$ typically reach their maximum in the vicinity of specific values of $k_z \sim \pi n/h, n \subset$ integer, which is a consequence of the fundamental properties of the Fourier transform. Moreover, for purely real dielectric profiles, if the field $\mathbf{E}_\omega$ is a solution of Maxwell's equations with frequency ω, then $\mathbf{E}^*_\omega$ is also a solution of

Maxwell's equations with the same frequency. This implies that if $E^{2D}_{k_t,k_z}(x,y)\exp(ik_z z)$ is a mode of a 2D photonic crystal, then $E^{*2D}_{k_t,k_z}(x,y)\exp(-ik_z z)$ is also a mode, and, therefore, $E^{2D}_{k_t,-k_z}(x,y) = E^{*2D}_{k_t,k_z}(x,y)$. These two observations allow us to rewrite (7.88) as:

$$E^{slab}_{k_t}(x,y,z) = \int_0^{+\infty} dk_z \left[A(k_z)E^{2D}_{k_t,k_z}(x,y)\exp(ik_z z) + A(-k_z)E^{*,2D}_{k_t,k_z}(x,y)\exp(-ik_z z) \right]$$

$$\sim \frac{\pi}{h} \sum_{n=0}^{\infty} \left[A_n E^{2D}_{k_t,k^n_z}(x,y)\exp(ik^n_z z) + A_{-n} E^{*2D}_{k_t,k^n_z}(x,y)\exp(-ik^n_z z) \right],$$

$$k^n_z = \frac{\pi n}{h}. \qquad (7.89)$$

Expansion (7.89) implies that slab modes can be considered as properly symmetrized superpositions of the counter-propagating (in the $\hat{z}$ direction) modes of a 2D photonic crystal. Moreover, out-of-plane components of the wave vectors of such modes belong to a discrete spectrum of the form $k^n_z \sim \pi n/h, n \subset$ integer.

To further demonstrate this point, consider a photonic-crystal slab surrounded by a perfect conductor instead of air cladding. In this case, boundary conditions at the slab boundaries require a zero value of the electric field transverse components:

$$E^{slab}_{t,k_t}(x,y,0) = E^{slab}_{t,k_t}(x,y,h) = 0. \qquad (7.90)$$

Taking $A_{-n} = -iA_n$, and assuming that modal fields of a 2D photonic crystal are purely real $E^{2D}_{k_t,k_z} = E^{*2D}_{k_t,k_z}$ (this is typically a valid assumption when operating at the high symmetry points of a Brillouin zone, and when $|k_z| \ll |k_t|$), we can rewrite (7.89) as:

$$E^{slab}_{k_t}(x,y,z) \sim \sum_{n=0}^{\infty} A_n E^{2D}_{k_t,k^n_z}(x,y)\sin(k^n_z z), \quad k^n_z = \frac{\pi n}{h}, \qquad (7.91)$$

for which boundary conditions (7.90) are satisfied for any choice of the expansion coefficients A_n. Therefore, under the above mentioned assumptions, the eigenmodes of a photonic-crystal slab surrounded by the perfect reflector can be chosen as:

$$E^{slab}_{k_t,n}(x,y,z) \sim E^{2D}_{k_t,k^n_z}(x,y)\sin(k^n_z z), \quad k^n_z = \frac{\pi n}{h}. \qquad (7.92)$$

To better understand the slab modes in the representation (7.89) we first consider 2D photonic-crystal modes with a nonzero value of the off-plane wave-vector component $k_z \neq 0$. In particular, in Figs. 7.15(a) and (b) we plot the dispersion relation of the modes of a 2D photonic crystal with $k_z = \pi/a$. For comparison, in Fig. 7.15(a) the dashed lines present TM modes of a 2D photonic crystal with $k_z = 0$, while in Fig. 7.15(b) the dashed lines present TE modes of a 2D photonic crystal with $k_z = 0$. At the K symmetry point, $k_z = 0$ modes of a 2D photonic crystal (dashed lines) are labeled as TM_m or TE_m, where the higher value of index m corresponds to the higher value of the modal frequency. Note that the origin of all the $k_z \neq 0$ bands of a 2D photonic crystal (solid lines) can be traced back to either TM $k_z = 0$ bands of Fig. 7.15(a), or TE $k_z = 0$ bands of Fig. 7.15(b), which can be further ascertained by considering the corresponding

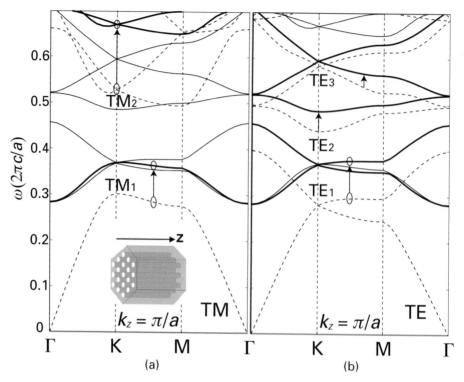

Figure 7.15 Band diagram of the modes of a 2D photonic crystal with $k_z = 2\pi/h$, $h = 2a$ (solid lines). Dashed lines in (a) correspond to the TM-polarized modes of a 2D photonic crystal with $k_z = 0$. Dashed lines in (b) correspond to the TE-polarized modes of a 2D photonic crystal with $k_z = 0$. Arrows indicate the related $k_z \neq 0$ and $k_z = 0$ modes.

field-intensity distributions. An important observation from Fig. 7.15 is that the $k_z \neq 0$ modes of a 2D photonic crystal can no longer be characterized as pure TE or TM, and, therefore, they are, strictly speaking, hybrid modes. Nevertheless, as seen from Fig. 7.15, such hybrid modes can still be thought of as predominantly TE- or TM-like, as their modal dispersion relations and field-intensity distributions can be easily associated with those of the pure TE- or TM-polarized modes with $k_z = 0$. Finally, in the limit when $k_z > |\mathbf{k}_t|$, the separation in frequency between the $k_z \neq 0$ bands and the corresponding $k_z = 0$ bands scales proportionally to k_z as:

$$\omega(\mathbf{k}_t, k_z) = n_{\text{eff}}^{-1}\sqrt{|\mathbf{k}_t|^2 + k_z^2} \underset{k_z > |\mathbf{k}_t|}{\approx} n_{\text{eff}}^{-1} k_z.$$

We conclude this section by demonstrating how dispersion relations of the modes of a photonic-crystal slab can be understood in terms of the dispersion relations of the $k_z = 0$ modes of a 2D photonic crystal. So far, we have established that a particular slab mode can be thought of as a superposition of the two counter-propagating (in the $\hat{\mathbf{z}}$ direction) modes of a 2D photonic crystal having an out-of-plane wave-vector component of the form $k_z^n \sim \pi n/h, n \subset$ integer. In other words, the slab mode along the $\hat{\mathbf{z}}$ direction is a standing wave with $\sim n$ nodes, while in the $\hat{x}\hat{y}$ plane it has a field distribution similar to that of the mode of a 2D photonic crystal having an out-of-plane wave-vector component k_z^n. Furthermore, the dispersion relation and field distribution of a particular 2D photonic crystal mode with $k_z \neq 0$ can be associated with those of a TE- or TM-polarized $k_z = 0$ mode of a 2D photonic crystal. Additionally, the dispersion relation of a 2D photonic-crystal mode with $k_z \neq 0$ can be approximated by shifting the dispersion relation of a corresponding $k_z = 0$ mode of a 2D photonic crystal upwards in frequency by the amount $\sim k_z^n$.

From these observations, it follows that dispersion relation of a slab mode can be put in correspondence with that of either the TM_m or the TE_m polarized $k_z = 0$ mode of a 2D photonic crystal, however shifted upward in frequency by the amount $\sim k_z^n \sim \pi n/h, n \subset$ integer. This fact allows us to introduce a double-index labeling scheme for the modes of a photonic-crystal slab. In particular, every mode of a photonic crystal slab can be labeled as either $\mathrm{TM}_{m,n}$ or $\mathrm{TE}_{m,n}$ depending on whether the mode is TM_m or TE_m-like (judged by the form of a dispersion relation and field distribution in the $\hat{x}\hat{y}$ plane). The index n indicates the order of the standing wave in the $\hat{z}$ direction, or, in other words, the order of the out-of-plane wave-vector component of the two counter-propagating 2D photonic crystal modes forming a standing wave. Moreover, when decreasing the thickness of a photonic-crystal slab, one expects shifting of the dispersion relations of the slab modes towards high frequencies as $\sim \pi n/h$.

In Fig. 7.16 the solid lines show the band diagrams of the even slab modes. For comparison, in Fig. 7.16(a) thick solid lines highlight the TM-like slab modes, while in Fig. 7.16(b) thick solid lines highlight the TE-like slab modes. The first subplots in Figs. 7.16(a) and (b), labeled $h = \infty$, present dispersion relations of the TM_m- and TE_m-polarized $k_z = 0$ modes of a 2D photonic crystal; the same dispersion relations are also shown in the other subplots as dotted lines. Arrows in the various subplots indicate the related modes of a slab and 2D photonic crystal as determined by the form of a dispersion relation, and the similarity of the modal field distributions in the $\hat{x}\hat{y}$ plane. Note that for the $\mathrm{TM}_{m,n}$ and $\mathrm{TE}_{m,n}$ modes with $n \geq 2$, as the slab thickness decreases, the dispersion relations of the slab modes shift towards high frequencies approximately as $\sim 1/h$. As the slab thickness continues to decrease, higher-order modes get pushed out into the radiation continuum of a cladding. To demonstrate this point, compare, for example, the band diagrams of the TM-like modes for the slab thicknesses $h = 2.0a$ and $h = 1.0a$ (see Fig. 7.16(a)). For $h = 2.0a$, there are three TM_1-like modes, the fundamental $\mathrm{TM}_{1,1}$ and two higher-order $\mathrm{TM}_{1,2}$ and $\mathrm{TM}_{1,3}$ modes. The modes were identified as TM_1-like by inspection of the modal field profiles (in the $\hat{x}\hat{y}$ plane) shown in Fig. 7.17. As expected, the number of nodes (along the $\hat{z}$ direction) in the electric field distributions of modes $\mathrm{TM}_{1,n}$ increases with the value of index n. When the slab thickness is reduced to $h = 1.0a$, the $\mathrm{TM}_{1,3}$ mode is pushed completely into the radiation continuum of a cladding, while $\mathrm{TM}_{1,2}$ is pushed upwards towards the cladding light cone.

Finally, we note that the band diagram of even slab modes does not contain the fundamental $\mathrm{TE}_{1,1}$ mode. This is because the $\mathrm{TE}_{1,1}$ mode is directly analogous to the TE_1-polarized $k_z = 0$ mode of a 2D photonic crystal; in the $z = 0$ plane, the field distribution of the $\mathrm{TE}_{1,1}$ mode is TE-like, which is incompatible with symmetry (7.86) of

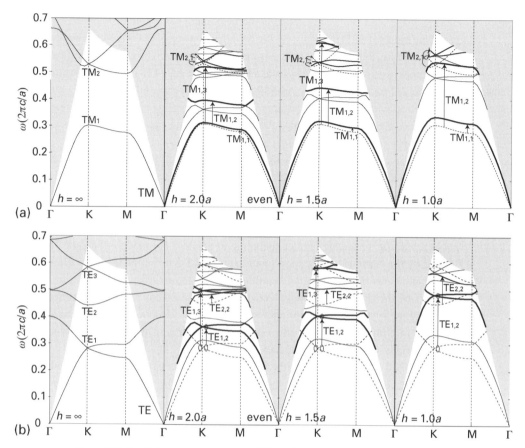

Figure 7.16 Band diagrams of the even slab modes for various thicknesses of a photonic-crystal slab. (a) Thick solid lines highlight the TM-like slab modes, while dotted lines show the TM-polarized $k_z = 0$ modes of a 2D photonic crystal. (b) Thick solid lines highlight the TE-like slab modes, while dotted lines show the TE-polarized $k_z = 0$ modes of a 2D photonic crystal. In all the subplots, arrows indicate the related modes of a slab and a 2D photonic crystal.

even slab modes. There is, however, no contradiction in the fact that $TE_{m,n}$ modes with $n \geq 2$ are present in the band diagram of even modes as $TE_{m,n}$ modes are analogous to the TE_m, $k_z \neq 0$ modes of a 2D photonic crystal, which are not pure TE modes (hybrid modes). In a similar manner, the band diagram of the odd modes does not contain the fundamental $TM_{1,1}$ mode, while it can contain the $TM_{m,n}$ modes with $n \geq 2$.

7.3.4   Modes of a photonic-crystal slab waveguide

Photonic-crystal slab waveguides are line defects in otherwise perfectly periodic photonic-crystal slabs. Many of the modal properties of photonic-crystal slab waveg-uides can be rationalized in a manner similar to that used in understanding modal properties of 2D photonic-crystal waveguides considered in Section 6.6. Slab waveg-uides exhibit one-dimensional discrete translational symmetry along a certain direction

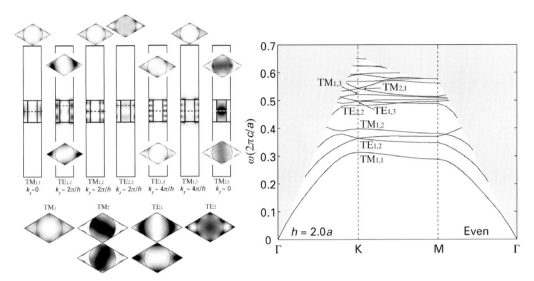

Figure 7.17 Band diagram of photonic crystal slab modes (even), and modal electric field intensity distributions at the K symmetry point. For comparison, modal field intensity distributions of the TE- and TM-polarized $k_z = 0$ modes of a 2D photonic crystal are also presented.

confined to the plane of a slab. A particular implementation of a photonic-crystal slab waveguide is presented in Fig. 7.18. There, the waveguide is formed by a single row of smaller diameter holes of radius r_d aligned along the $\hat{\mathbf{x}}$ direction of a hexagonal photonic-crystal slab defined previously in Fig. 7.13. We assume that the dielectric index of the waveguide holes is the same as that of a cladding. In Fig. 7.18(a) several unit cells in the $\hat{\mathbf{x}}\hat{\mathbf{y}}$ plane of a photonic-crystal slab waveguide are shown. The waveguide vertical cross-section (z plane) along the longer diagonal of a single lattice cell (dotted line in Fig. 7.18(a)) is shown in Fig. 7.18(b). The basis vector defining waveguide periodicity is $\bar{a}_1 = a \cdot (1, 0, 0)$, while the corresponding reciprocal lattice vector is $\bar{b}_1 = 2\pi/a \cdot (1, 0, 0)$.

As in the case of photonic crystal slabs, a standard way of computing eigenmodes of a photonic-crystal slab waveguide is by using the plane-wave expansion method in the supercell approximation. For a slab waveguide the only discrete translational symmetry remaining is the one along the direction of the waveguide (the $\hat{\mathbf{x}}$ direction in Fig. 7.18(a)). Discrete translational symmetry in the other two directions is broken, and a supercell approximation is used along these directions. The modes most suitable for being computed by the supercell approximation have to be strongly localized along the $\hat{\mathbf{y}}$ and $\hat{\mathbf{z}}$ directions. Generally, field localization for such modes in $\hat{\mathbf{z}}$ direction is caused by the total internal reflection on the boundary between the high-refractive-index slab and the low-refractive-index cladding. Localization in the $\hat{\mathbf{y}}$ direction (in the vicinity of a waveguide core) typically results from the bandgap of a photonic-crystal slab. The supercell used in a simulation of such modes (dark gray in Fig. 7.18(a)) typically consists of a single lattice cell in the $\hat{\mathbf{x}}$ direction, and multiple cells in the $\hat{\mathbf{y}}$ direction to

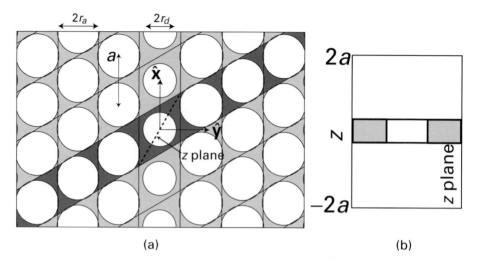

Figure 7.18 Example of a photonic-crystal slab waveguide. (a) The slab waveguide is formed by a row of smaller holes aligned along the $\hat{\mathbf{x}}$ direction in a hexagonal lattice of larger diameter holes. (b) In the vertical direction (z plane), the photonic-crystal slab has a finite thickness. The dielectric constant of the cladding is the same as the one of the holes.

ensure considerable modal field decay towards the horizontal cell boundaries. Similarly, the supercell size in the $\hat{\mathbf{z}}$ direction is chosen to ensure considerable modal field decay towards the vertical cell boundaries (Fig. 7.18(b)). With this choice of the supercell geometry, the eigenmodes of an artificially periodic three-dimensional supercell structure approximate the eigenmodes of a photonic crystal slab waveguide well. From Section 2.4.5 (the subsection on 3D discrete translational symmetry) it follows that eigenmodes of a three-dimensional system exhibiting discrete translational symmetries in all directions have the form (2.128) and can be labeled by a conserved wave vector $\mathbf{k}$ confined to a three-dimensional first Brillouin zone. However, the k_y and k_z components of a wave vector are confined to the intervals $-\pi/(aN_y) < k_y < \pi/(aN_y)$, $-\pi/(aN_z) < k_z < \pi/(aN_z)$, and, therefore, tend to zero when supercell sizes along $\hat{\mathbf{y}}$ (aN_y) and $\hat{\mathbf{z}}$ (aN_z) directions increase. Thus, in practical supercell simulations of a photonic-crystal slab waveguide shown in Fig. 7.18, the modal wave vector is typically chosen in the form $\mathbf{k} = (k_x, 0, 0)$, where $\mathbf{k}$ is then confined to a one-dimensional first Brillouin zone $-\pi/a < k_x < \pi/a$.

Figure 7.19 shows a band diagram of the modes of a photonic-crystal slab waveguide with the following parameters: $h = 0.6a$, $r_d = 0.35a$, $\varepsilon_b = \sqrt{12}$, $r_a = 0.45a$, $\varepsilon_a = 1$. Owing to time-reversal symmetry, the dispersion relations of the waveguide modes are presented only in the positive half of the first Brillouin zone $0 < k_x < \pi/a$. The waveguide modes were computed using MPB implementation of the plane-wave method [14] using a supercell with seven periods in $\hat{\mathbf{y}}$ direction and size $4a$ in $\hat{\mathbf{z}}$ direction. The light gray shading presents a light cone of the cladding radiation continuum described as $\omega \geq k_x/\sqrt{\varepsilon_a}$. The modes located inside the light cone are delocalized in the cladding, and, therefore, are not confined to the slab. Thus, truly guided slab waveguide modes are

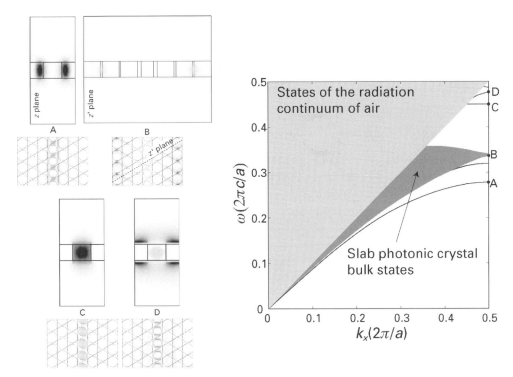

Figure 7.19 Dispersion relations and field distribution profiles of the modes of a photonic crystal slab waveguide of Fig. 7.18. States of the radiation continuum of air cladding are shown in light gray. States bound to the slab and delocalized across the photonic-crystal slab are shown in dark gray.

found only below the cladding light cone. Another interesting feature of the band diagram in Fig. 7.19 is the presence of a continuum of photonic crystal slab modes shown as dark gray. The modes of the slab continuum, although confined to the slab in $\hat{z}$ direction, are, however, delocalized throughout the photonic crystal cladding in the $\hat{x}\hat{y}$ plane. The field distribution in one of such modes is presented in subplot (B). Modes (C) and (D) are the bandgap-guided waveguide modes with most of their electric field concentrated in the air holes. Mode (D) is more strongly delocalized in the $\hat{z}$ direction than mode (C), as it is located closer to the light line of the cladding. Finally, mode (A) is concentrated mostly in the thicker veins of a slab material in the vicinity of a waveguide core, and, therefore, is confined by the total internal reflection in both the $\hat{y}$ and $\hat{z}$ directions.

7.4 Problems

7.1 Eigenvalues of a Helmholtz equation

By truncating the infinite system (7.46) to a finite one and turning it into a matrix eigenvalue problem, calculate the eigenvalues of (7.39) for $k_1 = k_2 = 0$, $E_0 = 15$, $I_0 = 6$, and compare them with those in Fig. 7.5 at the Γ point.

7.2   Orthogonality relation between Bloch modes

Prove the orthogonality relation (7.50) between Bloch modes:

$$\int_{-\infty}^{\infty}\int_{-\infty}^{\infty} B_m^*(x,y;k_1,k_2)B_n(x,y;\hat{k}_1,\hat{k}_2)dxdy = (2\pi)^2\delta(\hat{k}_1-k_1)\delta(\hat{k}_2-k_2)\delta_{m,n},$$

(7.93)

assuming normalization (7.51) when $m = n$.

(a) Rewrite the left-hand side of (7.50) as:

$$\sum_{j=-\infty}^{\infty} e^{i(\hat{k}_1-k_1)\pi j} \sum_{j=-\infty}^{\infty} e^{i(\hat{k}_2-k_2)\pi j} \int_0^\pi dx \int_0^\pi dy\, G_m^*(x,y;k_1,k_2)G_n(x,y;\hat{k}_1,\hat{k}_2).$$

(7.94)

(b) Demonstrate that:

$$\sum_{j_1=-N}^{+N} e^{i(\hat{k}-k)j\pi} = \frac{\sin\left[(\hat{k}-k)\pi(N+1/2)\right]}{\sin\left[(\hat{k}-k)\pi/2\right]}.$$

(7.95)

(c) By taking the limit $N\to\infty$ in the above formula and using the well-known result $\lim_{N\to\infty}\sin(\alpha(N+1/2))/\sin(\alpha/2)=2\pi\,\delta(\alpha)$ show that:

$$\sum_{j=-\infty}^{\infty} e^{i(\hat{k}-k)\pi j} = 2\delta(k-\hat{k}).$$

(7.96)

(d) Finally, show the orthogonality relation (7.50) using the above results and the fact that the Bloch modes in question are solutions of a Hermitian eigenvalue problem.

7.3   Computation of a defect mode localized at a strong defect in a photonic lattice

Implement numerically the iteration method (7.69)–(7.71) for computing the defect mode in Fig. 7.8(a).

References

[1] A. Argyros, T. A. Birks, S. G. Leon-Saval, *et al.* Photonic bandgap with an index step of one percent, *Opt. Express* **13** (2005), 309–314.
[2] J. W. Fleischer, M. Segev, N. K. Efremidis, and D. N. Christodoulides. Observation of two-dimensional discrete solitons in optically induced nonlinear photonic lattices, *Nature* **422** (2003), 147–150.
[3] Z. Chen and J. Yang. Optically-induced reconfigurable photonic lattices for linear and nonlinear control of light. In *Nonlinear Optics and Applications* (Kerala, India: Research Signpost, 2007).
[4] A. Szameit, J. Burghoff, T. Pertsch, *et al.* Two-dimensional soliton in cubic fs laser written waveguide arrays in fused silica, *Opt. Express* **14** (2006), 6055–6062.

[5] H. Eisenberg, Y. Silberberg, R. Morandotti, A. Boyd, and J. Aitchison. Discrete spatial optical solitons in waveguide arrays, *Phys. Rev. Lett.* **81** (1998), 3383–3386.

[6] R. Iwanow, R. Schieck, G. Stegeman, *et al.* Observation of discrete quadratic solitons, *Phys. Rev. Lett.* **93** (2004), 113902.

[7] A. Yariv. *Quantum Electronics*, 3rd edition (New York: Wiley, 1989).

[8] N. K. Efremidis, S. Sears, and D. N. Christodoulides. Discrete solitons in photorefractive optically induced photonic lattices, *Phys. Rev. E* **66** (2002), 046602.

[9] I. Makasyuk, Z. Chen, and J. Yang. Band-gap guidance in optically induced photonic lattices with a negative defect, *Phys. Rev. Lett.* **96** (2006), 223903.

[10] F. Fedele, J. Yang, and Z. Chen. Properties of defect modes in one-dimensional optically induced photonic lattices, *Stud. Appl. Math.* **115** (2005), 279.

[11] J. Yang and T. I. Lakoba. Universally-convergent squared-operator iteration methods for solitary waves in general nonlinear wave equations, *Stud. Appl. Math.* **118** (2007), 153–197.

[12] J. Wang, J. Yang, and Z. Chen. Two-dimensional defect modes in optically induced photonic lattices, *Phys. Rev. A* **76** (2007), 013828.

[13] S. G. Johnson and J. D. Joannopoulos. *Photonic Crystals: The Road from Theory to Practice* (Boston, MA: Springer, 2002).

[14] S. G. Johnson and J. D. Joannopoulos. Block-iterative frequency-domain methods for Maxwell's equations in a planewave basis, *Opt. Express* **8** (2001), 173–190.

8 Nonlinear effects and gap–soliton formation in periodic media

In the previous chapter we have demonstrated that defects in the otherwise periodic photonic-crystal lattice can localize light. In this chapter we will show that if material of a photonic crystal is nonlinear, light localization can be achieved even without any structural or material defects via the mechanism of self-localization.

From Section 7.2.1, we have established that if the refractive index changes little over a length scale of one optical wavelength, then under the slowly varying amplitude approximation, the propagation of a linearly polarized light beam can be described by the following Schrödinger equation:

$$i\frac{\partial U}{\partial z} + \frac{1}{2k}\nabla_\perp^2 U + \frac{k\Delta n}{n_b}U = 0. \tag{8.1}$$

Here U is the envelope function of the light beam, n_b is the constant background refractive index of the medium, $\nabla_\perp^2 = \partial^2/\partial x^2 + \partial^2/\partial y^2$ is the transverse Laplace operator, $k = \omega n_b/c$, ω is the frequency of light, and Δn is the index variation of the medium.

In photorefractive crystals, the nonlinear response of the medium to light is of saturable type, [1] i.e.:

$$\Delta n \propto -\frac{E_0}{1 + I_L + I}, \tag{8.2}$$

where I_L is the intensity function of the optically induced periodic lattice in the medium, $I = |U|^2$ is the intensity of light, and E_0 is the applied dc field. When E_0 is positive, the nonlinear response is of the focusing type, i.e., the refractive index is larger in higher light-intensity areas. When E_0 is negative, the nonlinear response is of the defocusing type.

In a Kerr nonlinear medium, the index variation Δn can be written as:

$$\Delta n = -V(x, y) + n_2 I, \tag{8.3}$$

where $-V$ is the index variation of the medium, $I = |U|^2$ is the beam intensity, and n_2 is the Kerr coefficient. When $n_2 > 0$, which is the case for most optical materials, the nonlinearity is of the focusing type, while when $n_2 < 0$, the nonlinearity is of the defocusing type. Physical examples of this Kerr model include PCF fibers with weak index variation [2] and laser-written waveguides, [3] and planar-etched waveguide arrays on AlGaAs substrates (see also Section 7.3) [4,5]. Substituting (8.3) into (8.1) and introducing some variable normalizations, we can get the dimensionless guiding equation for light propagation in a Kerr nonlinear medium with medium-index variations:

$$i\frac{\partial U}{\partial z} + \nabla_\perp^2 U - VU + \sigma|U|^2 U = 0, \tag{8.4}$$

where $\sigma = \text{sgn}(n_2)$ represents the sign of nonlinearity.

It is interesting to note that in the mean-field approximation, the nonlinear Gross–Pitaevskii equation describing the dynamics of a Bose–Einstein condensate loaded into an optical lattice is equivalent to the above nonlinear Schrödinger (NLS) equation (8.4). [6,7] In the Bose–Einstein condensate, U represents the macroscopic condensate wave-function, the periodic potential V is formed by an optical lattice, z is time, and $\sigma = \pm 1$ represents the sign of the s-wave scattering length of the atoms in the condensate. Therefore, the theoretical results for (8.4) in optics are also applicable to Bose–Einstein condensates.

8.1 Solitons bifurcated from Bloch bands in 1D periodic media

In the 1D case, the NLS equation (8.4) describing light propagation in a Kerr nonlinear medium with a periodic index variation is:

$$iU_z + U_{xx} - V(x)U + \sigma|U|^2 U = 0, \tag{8.5}$$

where $\sigma = \pm 1$. For simplicity, we take the index variation to be:

$$V(x) = V_0 \sin^2(x). \tag{8.6}$$

Soliton solutions of (8.5) are sought in the form:

$$U(x;z) = u(x)e^{-i\mu z}, \tag{8.7}$$

where μ is the propagation constant, and the amplitude function $u(x)$ satisfies the equation:

$$u_{xx} - V(x)u + \mu u + \sigma|u|^2 u = 0. \tag{8.8}$$

8.1.1 Bloch bands and bandgaps

When the function $u(x)$ is small, (8.8) becomes a linear equation:

$$u_{xx} - V(x)u + \mu u = 0. \tag{8.9}$$

Solutions of this linear equation are Bloch modes, labeled by the corresponding propagation constant μ. This 1D equation is equivalent to the Mathieu equation, and we seek its solution in the form:

$$p(x;\mu(k)) = e^{ikx}\tilde{p}(x;\mu(k)), \tag{8.10}$$

where $\tilde{p}(x;\mu(k))$ is periodic with the same period π as the potential $V(x)$, and $\mu(k)$ is the diffraction relation. The diffraction diagram for (8.6), (8.9) is shown in Fig. 8.1(a)

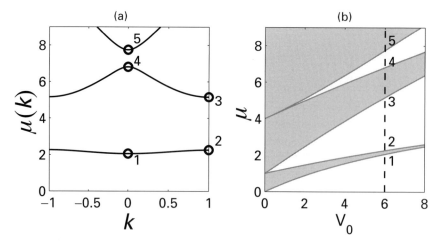

Figure 8.1 (a) Diffraction curves of the 1D equation (8.9) with $V_0 = 6$; (b) Projected band diagram of the Bloch bands (shaded) and bandgaps at various values of potential levels V_0. Circled points marked by the numbers 1–5 in plots (a) and (b) lie at the edges of the Bloch bands. Bloch modes at the edges 1–4 are displayed in Fig. 8.2.

for $V_0 = 6$. The projected bandgap structure of the 1D equation (8.9) at various values of V_0 is shown in Fig. 8.1(b). Notice that in a 1D case, bandgaps open for any nonzero value of V_0; in addition, the number of bandgaps is infinite. Modal distributions $p(x; \mu_n)$ of the Bloch modes at the edges of the first two Bloch bands (points $1 \leq n \leq 4$ in Fig. 8.1(a)) are displayed in Fig. 8.2. Notice that these Bloch waves are real-valued.

The above Bloch solutions are the solutions of nonlinear equation (8.8) in the approximation of infinitesimal amplitudes. When the amplitudes of these solutions increase, these Bloch solutions may self-localize and form solitons. The propagation constant μ describing such solutions would then move from the band edges into the bandgaps. In the following section, we employ a multiscale perturbation method to analyze how solitons bifurcate from the Bloch solutions at the band edges.

8.1.2   Envelope equations of Bloch modes

In this section, we develop an asymptotic theory to analyze small-amplitude solitons bifurcating from the Bloch waves near the band edges, and derive their envelope equations. This analysis is similar to that developed in [8].

Let us consider point μ_0 at the band edge, with the corresponding Bloch mode $p(x; \mu_0)$ in the form (8.10). As before, $L = \pi$ is the period of the potential function $V(x)$. Notice that $p(x; \mu_0)$ is periodic in x with period L or $2L$ as the band-edge point μ_0 corresponds to either $k = 0$ or $k = 1$ (see Fig. 8.1). Thus:

$$p(x + L) = \pm p(x); \quad p(x + 2L) = p(x), \qquad (8.11)$$

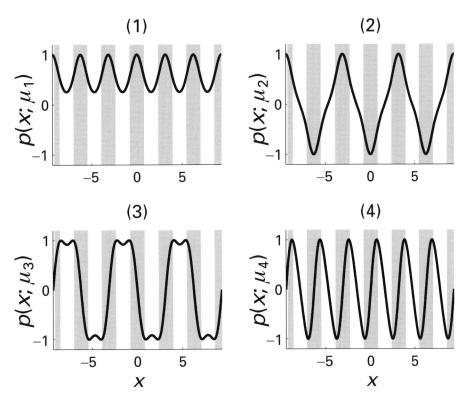

Figure 8.2 One-dimensional Bloch wave solutions of (8.9) (with $V_0 = 6$) at the four edges of the Bloch bands marked from 1 to 4 in Fig. 8.1(b). Shaded regions represent the lattice sites (regions of low values of $V(x)$ potential).

When solution $u(x)$ of (8.8) is infinitesimally small, the square term in the equation can be neglected and $u(x) \sim p(x)$. When $u(x)$ is small but not infinitesimal, we can expand it into a multiscale perturbation series:

$$u = \varepsilon u_0 + \varepsilon^2 u_1 + \varepsilon^3 u_2 + \cdots, \tag{8.12}$$

$$\mu = \mu_0 + \eta \varepsilon^2, \tag{8.13}$$

where

$$u_0 = A(X)p(x), \tag{8.14}$$

$\eta = \pm 1$, and $X = \varepsilon x$ is the slow spatial variable of the envelope function $A(X)$. Substituting the above expansions into (8.8), to the order of $O(\varepsilon)$ the equation is automatically satisfied. To the order of $O(\varepsilon^2)$, the equation for u_1 is:

$$u_{1xx} - V(x)u_1 + \mu_0 u_1 = -2\frac{\partial^2 u_0}{\partial x \partial X}. \tag{8.15}$$

The corresponding homogeneous equation to (8.15) is similar to (8.9), and, therefore, has a periodic solution $p(x)$. For the inhomogeneous equation (8.15) to admit a solution

u_1 periodic with respect to its fast variable x, the following Fredholm condition must be satisfied (see Problem 8.1):

$$\int_0^{2L} p(x)\frac{\partial^2 u_0}{\partial x \partial X}dx = 0.$$ (8.16)

Here, the integration length is $2L$ rather than L since the homogeneous solution $p(x)$ may be periodic with a period $2L$ according to (8.11).

Recalling the form (8.14) of a u_0 solution, it is easy to check that the Fredholm condition (8.16) is satisfied automatically. We now look for the solution of (8.15) in the form:

$$u_1 = \frac{dA}{dX}v(x),$$ (8.17)

where $v(x)$ is a periodic solution of the equation:

$$v_{xx} - V(x)v + \mu_0 v = -2p_x.$$ (8.18)

At $O(\varepsilon^3)$, the equation for u_2 is:

$$u_{2xx} + u_{2yy} - V(x)u_2 + \mu_0 u_2 = -\left(2\frac{\partial^2 u_1}{\partial x \partial X} + \frac{d^2 u_0}{dX^2} + \eta u_0 + |u_0|^2 u_0\right).$$ (8.19)

Substituting the formulas (8.14) and (8.17) for u_0 and u_1 into this equation, we get:

$$-\{u_{2xx} - V(x)u_2 + \mu_0 u_2\}$$
$$= \frac{d^2 A}{dX^2}[2v'(x) + p(x)] + p^3(x)|A|^2 A + \eta Ap(x).$$ (8.20)

Before applying the Fredholm condition to this inhomogeneous equation, we note the following identity (see Problem 8.2):

$$\int_0^{2L} [2v'(x) + p(x)]p(x)dx = D\int_0^{2L} p^2(x)dx,$$ (8.21)

where

$$D \equiv \frac{1}{2}\frac{d^2\mu}{dk^2}\bigg|_{\mu=\mu_0}.$$ (8.22)

Identity (8.21) can be verified by expanding the solution (8.10) of (8.9) into the power series of $k - k_0$ around the edge of the Bloch band $\mu = \mu_0(k = k_0)$, pursuing the expansion to the second order in $k - k_0$, and utilizing the Fredholm condition. Using this identity and (8.11), the Fredholm condition for (8.20) leads to the following NLS equation for the envelope function A:

$$D\frac{d^2 A}{dX^2} + \eta A + \sigma\alpha |A|^2 A = 0,$$ (8.23)

where

$$\alpha = \frac{\displaystyle\int_0^{2L} p^4(x)dx}{\displaystyle\int_0^{2L} p^2(x)dx} > 0. \tag{8.24}$$

Localized solutions for the NLS equation (8.23) exist only when $\mathrm{sgn}(D) = \mathrm{sgn}(\sigma)$, which means that lattice solitons exist under focusing nonlinearity ($\sigma = 1$) near band edges with positive diffraction coefficients D, and they can also exist under defocusing nonlinearity ($\sigma = -1$) near band edges with negative diffraction coefficients. The existence of solitons under a defocusing nonlinearity is a distinctive phenomenon of periodic media which cannot occur in homogeneous (bulk) media. When $\mathrm{sgn}(D) = \mathrm{sgn}(\sigma)$, soliton solutions require $\mathrm{sgn}(\sigma) = -\mathrm{sgn}(D)$, which simply means that the propagation constant of the soliton lies inside the bandgaps of (8.9), as one would normally expect. Under these conditions, the soliton solution of (8.23) is

$$A(X) = \sqrt{2|\alpha|}\,\mathrm{sech}\,\frac{X}{\sqrt{|D|}}. \tag{8.25}$$

8.1.3 Locations of envelope solitons

The envelope equation (8.23) is translation-invariant. In particular, any spatial translation $A(X - X_0)$ of a solution $A(X)$ would still be a solution of (8.23) for any constant X_0. However, only when X_0 takes some special values can the perturbation series (8.12) truly satisfy the original equation (8.8). The reason is that X_0 must satisfy a certain constraint. This constraint is exponentially small in ε, thus it can not be captured in the power series expansions of (8.12) and needs to be calculated separately.

First we derive this constraint for the envelope solution. Multiplying (8.8) by the complex conjugate of u_x, adding its conjugate equation, and integrating from $-\infty$ to $+\infty$, we get the following condition:

$$\int_{-\infty}^{\infty} V'(x)|u(x)|^2\,dx = 0. \tag{8.26}$$

Substituting the perturbation expansion (8.12) of solution $u(x)$ into the above equation, this condition to the leading order of ε becomes:

$$I(x_0) = \varepsilon^2 \int_{-\infty}^{\infty} V'(x)p^2(x)|A(X - X_0)|^2\,dx = 0, \tag{8.27}$$

where $X_0 = \varepsilon x_0$ is the center position of the envelope solution A. Since $V'(x)$ is antisymmetric and $p^2(x)$ is symmetric, and both are periodic with period L, the function $V'(x)p^2(x)$ has the following Fourier series expansion:

$$V'(x)p^2(x) = \sum_{m=1}^{\infty} c_m \sin(2\pi mx/L). \tag{8.28}$$

When the above Fourier expansion is substituted into (8.27), every Fourier mode in this expansion creates an exponentially small term in ε, and the exponential rate of decay of these terms is larger for higher values of index m. Keeping only the leading-order term obtained from the Fourier mode with $m= 1$, (8.27) is approximated as:

$$I(x_0) = \varepsilon^2 \int_{-\infty}^{\infty} c_1 \, |A(X - X_0)|^2 \sin(2\pi x/L)dx = 0. \tag{8.29}$$

Recalling the form of a solution (8.25) for $A(X)$, the above constraint can be simplified as:

$$I(x_0) = W_1 \sin(2\pi x_0/L) = 0, \tag{8.30}$$

where

$$W_1 \equiv \varepsilon^2 \int_{-\infty}^{\infty} c_1 \, |A(X)|^2 \cos(2\pi x/L)dx \neq 0. \tag{8.31}$$

Notice that W_1 is exponentially small in ε, thus the constraint (8.30) is also exponentially small. For (8.30) to hold, one must have

$$\sin(2\pi x_0/L) = 0. \tag{8.32}$$

Thus, the envelope solution A can only be centered at two locations:

$$x_0 = 0, \text{ or } L/2. \tag{8.33}$$

In the case of $x_0 = 0$, the resulting lattice soliton has its peak amplitude at a lattice site (minimum of potential $V(x)$), and is called an on-site lattice soliton. In the case of $x_0 = L/2$, the resulting lattice soliton has its peak amplitude in between two lattice sites (maximum of potential $V(x)$), and is called an off-site lattice soliton.

8.1.4   Soliton families bifurcated from band edges

The above asymptotic analysis predicts two families of lattice solitons bifurcating from each Bloch-band edge, and it also gives leading-order expressions for these solitons when their amplitudes are small (the weakly nonlinear case). As the soliton amplitudes increase (becoming strongly nonlinear), this asymptotic analysis starts to break down. In such cases, numerical methods are needed to determine the true profiles of soliton solutions. In this section, we numerically determine soliton families bifurcated from the edge points of Bloch bands. The purpose is not only to confirm the asymptotic theory, but also to obtain lattice soliton profiles when their amplitudes become large (strongly localized). The numerical method we used is a modified squared-operator iteration method developed in [9].

Numerically, we indeed find two families of lattice solitons (on-site and off-site, respectively) bifurcating from each band edge. The bifurcations of the two soliton families from the edge points μ_1 and μ_2 of the first Bloch band are displayed in Figs. 8.3 and 8.4. Figure 8.3 shows soliton power curves with the power defined as $P = \int_{-\infty}^{\infty} |u(x)|^2 dx$, and Fig. 8.4 presents typical soliton profiles at points near and far away from the band edges. From Fig. 8.3, we see that the two power curves bifurcate from each band edge, with the lower curve for on-site solitons, and the upper curve for off-site solitons. For the same value of propagation constant, off-site solitons have higher powers than on-site

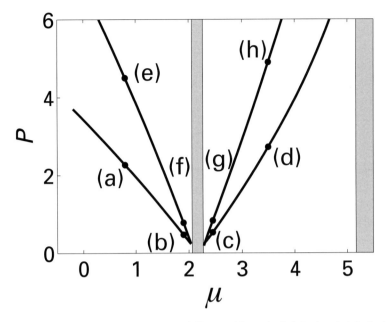

Figure 8.3 Power diagrams of solitons bifurcated from the left (μ_1) and right (μ_2) edges of the first Bloch band under focusing and defocusing nonlinearities, respectively. Upper curves: off-site solitons; lower curves: on-site solitons. Soliton profiles at the marked points are shown in Fig. 8.4.

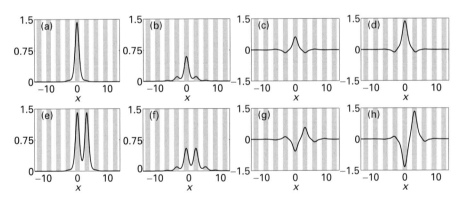

Figure 8.4 Soliton profiles at the letter-marked points in Fig. 8.3(a). (a, b, e, f) Solitons in the semi-infinite gap under focusing nonlinearity. (c, d, g, h) Solitons in the first gap under defocusing nonlinearity. The vertical gray stripes represent lattice sites. Upper row: on-site solitons; lower row: off-site solitons.

ones. It is important to note that the two soliton families in the semi-infinite gap exist under focusing nonlinearity ($\sigma = 1$), and the other two soliton families in the first gap exist under defocusing nonlinearity ($\sigma = -1$). These facts are consistent with the theoretical results derived above, in view that the dispersion coefficient D is positive at edge point μ_1 and negative at edge point μ_2.

Now we examine soliton profiles in these solution families. Inside the semi-infinite gap (under a focusing nonlinearity), solitons are all positive, which can be seen in Fig. 8.4(a), (b), (e), and (f). On-site solitons (see Fig. 8.4(a, b)) have a dominant hump residing at a lattice site, while off-site solitons (see Fig. 8.4(e, f)) have two dominant humps residing at two adjacent lattice sites. Near the band edge μ_1, these solitons develop in-phase tails on their sides, and their amplitudes and power decrease. When $\mu \to \mu_1$, these solitons approach the Bloch wave of Fig. 8.2(a) with infinitesimal amplitude. On-site solitons have been observed in etched-waveguide arrays. [4] In the first bandgap (under defocusing nonlinearity), adjacent peaks of solitons are out of phase. On-site solitons (see Fig. 8.4(c, d)) have a dominant hump at a lattice site, flanked by out-of-phase tails. Off-site solitons (see Fig. 8.4(g, h)) have two out-of-phase dominant humps at two adjacent lattice sites. The out-of-phase structure of neighboring peaks in these gap solitons originates from that of the Bloch wave at μ_2 (see Fig. 8.2(b)) from where these gap solitons bifurcate. Solitons bifurcated from higher bands have more complex spatial structures, and will not be shown here.

Regarding the linear stability of these lattice solitons, it has been shown in [8] that near the edges of Bloch bands, on-site solitons are linearly stable, while off-site solitons are linearly unstable (with unstable eigenvalues being real). Away from band edges, additional instabilities (with complex eigenvalues) can also arise.

8.2   Solitons bifurcated from Bloch bands in 2D periodic media

In this section, we study lattice solitons in two spatial dimensions. Many of the results for the 1D case above can be carried over to the 2D case. In such cases, our description will be brief. However, many new phenomena also arise in 2D, such as a much wider array of lattice-soliton structures, which often have no counterparts in 1D. Such results will be described in more detail.

To derive 2D soliton solutions, we first rewrite the dimensionless guiding equation (8.4) for light propagation in 2D periodic media as:

$$iU_z + U_{xx} + U_{yy} - V(x, y)U + \sigma|U|^2U = 0. \tag{8.34}$$

Here $V(x, y)$ is the periodic lattice potential, and $\sigma = \pm 1$ is the sign of the Kerr nonlinearity. For simplicity, the 2D lattice potential in (8.34) is taken as:

$$V(x, y) = V_0(\sin^2 x + \sin^2 y), \tag{8.35}$$

which is analogous to a 1D lattice potential of (8.6). This potential is separable, which makes the theoretical analysis easier. Similar analysis can be repeated for other types of periodic potentials and nonlinearities with minimal changes. Soliton solutions of (8.34) are sought in the form:

$$U(x, y; z) = u(x, y)e^{-i\mu z}, \tag{8.36}$$

where the amplitude function $u(x, y)$ satisfies the following equation:

$$u_{xx} + u_{yy} - [F(x) + F(y)]u + \mu u + \sigma |u|^2 u = 0, \tag{8.37}$$

$$F(x) = V_0 \sin^2 x, \tag{8.38}$$

and μ is a propagation constant.

The focus of the remaining presentation is to determine solitons of (8.37). To do so, we first need to understand Bloch bands and bandgaps of this 2D equation.

8.2.1   Two-dimensional Bloch bands and bandgaps of linear periodic systems

When function $u(x, y)$ is infinitesimal, (8.37) becomes a linear equation:

$$u_{xx} + u_{yy} - [F(x) + F(y)]u + \mu u = 0. \tag{8.39}$$

Since the periodic potential in this equation is separable, its Bloch solutions and Bloch bands can be constructed from solutions of a 1D equation. Specifically, the 2D Bloch solution $u(x, y)$ of (8.39) and its propagation constant μ can be presented in a separable form:

$$u(x, y) = p(x; \mu_a)p(y; \mu_b), \quad \mu = \mu_a + \mu_b, \tag{8.40}$$

where $p(x; \mu)$ is a solution of the 1D equation (8.9).

Using the 1D results of Section 8.1 and (8.40) connecting the 1D and 2D Bloch solutions, we can construct the diffraction surfaces and bandgap structures of the 2D problem (8.39). In particular, the 2D Bloch-mode solutions are of the form:

$$u(x, y) = e^{ik_x x + ik_y y} \tilde{p}[x; \mu(k_x)]\tilde{p}[y; \mu(k_y)], \tag{8.41}$$

where

$$\mu = \mu(k_x) + \mu(k_y), \tag{8.42}$$

is the 2D diffraction relation over the first Brillouin zone $-1 \le k_x, k_y \le 1$.

This diffraction relation for $V_0 = 6$ is shown in Fig. 8.5(a). This figure contains a number of diffraction surfaces whose μ values make up the Bloch bands. Between these surfaces, two bandgaps can be seen. At other V_0 values, the 2D bandgap structure is summarized in Fig. 8.5(b). This figure reveals that, unlike in the 1D case, there is only a finite number of bandgaps in the 2D problem at a given V_0 value. In addition, bandgaps appear only when V_0 is above a certain threshold. As the lattice potential V_0 increases, so does the number of bandgaps. At $V_0 = 6$ one observes two bandgaps which are also clearly visible in Fig. 8.5(a). The edges of Bloch bands at this V_0 value are marked in Fig. 8.5(b) as A, B, C, D, E, respectively.

The positions of the band edges within the first Brillouin zone are important, as they determine the symmetry properties of the corresponding Bloch modes. In the first Brillouin zone, band edges A, B in Fig. 8.5(b) are located at the Γ and M points respectively, where only a single Bloch wave exists. The band edges C and D, however, are located at X and X′ points, where two linearly independent Bloch solutions exist. The

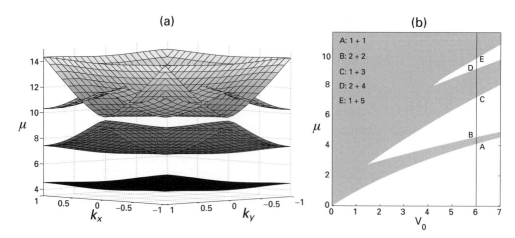

Figure 8.5 (a) Diffraction surfaces of the 2D equation (8.39) with $V_0 = 6$. (b) The 2D bandgap structure for various values of V_0. A, B, C, D, E mark the edges of Bloch bands at $V_0 = 6$. The labels are explained in the text.

Bloch modes at these edges are $u(x, y) = p(x; \mu_1)p(y; \mu_1)$ for A, $p(x; \mu_2)p(y; \mu_2)$ for B, $p(x; \mu_1)p(y; \mu_3)$ and $p(y; \mu_1)p(x; \mu_3)$ for C, $p(x; \mu_2)p(y; \mu_4)$ and $p(y; \mu_2)p(x; \mu_4)$ for D, where functions $p(x; \mu_k)$ are shown in Fig. 8.2. For convenience, we denote point 'A' as '$1 + 1$', 'B' as '$2 + 2$', 'C' as '$1 + 3$', and 'D' as '$2 + 4$'. The Bloch modes at 'A, B' are very similar to those in Fig. 7.5(A, B). These solutions have the symmetry of $u(x, y) = u(y, x)$. Adjacent peaks in the Bloch mode of 'A' are in-phase, while those of 'B' are out-of-phase. The Bloch modes $p(y; \mu_1)p(x; \mu_3)$ and $p(x; \mu_2)p(y; \mu_4)$ at 'C, D' are very similar to those in Fig. 7.6(C, D). These modes do not have the symmetry of $u(x, y) = u(y, x)$. The degenerate modes $p(y; \mu_1)p(x; \mu_3)$ and $p(x; \mu_2)p(y; \mu_4)$ at these points are a 90° rotation of those shown in Fig. 7.6(C, D). The presence of several linearly independent Bloch modes at a band edge is a new feature in two spatial dimensions, and it has important implications for soliton bifurcations as established in the next section.

8.2.2   Envelope equations of 2D Bloch modes

In this section, we study bifurcations of small-amplitude soliton packets from the edges of 2D Bloch bands, and derive their envelope equations. The main difference between the 2D and 1D cases here is that, because there can be two linearly independent Bloch modes at a 2D band edge, solitons can bifurcate from a linear combination of them. This leads to coupled envelope equations for the two Bloch modes and a wide variety of soliton structures that have no counterparts in 1D.

Let us consider a doubly degenerate Bloch mode at a 2D band edge. At such an edge one can write $\mu_0 = \mu_{0,1} + \mu_{0,2}$, where $\mu_{0,n}(n = 1, 2)$ are the 1D band edges with $\mu_{0,1} \neq \mu_{0,2}$. The corresponding Bloch modes are $p_1(x)p_2(y)$ and $p_1(y)p_2(x)$ with $p_n(x) = p(x; \mu_{0,n})$. When the solution $u(x, y)$ of (8.37) is infinitesimal, this solution (at the band edge) is a linear superposition of these two Bloch modes. When $u(x, y)$ is small but not infinitesimal, we can expand solution $u(x, y)$ of (8.37) into a multiscale perturbation series:

$$u = \varepsilon u_0 + \varepsilon^2 u_1 + \varepsilon^3 u_2 + \cdots, \qquad (8.43)$$

$$\mu = \mu_0 + \eta \varepsilon^2, \qquad (8.44)$$

where

$$u_0 = A_1(X, Y) p_1(x) p_2(y) + A_2(X, Y) p_2(x) p_1(y), \qquad (8.45)$$

$\eta = \pm 1$, and $X = \varepsilon x$, $Y = \varepsilon y$ are the slow-varying spatial scales of envelope functions A_1 and A_2. The derivation of envelope equations for A_1 and A_2 below is analogous to that for 1D, and thus will only be sketched. Substituting the above expansions into (8.37), the equation at $O(\varepsilon)$ is automatically satisfied. At $O(\varepsilon^2)$, the equation for u_1 is:

$$u_{1xx} + u_{1yy} - [F(x) + F(y)]u_1 + \mu_0 u_1 = -2 \left(\frac{\partial^2 u_0}{\partial x \partial X} + \frac{\partial^2 u_0}{\partial y \partial Y} \right). \qquad (8.46)$$

One can verify that the following expression is a solution of (8.46):

$$u_1 = \frac{\partial A_1}{\partial X} v_1(x) p_2(y) + \frac{\partial A_1}{\partial X} v_2(y) p_1(x)$$

$$+ \frac{\partial A_2}{\partial X} v_2(x) p_1(y) + \frac{\partial A_2}{\partial X} v_1(y) p_2(x), \qquad (8.47)$$

where $v_n(x)$ is a periodic solution of equation:

$$v_{n,xx} - F(x) v_n + \omega_{0,n} v_n = -2 p_{n,x}, \ n = 1, 2. \qquad (8.48)$$

At $O(\varepsilon^3)$, the equation for u_2 becomes:

$$u_{2xx} + u_{2yy} - [F(x) + F(y)]u_2 + \mu_0 u_2$$

$$= - \left(2\frac{\partial^2 u_1}{\partial x \partial X} + 2\frac{\partial^2 u_1}{\partial y \partial Y} + \frac{\partial^2 u_0}{\partial X^2} + \frac{\partial^2 u_0}{\partial Y^2} + \eta u_0 + |u_0|^2 u_0 \right). \qquad (8.49)$$

Substituting (8.45) and (8.47) for u_0 and u_1 into the right-hand side of (8.49) and utilizing the Fredholm conditions, the following coupled nonlinear equations for the envelope functions A_1 and A_2 are obtained [10]:

$$D_1 \frac{\partial^2 A_1}{\partial X^2} + D_2 \frac{\partial^2 A_1}{\partial Y^2} + \eta A_1 + \sigma \left[\alpha |A_1|^2 A_1 + \beta \left(\overline{A}_1 A_2^2 + 2A_1 |A_2|^2 \right) \right.$$

$$\left. + \gamma (|A_2|^2 A_2 + \overline{A}_2 A_1^2 + 2A_2 |A_1|^2) \right] = 0, \qquad (8.50)$$

$$D_2 \frac{\partial^2 A_2}{\partial X^2} + D_1 \frac{\partial^2 A_2}{\partial Y^2} + \eta A_2 + \sigma \left[\alpha |A_2|^2 A_2 + \beta \left(\overline{A}_2 A_1^2 + 2A_2 |A_1|^2 \right) \right.$$

$$\left. + \gamma (|A_1|^2 A_1 + \overline{A}_1 A_2^2 + 2A_1 |A_2|^2) \right] = 0, \qquad (8.51)$$

where $D_k = 1/2 \cdot d^2\omega/dk^2|_{\omega=\omega_k}$ and

$$\alpha = \frac{\displaystyle\int_0^{2L}\int_0^{2L} p_1^4(x)p_2^4(y)dxdy}{\displaystyle\int_0^{2L}\int_0^{2L} p_1^2(x)p_2^2(y)dxdy}, \tag{8.52}$$

$$\beta = \frac{\displaystyle\int_0^{2L}\int_0^{2L} p_1^2(x)p_2^2(y)p_1^2(y)p_2^2(x)dxdy}{\displaystyle\int_0^{2L}\int_0^{2L} p_1^2(x)p_2^2(y)dxdy}, \tag{8.53}$$

$$\gamma = \frac{\displaystyle\int_0^{2L}\int_0^{2L} p_1^3(x)p_2(x)p_2^3(y)p_1(y)dxdy}{\displaystyle\int_0^{2L}\int_0^{2L} p_1^2(x)p_2^2(y)dxdy}. \tag{8.54}$$

Note that α and β are always positive, but γ may be positive, negative, or zero.

At the band edge where only a single Bloch mode exists (points A and B in Fig. 8.5(b)), the envelope equations are simpler. In this case, a single Bloch mode has the form $p_1(x)p_1(y)$ with $\mu_0 = 2\mu_{0,1}$, where $p_1(x) = p(x; \mu_{0,1})$, and $\mu_{0,1}$ is a 1D band edge. The leading-order solution $u_0(x, y)$ in (8.43) now becomes $A_1(X, Y)p_1(x)p_1(y)$, and the envelope equation for $A_1(X, Y)$ can be readily found to be:

$$D_1\left(\frac{\partial^2 A_1}{\partial X^2} + \frac{\partial^2 A_1}{\partial Y^2}\right) + \eta A_1 + \sigma\alpha_0|A_1|^2 A_1 = 0, \tag{8.55}$$

where D_1 is defined in the same way as in (8.22) and:

$$\alpha_0 = \left(\frac{\displaystyle\int_0^{2L} p_1^4(x)dx}{\displaystyle\int_0^{2L} p_1^2(x)dx}\right)^2. \tag{8.56}$$

Similarly to the 1D case, the envelope solitons (A_1, A_2) of (8.50), (8.51) must be located at certain special positions in the lattice owing to constraints that are the counterparts of (8.26) for the 1D case. Following similar calculations as in 1D, we can show that if $|A_1|^2$, $|A_2|^2$ and $A_1\overline{A}_2 + \overline{A}_1 A_2$ are symmetric in X and Y about the center position $(\varepsilon x_0, \varepsilon y_0)$, then (x_0, y_0) can only be located at four possible positions:

$$(x_0, y_0) = (0, 0), (0, L/2), (L/2, 0), (L/2, L/2), \tag{8.57}$$

where $L = \pi$ is the lattice period.

Envelope equations (8.50)–(8.51) and (8.55) show that soliton bifurcations are possible only when the diffraction coefficients D_1, D_2 and the nonlinearity coefficient σ are of the same sign. For instance, at points A, C and E in Fig. 8.5 where $D_1 > 0$ and $D_2 > 0$, solitons bifurcate out only when $\sigma > 0$, i.e., under focusing nonlinearity, but not under defocusing nonlinearity $(\sigma < 0)$. The situation is the opposite at points B and D.

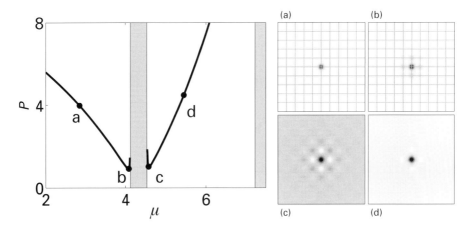

Figure 8.6 Left: power diagrams of solitons bifurcated from the left and right edge points 'A, B' of the first Bloch band under focusing and defocusing nonlinearities respectively. Soliton profiles at the marked points are shown on the right.

8.2.3 Families of solitons bifurcated from 2D band edges

Envelope equations (8.50)–(8.51) and (8.55) admit various types of solutions, and each envelope solution generates four families of lattice solitons corresponding to the four envelope locations (8.57). Thus, a large number of soliton families can bifurcate from each edge of a 2D Bloch band. In this section, we will only discuss a few such solution families whose envelopes are located at $(x_0, y_0) = (0, 0)$.

At the band edge A in Fig. 8.5(b), the envelope equation is the scalar 2D NLS equation (8.55) with $D_1 > 0$. It admits a real and radially symmetric Gaussian-like solution, whose corresponding leading-order analytical solution $u_0(x, y)$ is a nodeless soliton packet. At $\mu = 4.086$, which is slightly below the edge A (see point b in Fig. 8.6), this analytical solution looks almost the same as the true solution shown in Fig. 8.6(b). It has one main peak at a lattice site, flanked by in-phase tails on all four sides. From this solution, a family of lattice solitons bifurcates out. This soliton family resides inside a semi-infinite bandgap, and its power curve is displayed in Fig. 8.6 (left). The soliton power is defined as $P = \int_{-\infty}^{\infty} \int_{-\infty}^{\infty} |u(x, y)|^2 dx dy$. This curve is nonmonotonic. It has a nonzero minimum value, below which solitons do not exist. This contrasts the 1D case where solitons exist at all power levels (see Fig. 8.3). When μ moves away from the band edge, the soliton becomes more localized, and its tails gradually disappear (see Fig. 8.6(a)). These focusing lattice solitons have been observed in [1,11]. Regarding the linear stability of these solitons, it has been shown in [12] that near the band edge where $P'(\mu) > 0$, the soliton is linearly unstable. Under perturbations, it either decays away, or self-focuses into a localized bound state. Away from the band edge where $P'(\mu) < 0$, the soliton is linearly stable. This stability behavior can be readily explained by the Vakhitov–Kolokolov stability criterion. [13]

At the band edge B, the envelope equation is a scalar 2D NLS equation (8.55) with $D_1 < 0$, which admits a Gaussian-like ground-state solution under defocusing

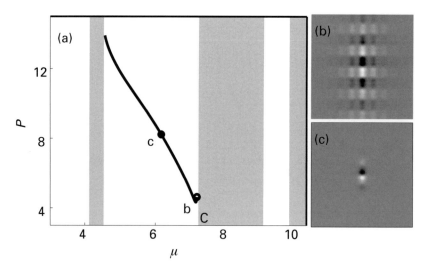

Figure 8.7 (a) Power diagram of the family of single-Bloch-wave solitons bifurcated from the edge point C (under focusing nonlinearity). Soliton profiles at marked points are displayed in (b, c) (after [10], © 2007 APS).

nonlinearity. At $\mu = 4.574$, which is slightly above this band edge (see point c in Fig. 8.6), the analytical solution $u_0(x, y)$ is almost the same as a true solution shown in Fig. 8.6(c). It has one main peak at a lattice site, flanked by out-of-phase tails on all sides. From this solution, a family of lattice solitons bifurcates out in the first bandgap, whose power curve is displayed in Fig. 8.6 (left). This curve is also nonmonotonic with a nonzero minimum value. When μ moves away from the band edge, the soliton becomes more localized (see Fig. 8.6(d)). These defocusing lattice solitons have been observed in [1,14].

At the band edge C, the envelope equations are (8.50)–(8.51) with $D_1, D_2 > 0, D_1 \neq D_2$, and $\gamma = 0$. This coupled system admits several types of envelope solutions under focusing nonlinearity. One of them is $A_1 \neq 0, A_2 = 0$. In this case, the A_1 equation is a single 2D NLS equation with different dispersion coefficients along the X and Y directions. Thus, it admits an elliptical envelope soliton like a stretched Gaussian function. At $\mu = 7.189$, which is slightly below the edge C (see point b in Fig. 8.7(a)), the leading-order analytical solution $u_0(x, y)$ is almost the same as a true solution plotted in Fig. 8.7(b). This soliton is narrower along the x direction, and broader along the y direction. Solutions at adjacent lattice sites are in-phase along x and out-of-phase along y. Since this solution contains only a single Bloch mode ($A_2 = 0$), we call it a single-Bloch-mode soliton. Far away from the edge point C, the soliton in this solution family becomes a strongly localized dipole mode aligned along the y direction. Note that the two peaks of this dipole reside in a single lattice site. These solitons have been observed in [15] (where they were called reduced-symmetry solitons).

At the band edge C, envelope equations (8.50)–(8.51) also admit other solutions. One of them is where $A_1 > 0, A_2 > 0$. In this case, the envelope solutions are both real and positive, and they are shown in Fig. 8.8(b, c). They are both ellipse-shaped but

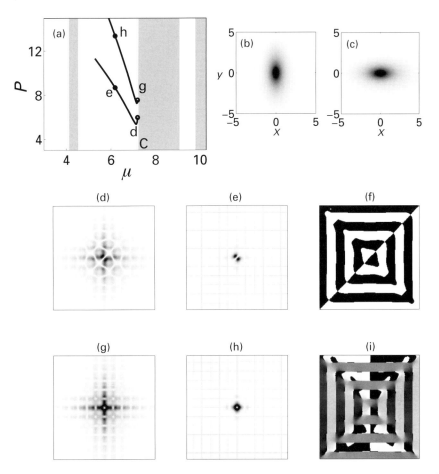

Figure 8.8 (a) Power diagrams of the diagonal-dipole and vortex soliton families bifurcated from the edge point C (under focusing nonlinearity). (b,c) Envelope solutions $A_1 > 0$ (left) and $A_2 > 0$ (right) at point C. (d,f) Amplitude ($|u|$) and phase (black is π, white is $-\pi$) of the diagonal-dipole soliton at point 'd' in (a). (e) Amplitude of the diagonal-dipole soliton at point 'e' in (a). (g, i) Amplitude ($|u|$) and phase of the vortex soliton at point 'g' in (a). (h) Amplitude of the vortex soliton at point 'h' in (a) (after [10], © 2007 APS).

stretched along opposite directions. The corresponding leading-order analytical solution $u_0(x, y)$ is almost the same as the true solution plotted in Fig. 8.8(d, f) for $\mu = 7.189$ slightly below the edge C. The central region of this soliton is an out-of-phase dipole aligned along a diagonal lattice direction, with its two peaks residing inside a single lattice site. The outer region of this soliton is aligned along the horizontal x and vertical y directions. When μ moves away from the edge C, the soliton becomes strongly localized into a diagonal dipole with very weak tails (see Fig. 8.8(e)), and its power increases (see Fig. 8.8(a)). The phase structure of the soliton remains roughly the same as in Fig. 8.8(f).

Another envelope solution admitted at edge C is the one for which $A_1 > 0$, $A_2 = i\hat{A}_2$, $\hat{A}_2 > 0$. In this case, the envelope of one Bloch wave is real, while that of the other

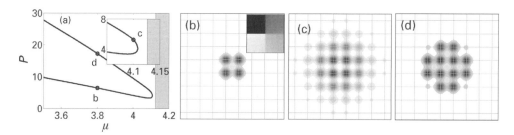

Figure 8.9 A family of the off-site vortex solitons of (8.37) under focusing nonlinearity. (a) Power curve. (b, c, d) Amplitude profiles ($|u|$) of the three vortex solitons at locations marked by letters in (a). The inset in (b) shows the phase distribution.

Bloch wave is purely imaginary, with a relative $\pi/2$ phase delay between the two modes. The envelope functions A_1 and $\hat{A}_2$ are found to be very similar to A_1 and A_2 shown in Fig. 8.8(b, c). The leading-order analytical solution $u_0(x, y)$ for these envelope solutions is almost the same as the true solution displayed in Fig. 8.8(g, i). This soliton looks quite different from the diagonal-dipole soliton in Fig. 8.8(d, f). The most significant difference is that, when winding around the lattice centers (i.e., points $x = m\pi$, $y = n\pi$ with m, n being integers), the phase of the soliton increases or decreases by 2π. In other words, the solution around each lattice site has a vortex structure. When μ moves away from edge C, the vortex becomes strongly localized (see Fig. 8.8(h)), while its phase structure remains roughly the same as Fig. 8.8(i). These gap vortices have been observed. [16] It is important to notice that these vortices are largely confined at the central lattice site, and are fundamentally different from the vortex solitons located at the four adjacent lattice sites, as was reported in [12].

At higher band edges D, E of Fig. 8.5, further novel gap soliton structures will arise, [10] but they will not be discussed here.

8.3   Soliton families not bifurcated from Bloch bands

In addition to the above lattice solitons which bifurcate from edges of Bloch bands, there are also solitons of (8.37) that do not bifurcate from the band edges. Examples include the 1D and 2D dipole solitons residing in the semi-infinite gap. [17] We conclude this chapter by describing certain vortex solitons that do not bifurcate from band edges. Such solitons have attracted considerable interest in optics in recent years.

Under focusing nonlinearity ($\sigma = 1$, $V_0 = 6$), a family of vortex solitons is found for this equation in the semi-infinite gap. The power curve of this vortex family is displayed in Fig. 8.9(a). Unlike the power curves of all previous soliton families, this power curve does not reach the Bloch band, indicating that this soliton family does not bifurcate from the edge of a Bloch band. At point a on the power curve, the intensity and phase of the soliton are displayed in Fig. 8.9(b). The phase plot shows that when winding around the soliton center, the phase increases by 2π, signaling that this soliton is a vortex with charge one. The intensity plot shows that this vortex has four main peaks located at

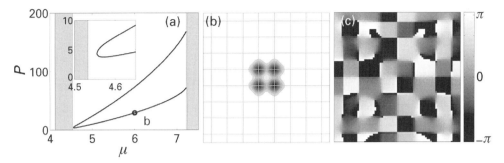

Figure 8.10 A family of the off-site vortex solitons in (8.37) under defocusing nonlinearity. (a) Power curve. (b, c) Intensity and phase of a vortex soliton at point 'b' marked in (a).

adjacent lattice sites forming a square pattern, and the center of this vortex is between lattice sites (off-site vortex soliton). This vortex was first predicted in [12], and was subsequently observed. [18,19] At other positions of the power curve, while the phase distribution of the vortex shows little changes, the intensity distribution changes drastically. For instance, at point c near the Bloch band, the vortex spreads out significantly, while at point d of the upper branch, the vortex becomes a twelve-peaked structure (see Fig. 8.9(d)). This drastic change of vortex-soliton profiles within the same solution family is quite surprising.

When the nonlinearity is self-defocusing, vortex solitons of (8.37) also exist, but in the first bandgap. A family of the off-site defocusing vortices is shown in Fig. 8.10. This defocusing vortex family also does not bifurcate from the Bloch band. Its power curve has a slanted U-shape like the previous case, but tilted in the opposite direction (see Fig. 8.10(a)). A vortex at point b of the power curve is displayed in Fig. 8.10(b, c). The intensity distribution of this vortex (see Fig. 8.10(b)) resembles that in Fig. 8.9(b) of the focusing case, but its phase structure is more complicated (see Fig. 8.10(c)). This defocusing vortex was first reported in [20]. When μ moves from the lower power branch to the upper one, the vortex profile undergoes drastic changes as well.

8.4   Problems

8.1   Fredholm condition

Show that if $p(x)$ is a periodic solution with period $2L$ of a homogeneous equation $p_{xx} + (\mu_0 - V(x)) p = 0$, then the periodic solution $u(x)$ with the same period $2L$ of a nonhomogeneous equation $u_{xx} + (\mu_0 - V(x)) u = f(x)$ is orthogonal to the solution of a homogeneous equation:

$$\int_0^{2L} p(x) f(x) dx = 0.$$

In your derivations assume that both $V(x)$ and $f(x)$ are periodic with a period $2L$.

(Hint: use the fact that operator $\hat{H} = \partial^2/\partial x^2 + (\mu_0 - V(x))$ has the following property $\langle p| \hat{H} |u\rangle = \int_0^{2L} u(x)\hat{H}p(x)\mathrm{d}x = \langle u| \hat{H}|p\rangle$).

8.2 Prove Identity (8.21)

In the derivation of the envelope equation (8.23), the following identity was used:

$$\int_0^{2L} [2v'(x) + p(x)]p(x)\mathrm{d}x = D \int_0^{2L} p^2(x)\mathrm{d}x.$$

Here $v(x)$ is a periodic solution of (8.18), $p(x)$ is the Bloch wave at a band edge μ_0, and $D = \mu''(k)/2$ at $\mu = \mu_0$. Prove this identity by the following steps:

(a) First rewrite the solution (8.10) as $p(x;\mu) = e^{i(k-k_0)x}q(x;\mu(k))$, where k_0 is the corresponding wavenumber of the band edge μ_0, and $q(x;\mu)$ is a periodic function with period $2L$.
(b) Derive the $q(x;\mu)$ equation by substituting $p(x;\mu)$ into (8.9).
(c) Expand $q(x;\mu)$ and $\mu(k)$ into the power series with respect to $k - k_0$ up to $O((k - k_0)^2)$.
(d) Using expansion (c), find the solution $q(x;\mu)$ to the order $O((k - k_0)^2)$. By using the Fredholm condition derive the identity (8.21).

8.3 Center coordinates for the 2D solitons

Prove that for two-dimensional Bloch-wave packets (8.45), if $|A_1|^2$, $|A_2|^2$ and $A_1\overline{A_2} + \overline{A_1}A_2$ are symmetric in X and Y about the center position $(\varepsilon x_0, \varepsilon y_0)$, then (x_0, y_0) can only be located at four possible positions $(0, 0)$, $(0, L/2)$, $(L/2, 0)$, $(L/2, L/2)$, where L is the lattice period.

Hint: first derive the constraints analogous to (8.26) of the 1D case; then substitute the leading order term of $u(x, y)$ into these constraints and simplify.

References

[1] J. W. Fleischer, M. Segev, N. K. Efremidis, and D. N. Christodoulides. Observation of two-dimensional discrete solitons in optically induced nonlinear photonic lattices, *Nature* **422** (2003), 147–150.
[2] A. Argyros, T. A. Birks, S. G. Leon-Saval, *et al.* Photonic bandgap with an index step of one percent, *Opt. Express* **13** (2005), 309.
[3] A. Szameit, J. Burghoff, T. Pertsch, *et al.* Two-dimensional soliton in cubic fs laser written waveguide arrays in fused silica, *Opt. Express* **14** (2006), 6055–6062.
[4] H. Eisenberg, Y. Silberberg, R. Morandotti, A. Boyd, and J. Aitchison. Discrete spatial optical solitons in waveguide arrays, *Phys. Rev. Lett.* **81** (1998), 3383–3386.
[5] R. Iwanow, R. Schieck, G. Stegeman, *et al.* Observation of discrete quadratic solitons, *Phys. Rev. Lett.* **93** (2004), 113902.

[6] F. Dalfovo, S. Giorgini, L. P. Pitaevskii, and S. Stringari. Theory of Bose–Einstein condensation in trapped gases, *Rev. Mod. Phys.* **71** (1999), 463–512.

[7] K. Xu, Y. Liu, J. R. Abo-Shaeer, *et al.* Sodium Bose–Einstein condensates in an optical lattice, *Phys. Rev. A* **72** (2005), 043604.

[8] D. E. Pelinovsky, A. A. Sukhorukov, and Y. S. Kivshar. Bifurcations and stability of gap solitons in periodic potentials, *Phys. Rev. E* **70** (2004), 036618.

[9] J. Yang and T. I. Lakoba. Universally-convergent squared-operator iteration methods for solitary waves in general nonlinear wave equations, *Stud. Appl. Math.* **118** (2007), 153–197.

[10] Z. Shi and J. Yang. Solitary waves bifurcated from Bloch-band edges in two-dimensional periodic media, *Phys. Rev. E* **75** (2007), 056602.

[11] H. Martin, E. D. Eugenieva, Z. Chen, and D. N. Christodoulides. Discrete solitons and soliton-induced dislocations in partially coherent photonic lattices, *Phys. Rev. Lett.* **92** (2004), 123902.

[12] J. Yang and Z. Musslimani. Fundamental and vortex solitons in a two-dimensional optical lattice, *Opt. Lett.* **28** (2003), 2094–2096.

[13] N. G. Vakhitov and A. A. Kolokolov, Stationary solutions of the wave equation in the medium with nonlinearity saturation, *Izv Vyssh. Uchebn. Zaved. Radiofiz.* **16** (1973), 1020–1028.

[14] C. Lou, X. Wang, J. Xu, Z. Chen, and J. Yang. Nonlinear spectrum reshaping and gap-soliton-train trapping in optically induced photonic structures, *Phys. Rev. Lett.* **98** (2007), 213903.

[15] R. Fischer, D. Trager, D. N. Neshev, *et al.* Reduced-symmetry two-dimensional solitons in photonic lattices, *Phys. Rev. Lett.* **96** (2006), 023905.

[16] G. Bartal, O. Manela, O. Cohen, J. W. Fleischer, and M. Segev. Observation of second-band vortex solitons in 2D photonic lattices, *Phys. Rev. Lett.* **95** (2005), 053904.

[17] J. Yang, I. Makasyuk, A. Bezryadina, and Z. Chen. Dipole and quadrupole solitons in optically-induced two-dimensional photonic lattices: theory and experiment, *Stud. Appl. Math.* **113** (2004), 389–412.

[18] D. N. Neshev, T. J. Alexander, E. A. Ostrovskaya, *et al.* Observation of discrete vortex solitons in optically induced photonic lattices, *Phys. Rev. Lett.* **92** (2004), 123903.

[19] J. W. Fleischer, G. Bartal, O. Cohen, *et al.* Observation of vortex-ring discrete solitons in 2D photonic lattices, *Phys. Rev. Lett.* **92** (2004), 123904.

[20] E. A. Ostrovskaya and Y. S. Kivshar. Matter-wave gap vortices in optical lattices, *Phys. Rev. Lett.* **93** (2004), 160405.

Problem solutions

Chapter 2

Solution of Problem 2.1

(a) The solution in the half space $z \geq 0$ can be presented in terms of a superposition of plane waves (2.19):

$$\mathbf{H}(\mathbf{r}, t) = \int_{-\infty}^{+\infty} dk_y\, H(k_y)\hat{\mathbf{x}} \exp(\mathrm{i}k_y y + \mathrm{i}k_z z - \mathrm{i}\omega t), \qquad (\text{P2.1.1})$$

where $k_z = \sqrt{(\omega/c)^2 - k_y^2}$, and $k_y \subset (-\infty, +\infty)$. Note that $k_y \subset [-\omega/c, \omega/c]$ describe waves that are delocalized in $\hat{\mathbf{z}}$, while $k_y \subset (-\infty, -\omega/c) \cup (\omega/c, +\infty)$ describe waves that are evanescent (exponentially decaying) in $\hat{\mathbf{z}}$. To find the expansion coefficients $H(k_y)$, we have to match (P2.1.1) with a boundary condition at $z = 0$. In particlar:

$$\int_{-\infty}^{+\infty} dk_y\, H(k_y)\hat{\mathbf{x}} \exp(\mathrm{i}k_y y) = \begin{cases} H_0\hat{\mathbf{x}}, & -a/2 < y < a/2 \\ 0, & \text{elsewhere.} \end{cases} \qquad (\text{P2.1.2})$$

Using the Fourier transform of a step function, we can readily find the expansion coefficients $H(k_y) = H_0 \sin(k_y a/2)/\pi k_y$. After the substitution of these coefficients into (P2.1.1) we get the complete solution:

$$\mathbf{H}(\mathbf{r}, t) = H_0\hat{\mathbf{x}} \exp(-\mathrm{i}\omega t) \int_{-\infty}^{+\infty} dk_y \frac{\sin(k_y a/2)}{\pi k_y} \exp(\mathrm{i}k_y y + \mathrm{i}\sqrt{(\omega/c)^2 - k_y^2}z). \quad (\text{P2.1.3})$$

(b) In the case of a subwavelength slit $a \ll \lambda$ and $z < \lambda$ we can limit expansion (P2.1.3) to the waves with $|k_y| < 2/a$ having almost constant expansion coefficients $H(k_y) \simeq H_0 a/2\pi$. Taking $y = 0$ we rewrite (P2.1.3) as:

$$\mathbf{H}(\mathbf{r}, t) \approx H_0\hat{\mathbf{x}} \frac{a}{2\pi} \exp(-\mathrm{i}\omega t) \int_{-\infty}^{+\infty} dk_y \exp\left(\mathrm{i}\sqrt{(\omega/c)^2 - k_y^2}z\right)$$

$$= \mathbf{H}_{\text{delocalized}}(\mathbf{r}, t) + \mathbf{H}_{\text{evanescent}}(\mathbf{r}, t)$$

$$= H_0\hat{\mathbf{x}}\frac{2a}{\lambda}\exp(-i\omega t)$$

$$\times\left[\int_0^1 d\xi\,\exp\left(i\sqrt{1-\xi^2}\frac{2\pi z}{\lambda}\right) + \int_1^{+\infty} d\xi\,\exp\left(-\sqrt{\xi^2-1}\frac{2\pi z}{\lambda}\right)\right], \quad \text{(P2.1.4)}$$

where finally:

$$\mathbf{H}_{\text{delocalized}}(\mathbf{r}, t) = H_0\hat{\mathbf{x}}\frac{2a}{\lambda}\exp(-i\omega t)\int_0^1 d\xi\,\exp\left(i\sqrt{1-\xi^2}\frac{2\pi z}{\lambda}\right)$$

$$\mathbf{H}_{\text{evanescent}}(\mathbf{r}, t) = H_0\hat{\mathbf{x}}\frac{2a}{\lambda}\exp(-i\omega t)\int_1^{+\infty} d\xi\,\exp\left(-\sqrt{\xi^2-1}\frac{2\pi z}{\lambda}\right)$$

$$\underset{z<\lambda}{\simeq} H_0\hat{\mathbf{x}}\frac{a}{\pi z}\exp\left(-\frac{2\pi z}{\lambda}\right)\exp(-i\omega t). \quad \text{(P2.1.5)}$$

Solution of Problem 2.2

$$\left.\frac{d\omega(k_z)}{dk_z}\right|_{\frac{\pi}{a}} = \lim_{\delta\to 0}\frac{\omega\left(\frac{\pi}{a}\right) - \omega\left(\frac{\pi}{a} - \delta\right)}{\delta}\underset{\omega(k_z)=\omega\left(k_z-\frac{2\pi}{a}\right)}{=}\lim_{\delta\to 0}\frac{\omega\left(-\frac{\pi}{a}\right) - \omega\left(-\frac{\pi}{a} - \delta\right)}{\delta}$$

$$\underset{\omega(k_z)=\omega(-k_z)}{=}\lim_{\delta\to 0}\frac{\omega\left(\frac{\pi}{a}\right) - \omega\left(\frac{\pi}{a} + \delta\right)}{\delta} = -\lim_{\delta\to 0}\frac{\omega\left(\frac{\pi}{a} + \delta\right) - \omega\left(\frac{\pi}{a}\right)}{\delta}$$

$$= -\left.\frac{d\omega(k_z)}{dk_z}\right|_{\frac{\pi}{a}} \rightarrow \left.\frac{d\omega(k_z)}{dk_z}\right|_{\frac{\pi}{a}} = 0. \quad \text{(P2.2.3)}$$

Solution of Problem 2.3

We start with a general form of a solution for a system exhibiting 1D continuous translational symmetry along the $\hat{\mathbf{z}}$ axis (see Section 2.4.3):

$$\mathbf{H}_{k_z}(\mathbf{r}) = \exp(ik_z z)\mathbf{U}_{k_z}(\mathbf{r}_t).$$

If, additionally, the system exhibits C_N discrete rotational symmetry, then $\hat{\mathbf{R}}_{(\hat{z},\theta_p)}\mathbf{H}_{k_z}(\mathbf{r})$ will be a set of degenerate solutions defined for $\theta_p = \frac{2\pi}{N}p, p = [0, N-1]$. The general solution expressed in the Cartesian coordinate system will be a linear combination of the degenerate solutions in the form:

$$\mathbf{H}(\mathbf{r})|_{\text{Cartesian}} = \sum_{p=0}^{N-1} A_p \hat{\mathbf{R}}_{(\hat{z},\theta_p)} \exp(ik_z z) \mathbf{U}_{k_z}(\mathbf{r}_t) \tag{P2.3.1}$$

$$= \exp(ik_z z) \sum_{p=0}^{N-1} A_p \Re_{(\hat{z},\theta_p)} \mathbf{U}_{k_z}(\Re_{(\hat{z},\theta_p)}^{-1} \mathbf{r}_t).$$

In the cylindrical coordinate system, (P2.3.1) can be re-written as:

$$\mathbf{H}_{k_z}(\rho, \theta, z) = \exp(ik_z z) \sum_{p=0}^{N-1} A_p \Re_{(\hat{z},\theta_p-\theta)} \mathbf{U}_{k_z}(\rho, \theta - \theta_p). \tag{P2.3.2}$$

An alternative form of a general solution (P2.3.2) exhibiting C_N discrete rotational symmetry in cylindrical coordinates is given by (2.136):

$$\mathbf{H}_m(\rho, \theta, z) = \exp(im\theta)\mathbf{U}_m(\rho, \theta, z)$$
$$\mathbf{U}_m\left(\rho, \theta + \frac{2\pi}{N}k, z\right) = \mathbf{U}_m(\rho, \theta, z), \tag{P2.3.3}$$
$$k - \text{integer}, \quad m = [0, \ldots, N-1].$$

One can verify that, by choosing $A_p = \exp(im\theta_p)$, (P2.3.2) also satisfies (P2.3.3). Indeed, after substitution of these expansion coefficients into (P2.3.2) we get:

$$\mathbf{H}_{k_z}(\rho, \theta, z) = \exp(ik_z z) \sum_{p=0}^{N-1} \exp(im\theta_p)\Re_{(\hat{z},\theta_p-\theta)}\mathbf{U}_{k_z}(\rho, \theta - \theta_p)$$

$$= \exp(ik_z z)\exp(im\theta) \sum_{p=0}^{N-1} \exp(im(\theta_p - \theta))\Re_{(\hat{z},\theta_p-\theta)}\mathbf{U}_{k_z}(\rho, \theta - \theta_p)$$

$$= \exp(ik_z z)\exp(im\theta)\mathbf{F}_{k_z}(\rho, \theta), \tag{P2.3.4}$$

where

$$\mathbf{F}_{k_z}(\rho, \theta) = \sum_{p=0}^{N-1} \exp(im(\theta_p - \theta))\Re_{(\hat{z},\theta_p-\theta)}\mathbf{U}_{k_z}(\rho, \theta - \theta_p). \tag{P2.3.5}$$

Moreover, we are going to demonstrate that:

$$\mathbf{F}_m(\rho, \theta + \theta_k) = \mathbf{F}_m(\rho, \theta),$$
$$\theta_k = \frac{2\pi}{N}k, \quad k - \text{integer}, \quad m = [0, \ldots, N-1]. \tag{P2.3.6}$$

To simplify further manipulations we define:

$$\varphi(\theta_p - \theta) = \exp(im(\theta_p - \theta))\Re_{(\hat{z},\theta_p-\theta)}\mathbf{U}_{k_z}(\rho, \theta - \theta_p). \tag{P2.3.7}$$

Then,

$$\mathbf{F}_m(\rho, \theta + \theta_k) = \sum_{p=0}^{N-1} \varphi(\theta_p - \theta_k - \theta)$$

$$= \sum_{p-k=-k}^{-1} \varphi(\theta_{p-k} - \theta) + \sum_{p-k=0}^{N-1-k} \varphi(\theta_{p-k} - \theta)$$

$$= \sum_{\substack{p'=p''=p-k \\ p'=-k}}^{-1} \varphi(\theta_{p'} - \theta) + \sum_{p''=0}^{N-1-k} \varphi(\theta_{p''} - \theta)$$

$$= \sum_{p'=-k}^{-1} \varphi\left(\theta_{p'} + \frac{2\pi}{N}N - \theta\right) + \sum_{p''=0}^{N-1-k} \varphi(\theta_{p''} - \theta)$$

$$= \sum_{\substack{p''=N+p' \\ \theta_{p'}+2\pi=\theta_{N+p'}}}^{N-1} \varphi(\theta_{p''} - \theta) + \sum_{p''=0}^{N-1-k} \varphi(\theta_{p''} - \theta)$$

$$= \sum_{p=0}^{N-1} \varphi(\theta_p - \theta) = \mathbf{F}_m(\rho, \theta, z). \tag{P2.3.8}$$

Finally, we conclude that the eigenstate of a system exhibiting 1D continuous translational symmetry plus C_N discrete rotational symmetry is characterized by the two conserved numbers

$$k_z \subset [0, \pi/a], \quad m = [0, \ldots, N-1],$$

and has a general form:

$$\mathbf{H}_{m,k_z}(\rho, \theta, z) = \exp(ik_z z) \exp(im\theta) \mathbf{F}_{m,k_z}(\rho, \theta)$$

$$\mathbf{F}_{m,k_z}(\rho, \theta + \theta_p) = \mathbf{F}_{m,k_z}(\rho, \theta), \text{ for any } \theta_p = \frac{2\pi}{N} p$$

$$p - \text{integer}, \quad m = [0, N-1] \tag{P2.3.9}$$

Solution of Problem 2.4

We start with a general form of a solution for the system exhibiting 1D discrete translational symmetry with period a along the $\hat{z}$ axis (see Section 2.4.5.1):

$$\mathbf{H}_{k_z}(\mathbf{r}) = \exp(ik_z z) \mathbf{U}_{k_z}(\mathbf{r})$$

$$\mathbf{U}_{k_z}(\mathbf{r}) = \mathbf{U}_{k_z}(\mathbf{r} + a\hat{z}N_1), \quad N_1 \subset \text{integer}. \tag{P2.4.1}$$

If, additionally, the system exhibits C_N discrete rotational symmetry, then $\hat{\mathbf{R}}_{(\hat{z},\theta_p)}\mathbf{H}_{k_z}(\mathbf{r})$ will be a set of degenerate solutions defined for $\theta_p = \frac{2\pi}{N}p, p = [0, N-1]$. The general solution expressed in the Cartesian coordinate system will be a linear combination of the degenerate solutions in the form:

$$\mathbf{H}(\mathbf{r})|_{\text{Cartesian}} = \sum_{p=0}^{N-1} A_p \hat{\mathbf{R}}_{(\hat{z},\theta_p)} \exp(ik_z z) \mathbf{U}_{k_z}(\mathbf{r})$$

$$= \exp(ik_z z) \sum_{p=0}^{N-1} A_p \Re_{(\hat{z},\theta_p)} \mathbf{U}_{k_z}(\Re_{(\hat{z},\theta_p)}^{-1} \mathbf{r}). \tag{P2.4.2}$$

In the cylindrical coordinate system, (P2.4.2) can be written as:

$$\mathbf{H}_{k_z}(\rho, \theta, z) = \exp(ik_z z) \sum_{p=0}^{N-1} A_p \mathfrak{R}_{(\hat{z},\theta_p - \theta)} \mathbf{U}_{k_z}(\rho, \theta - \theta_p, z). \tag{P2.4.3}$$

An alternative form of a general solution (P2.4.3) exhibiting C_N discrete rotational symmetry in cylindrical coordinates is given by (2.136):

$$\mathbf{H}_m(\rho, \theta, z) = \exp(im\theta)\mathbf{U}_m(\rho, \theta, z)$$

$$\mathbf{U}_m\left(\rho, \theta + \frac{2\pi}{N}k, z\right) = \mathbf{U}_m(\rho, \theta, z),$$

$$k - \text{integer}, \quad m = [0, \ldots, N-1]. \tag{P2.4.4}$$

Similarly to the previous problem one can verify that, by choosing $A_p = \exp(im\theta_p)$, (P2.4.3) becomes of the form (P2.4.4):

$$\mathbf{H}_{k_z}(\rho, \theta, z) = \exp(ik_z z)\exp(im\theta)\mathbf{F}_{k_z}(\rho, \theta, z), \tag{P2.4.5}$$

where

$$\mathbf{F}_{k_z}(\rho, \theta, z) = \sum_{p=0}^{N-1} \exp(im(\theta_p - \theta))\mathfrak{R}_{(\hat{z},\theta_p - \theta)} \mathbf{U}_{k_z}(\rho, \theta - \theta_p, z). \tag{P2.4.6}$$

Moreover, in a manner identical to the previous problem one can also demonstrate that:

$$\mathbf{F}_m(\rho, \theta + \theta_k, z) = \mathbf{F}_m(\rho, \theta, z),$$

$$\theta_k = \frac{2\pi}{N}k, \quad k - \text{integer}, \quad m = [0, \ldots, N-1]. \tag{P2.4.7}$$

From (P2.4.1) and (P2.4.6) it also follows that:

$$\mathbf{F}_{m,k_z}(\rho, \theta + \theta_p, z + aN_1) = \mathbf{F}_{m,k_z}(\rho, \theta, z); \quad N_1 \text{ an integer.} \tag{P2.4.8}$$

Indeed,

$$\mathbf{F}_{m,k_z}(\rho, \theta, z + aN_1) = \sum_{p=0}^{N-1} \exp(im(\theta_p - \theta))\mathfrak{R}_{(\hat{z},\theta_p - \theta)} \mathbf{U}_{k_z}(\rho, \theta - \theta_p, z + aN_1)$$

$$= \sum_{p=0}^{N-1} \exp(im(\theta_p - \theta))\mathfrak{R}_{(\hat{z},\theta_p - \theta)} \mathbf{U}_{k_z}(\rho, \theta - \theta_p, z)$$

$$= \mathbf{F}_{m,k_z}(\rho, \theta, z). \tag{P2.4.9}$$

Finally, we conclude that the eigenstate of a system exhibiting 1D discrete translational symmetry plus C_N discrete rotational symmetry is characterized by the two conserved numbers

$$k_z \subset [0, \pi/a], \quad m = [0, \ldots, N-1],$$

and has the general form:

$$\mathbf{H}_{m,k_z}(\rho,\theta,z) = \exp(ik_z z)\exp(im\theta)\mathbf{F}_{m,k_z}(\rho,\theta,z)$$

$$\mathbf{F}_{m,k_z}(\rho,\theta+\theta_p,z+aN_1) = \mathbf{F}_{m,k_z}(\rho,\theta,z),\ \text{for any}\ \theta_p = \frac{2\pi}{N}p,\ \text{and}\ N_1$$

$$p-\text{integer},\quad m = [0, N-1].\qquad\qquad\qquad (P2.4.10)$$

Solution of Problem 2.5

(a) This can be readily verified by using angular invariance of the fiber dielectric constant and expression of $\nabla\times$ in cylindrical coordinates:

$$\nabla\times\mathbf{H} = \hat{\rho}\left(\frac{1}{\rho}\frac{\partial H_z}{\partial\theta} - \frac{\partial H_\theta}{\partial z}\right) + \hat{\theta}\left(\frac{\partial H_\rho}{\partial z} - \frac{\partial H_z}{\partial\rho}\right) + \frac{\hat{z}}{\rho}\left(\frac{\partial(\rho H_\theta)}{\partial\rho} - \frac{\partial H_\rho}{\partial\theta}\right).$$

(b) From (P2.5.1) it follows that if $\mathbf{H}_m(\rho,\theta,z) = \exp(im\theta)\exp(ik_z z)(h_\rho^m(\rho),$ $h_\theta^m(\rho), h_z^m(\rho))$ is a solution, then $\mathbf{H}(\rho,\theta,z) = \exp(-im\theta)\exp(ik_z z)(h_\rho^m(\rho), -h_\theta^m(\rho),$ $h_z^m(\rho))$ is also a solution of the same frequency. The last expression, however, has the functional form of $\mathbf{H}_{-m}(\rho,\theta,z)$ owing to $\exp(-im\theta)$ dependence. Therefore, states with the opposite values of angular momenta $m, -m$ are degenerate. Finally, up to a constant, the fields of a mode with $-m$ angular momentum can be, thus, defined as:

$$\mathbf{H}_{-m}(\rho,\theta,z) = \exp(-im\theta)\exp(ik_z z)\left(h_\rho^m(\rho), -h_\theta^m(\rho), h_z^m(\rho)\right)$$

$$\mathbf{E}_{-m}(\rho,\theta,z) = \exp(-im\theta)\exp(ik_z z)\left(-e_\rho^m(\rho), e_\theta^m(\rho), -e_z^m(\rho)\right).\quad (P2.5.5)$$

(c) As suggested by (P2.5.4), the eigenstates of an angular reflection operator can be presented as a linear combination of the two degenerate modes:

$$\mathbf{H}(\mathbf{r}) = A\mathbf{H}_m(\rho,\theta,z) + B\mathbf{H}_{-m}(\rho,\theta,z)$$

$$= \exp(ik_z z)\begin{pmatrix} [A\exp(im\theta) + B\exp(-im\theta)]\,h_\rho^m(\rho) \\ [A\exp(im\theta) - B\exp(-im\theta)]\,h_\theta^m(\rho) \\ [A\exp(im\theta) + B\exp(-im\theta)]\,h_z^m(\rho) \end{pmatrix}.\quad (P2.5.6)$$

From the definition (P2.5.3) of a $\hat{\sigma}_\theta$ operator it follows that the modal polarization corresponding to the operator eigenvalue $\sigma = 1$ satisfies the following equation:

$$\hat{\sigma}_\theta\mathbf{H}(\mathbf{r}) = \mathbf{H}(\mathbf{r})$$

$$\left(H_\rho(\rho,-\theta,z), -H_\theta(\rho,-\theta,z), H_z(\rho,-\theta,z)\right)$$

$$= \left(H_\rho(\rho,\theta,z), H_\theta(\rho,\theta,z), H_z(\rho,\theta,z)\right).\quad (P2.5.7)$$

Substitution of (P2.5.6) into (P2.5.7) leads to the conclusion that $A = B$, and the following form of the modal fields (expressions for the electric fields are derived in a similar manner):

$$\mathbf{H}_{\sigma=1}(\mathbf{r}) = \exp(ik_z z)\left(\cos(m\theta)h_\rho^m(\rho),\ i\sin(m\theta)h_\theta^m(\rho),\ \cos(m\theta)h_z^m(\rho)\right)$$

$$\mathbf{E}_{\sigma=1}(\mathbf{r}) = \exp(ik_z z)\left(i\sin(m\theta)e_\rho^m(\rho),\ \cos(m\theta)e_\theta^m(\rho),\ i\sin(m\theta)e_z^m(\rho)\right).\quad (P2.5.8)$$

In a similar manner for the other polarization corresponding to $\sigma = -1$ we get:

$$\mathbf{H}_{\sigma=-1}(\mathbf{r}) = \exp(ik_z z) \left(i \sin(m\theta) h_\rho^m(\rho), \cos(m\theta) h_\theta^m(\rho), i \sin(m\theta) h_z^m(\rho) \right)$$
$$\mathbf{E}_{\sigma=-1}(\mathbf{r}) = \exp(ik_z z) \left(\cos(m\theta) e_\rho^m(\rho), i \sin(m\theta) e_\theta^m(\rho), \cos(m\theta) e_z^m(\rho) \right).$$

(P2.5.9)

(d) In the case when $m = 0$, (P2.5.8) and (P2.5.9) can be further simplified to define the so-called TE and TM polarizations:

$$\mathbf{H}_{\text{TE}}(\mathbf{r}) = \exp(ik_z z) \left(h_\rho^m(\rho), 0, h_z^m(\rho) \right);$$
$$\mathbf{E}_{\text{TE}}(\mathbf{r}) = \exp(ik_z z) \left(0, e_\theta^m(\rho), 0 \right),$$

(P2.5.10)

$$\mathbf{H}_{\text{TM}}(\mathbf{r}) = \exp(ik_z z) \left(0, h_\theta^m(\rho), 0 \right);$$
$$\mathbf{E}_{\text{TM}}(\mathbf{r}) = \exp(ik_z z) \left(e_\rho^m(\rho), 0, e_z^m(\rho) \right).$$

(P2.5.11)

Chapter 3

Solution of Problem 3.1

According to (3.25), the complete transfer matrix of the problem is $M = M_{\text{clad2,core}} M_{\text{clad1,core}}$. Effective refractive indices of guided modes can then be found by solving:

$$M_{2,2}(n_{\text{eff}}) = 0. \tag{P3.1.1}$$

Using definitions of individual transfer matrices for TE polarization, for example, we get:

$$M = \frac{1}{4} \begin{pmatrix} \left(1 + \dfrac{k_z^{\text{core}}}{k_z^{\text{clad}}}\right) \exp\left(ik_z^{\text{core}} a\right) \left(1 - \dfrac{k_z^{\text{core}}}{k_z^{\text{clad}}}\right) \exp\left(-ik_z^{\text{core}} a\right) \\ \left(1 - \dfrac{k_z^{\text{core}}}{k_z^{\text{clad}}}\right) \exp\left(ik_z^{\text{core}} a\right) \left(1 + \dfrac{k_z^{\text{core}}}{k_z^{\text{clad}}}\right) \exp\left(-ik_z^{\text{core}} a\right) \end{pmatrix} \times$$

$$\begin{pmatrix} \left(1 + \dfrac{k_z^{\text{clad}}}{k_z^{\text{core}}}\right) \left(1 - \dfrac{k_z^{\text{clad}}}{k_z^{\text{core}}}\right) \\ \left(1 - \dfrac{k_z^{\text{clad}}}{k_z^{\text{core}}}\right) \left(1 + \dfrac{k_z^{\text{clad}}}{k_z^{\text{core}}}\right) \end{pmatrix}.$$

(P3.1.2)

In fact, one can show that for both TE and TM polarizations, $M_{2,2}$ has the following universal form:

$$M_{2,2} = \frac{1}{4\tau} \left[(1+\tau)^2 \exp(-i\theta) - (1-\tau)^2 \exp(i\theta) \right], \tag{P3.1.3}$$

where definition of τ varies depending upon polarization in the following way:

$$\tau_{\text{TE}} = \frac{k_z^{\text{clad}}}{k_z^{\text{core}}}; \quad \tau_{\text{TM}} = \frac{k_z^{\text{clad}} \varepsilon_{\text{core}}}{k_z^{\text{core}} \varepsilon_{\text{clad}}}; \quad \theta = k_z^{\text{core}} a$$

$$k_z^{\text{clad}} = \frac{\omega}{c} \sqrt{n_{\text{clad}}^2 - n_{\text{eff}}^2}; \quad k_z^{\text{core}} = \frac{\omega}{c} \sqrt{n_{\text{core}}^2 - n_{\text{eff}}^2}. \tag{P3.1.4}$$

Substitution of (P3.1.3) into (3.25) leads to the following equation for the fundamental mode of a slab waveguide:

$$\exp(i\theta) = \frac{1+\tau}{1-\tau}. \tag{P3.1.5}$$

In the case of a low-refractive-index contrast and long-wavelength regime, (P3.1.1), (P3.1.3), (P3.1.4) could be greatly simplified. In particular, retaining only the leading order terms in $\delta n/n_0$, $\delta n_{\text{eff}}/n_0$ we get:

$$k_z^{\text{clad}} = \frac{\omega}{c}\sqrt{n_{\text{clad}}^2 - n_{\text{eff}}^2} \simeq i\frac{\omega}{c}\sqrt{2n_0\delta n_{\text{eff}}};$$

$$k_z^{\text{core}} = \frac{\omega}{c}\sqrt{n_{\text{core}}^2 - n_{\text{eff}}^2} \simeq \frac{\omega}{c}\sqrt{2n_0\left(\delta n - \delta n_{\text{eff}}\right)};$$

$$\left[\tau_{\text{TE}} = \frac{k_z^{\text{clad}}}{k_z^{\text{core}}}\right] \simeq \left[\tau_{\text{TM}} = \frac{k_z^{\text{clad}}\varepsilon_{\text{core}}}{k_z^{\text{core}}\varepsilon_{\text{clad}}}\right] \simeq i\sqrt{\frac{\delta n_{\text{eff}}}{\delta n - \delta n_{\text{eff}}}};$$

$$\theta = k_z^{\text{core}}a \simeq \frac{\omega a}{c}\sqrt{2n_0\left(\delta n - \delta n_{\text{eff}}\right)}. \tag{P3.1.6}$$

As $\tau_{\text{TE}} \simeq \tau_{\text{TM}}$, the solution of (P3.1.3) will be almost the same for both polarizations, thus confirming the near degeneracy of the TE and TM modes in the low-refractive-index-contrast regime. Furthermore, in the long-wavelength regime $\theta \ll 1$ and (P3.1.5) can be further simplified. Assuming that $\delta n_{\text{eff}} \ll \delta n$ we find:

$$\exp(i\theta) = \frac{1+\tau}{1-\tau}\underset{|\tau|\ll1,\theta\ll1}{\Rightarrow} 1 + i\theta + O\left(\theta^2\right)$$

$$= 1 + i\frac{2|\tau|}{1+|\tau|^2} + O\left(|\tau|^2\right) \Rightarrow \theta = \frac{2|\tau|}{1+|\tau|^2}. \tag{P3.1.7}$$

Finally, substituting (P3.1.6) into (P3.1.7) we find the effective refractive index of the fundamental slab mode:

$$\delta n_{\text{eff}} = \frac{n_0}{2}\left(\frac{\omega a}{c}\delta n\right)^2. \tag{P3.1.8}$$

From (P3.1.8) it follows that our assumption of $\delta n_{\text{eff}} \ll \delta n$ is correct when $\omega \ll c/(a\sqrt{n_0\delta n/2})$, which is consistent with our prior definition of the long-wavelength regime. Dependence of the refractive index is presented in Fig. S3.1.1

Solution of Problem 3.2

Part (a)
Mode 1 is within the light cone of a cladding, hence it is delocalized in air. Mode 1 is also inside a band of the states that are delocalized in the multilayer. Therefore, mode 1 is delocalized both in air and in the multilayer.

Mode 2 is within the light cone of a cladding, hence it is delocalized in air. Mode 2 is also inside a bandgap of a multilayer. Therefore, mode 2 is primarily delocalized in air and evanescent in the multilayer.

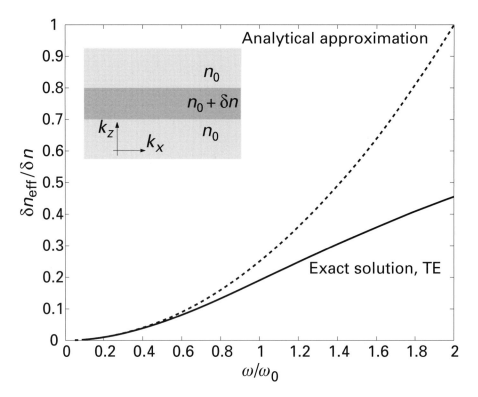

Figure S3.1.1 Effective refractive index of a fundamental slab mode $n_{\text{eff}} = n_0 + \delta n_{\text{eff}}$ as a function of normalized frequency ω/ω_0, $\omega_0 = c/(a\sqrt{2n_0\delta n})$. Parameters used $a = 1$; $\delta n = 0.01$; $n_{\text{clad}} = 2$.

Mode 3 is below the light cone of the cladding, hence it is evanescent in air. Mode 3 is also inside a band of the states that are delocalized in the multilayer. Therefore, mode 3 is mostly delocalized in the multilayer and evanescent in air.

Mode 4 is below the light cone of the cladding, hence it is evanescent in air. Mode 4 is also inside a bandgap of a multilayer. Therefore, mode 4 is evanescent both in air and in the multilayer, therefore it can only be localized at the cladding–multilayer interface.

Mode 5 is similar to mode 4 in the sense that it is evanescent both in air and in the multilayer, therefore it can only be localized on the cladding–multilayer interface. However, mode 5 is not in the bandgap of a multilayer, but is rather located below the light line of a low-index material of a multilayer. This signifies that mode 5 is evanescent in low-index layers and, therefore, it is primarily concentrated in the high-refractive-index layers in the vicinity of a cladding–multilayer interface.

Part (b)
Denoting M_{0h} to be the transfer matrix from the cladding to the high-index layer, (A_0, B_0) to be the field expansion coefficients in the cladding (see Fig. 3.2(b)), and (A_N, B_N),

$N \subset 1, \ldots$ to be the field expansion coefficients in the high-index layer of the Nth bilayer, similarly to (3.13) we can write:

$$\begin{pmatrix} A_N \\ B_N \end{pmatrix} = (M_{hlh})^{N-1} M_{0h} \begin{pmatrix} A_0 \\ B_0 \end{pmatrix}. \tag{P3.2.1}$$

Using the representation of a bilayer transfer matrix in terms of its eigenvalues and eigenvectors we get:

$$\begin{pmatrix} A_N \\ B_N \end{pmatrix} = V_{hlh} \Lambda_{hlh}^{N-1} V_{hlh}^{-1} M_{0h} \begin{pmatrix} A_0 \\ B_0 \end{pmatrix} = V_{hlh} \begin{pmatrix} \lambda_{<1}^{N-1} & 0 \\ 0 & \lambda_{>1}^{N-1} \end{pmatrix} V_{hlh}^{-1} M_{0h} \begin{pmatrix} A_0 \\ B_0 \end{pmatrix}. \tag{P3.2.2}$$

As mode 4 is located inside the multilayer bandgap, only the eigenvalue with magnitude smaller than one should be excited, therefore:

$$\begin{pmatrix} A_0 \\ B_0 \end{pmatrix} = M_{0h}^{-1} V_{hlh} \begin{pmatrix} \alpha \\ 0 \end{pmatrix}. \tag{P3.2.3}$$

Moreover, as mode 4 is located below the light line of the cladding, then $A_0 = 0$. Finally, we arrive at the following equation, the solution of which gives the propagation constant k_x of a guided surface state:

$$\begin{pmatrix} 0 \\ B_0 \end{pmatrix} = M_{0h}^{-1} V_{hlh} \begin{pmatrix} \alpha \\ 0 \end{pmatrix}$$

$$\text{defining } M_{0h}^{-1} V_{hlh} = \begin{pmatrix} a_{1,1}(k_x) & a_{1,2}(k_x) \\ a_{2,1}(k_x) & a_{2,2}(k_x) \end{pmatrix} \rightarrow a_{1,1}(k_x) = 0. \tag{P3.2.4}$$

Part (c)
(i) As established in part (a) points 1 and 2 describe modes evanescent in air and in the multilayer with the fields mostly concentrated at the cladding–multilayer interface. When going from point 1 towards point 2 one approaches the light line of air, therefore mode 2 is expected to be more delocalized in the air cladding. In particular, for the modes below the air light line, dependence of the modal fields in the air region can be represented as $\exp(-|k_z^0| \cdot |z - z_0|)$, defining penetration depth into the air cladding as $\sim 1/|k_z^0|$. As mode 2 is located closer to the light line of air than mode 1, then

$$|k_{z,\text{mode }1}^0| = \sqrt{k_{x,\text{mode }1}^2 - \omega^2 n_0^2} > |k_{z,\text{mode }2}^0| = \sqrt{k_{x,\text{mode }2}^2 - \omega^2 n_0^2},$$

and, therefore, the penetration depth of mode 1 into the air is smaller than that of mode 2. The field distributions of modes 1 and 2 are presented in Fig. S3.2.1.

(ii) As established in part (a) points 3 and 4 describe modes evanescent in air and in the low-refractive-index layers of a multilayer, with the fields mostly concentrated in the high-index layers at the cladding–multilayer interface. When going from point 3 towards point 4, one approaches a continuum of the states delocalized in the multilayer, therefore mode 4 is expected to be more delocalized in the multilayer than point 3. In particular, the number of bilayers of substantial penetration of the modal fields into the multi-layer is given by $N_{\text{attenuation}} = 1/\log(1/|\lambda_{<1}|)$. Near the band edge, the smallest eigen-value of a bilayer transfer matrix can be approximated as $1 - |\lambda_{<1}| \sim |\omega - \omega_{\text{band edge}}|^\alpha$, and therefore, $N_{\text{attenuation}} \sim 1/|\omega - \omega_{\text{band edge}}|^\alpha$. As mode 3 is locater further away from

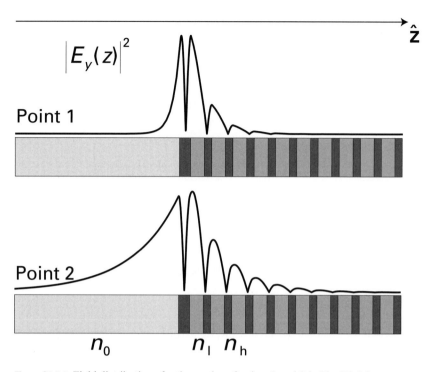

Figure S3.2.1 Field distributions for the modes of points 1 and 2 in Fig. P3.2.2.

the band edge than mode 4, then $N_{\text{attenuation}}^{\text{mode3}} < N_{\text{attenuation}}^{\text{mode4}}$, therefore one expects that mode 4 will be more delocalized into the multilayer than mode 3. The field distributions of modes 3 and 4 are presented in Fig. S3.2.2.

Part (d)
   (i) See Fig. S3.2.3.
   (ii) After the introduction of a periodic perturbation, mode 2 will end up in the continuum of the bulk states of a multilayer, therefore it will no longer be localized at the cladding–multilayer interface, but rather delocalized into the multilayer. Mode 2, however, will remain evanescent in the air cladding as even after folding into the first Brillouin zone, its dispersion relation is located below the light line of air.
   (iii) After introduction of a periodic perturbation, mode 3 will end up in the continuum of a multilayer bulk state located above the light line of air. Therefore it will no longer be localized at the cladding–multilayer interface, but rather delocalized both into the multilayer and into the air cladding.
   (iv) After introduction of a periodic perturbation, mode 5 will end up in the continuum of the bulk states of a multilayer, and located above the light line of air. Therefore, it will no longer be evanescent in the air cladding, but rather delocalized both into the multilayer and into the air cladding.

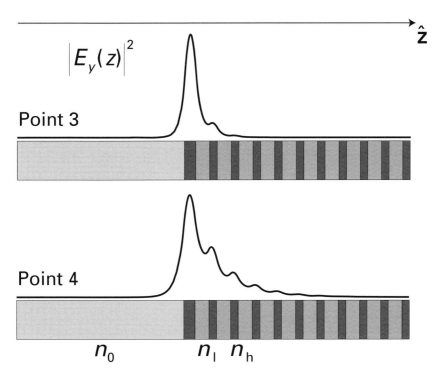

Figure S3.2.2 Field distributions for the modes of points 3 and 4 in Fig. P3.2.2.

Solution of Problem 3.3
We use (3.33), which determines the dispersion relation of the modes:

$$\cos(k_z a) = \cos(\phi_h)\cos(\phi_l) - \xi \sin(\phi_h)\sin(\phi_l), \qquad (P3.3.1)$$

and apply it at the edge $2p\omega_0 + \delta\omega$ of a bandgap centered around $2p$ even harmonic of ω_0. In this case, the solution of (P3.3.1) is $k_z = 0$, which transforms (P3.3.1) into:

$$1 = \cos(\phi_h)\cos(\phi_l) - \xi \sin(\phi_h)\sin(\phi_l). \qquad (P3.3.3)$$

Phases can be expressed in the following way:

$$\phi_h = \omega d_h n_h \underset{\substack{\omega=2p\omega_0+\delta\omega \\ d_h=d_h^0(1+\delta)}}{=} \omega_0 d_h^0 n_h (1+\delta)(2p+\delta\omega/\omega_0) \underset{\omega_0 d_h^0 n_h = \frac{\pi}{2}}{=} \pi (1+\delta)(p+\delta\omega/2\omega_0),$$

$$\phi_h = \pi (p + p\delta + \delta\omega/2\omega_o) = p\pi + \delta\phi_h$$

$$\delta\phi_h = \pi (\delta\omega/2\omega_o + p\delta), \qquad (P3.3.4)$$

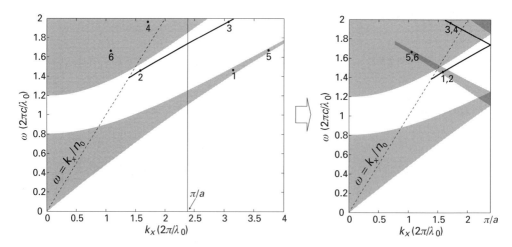

Figure S3.2.3 The band diagram of a periodic system with vanishingly small perturbation can be constructed from the band diagram of an unperturbed system by folding all the modal dispersion curves into the first Brillouin zone.

as well as:

$$\phi_l = \omega d_1 n_1 \Big|_{\substack{\omega=2p\omega_0+\delta\omega \\ d_1=d_1^0(1-\delta)}} = \omega_0 d_1^0 n_1 (1-\delta)(2p+\delta\omega\,\omega_0) \Big|_{\omega_0 d_1^0 n_1} = \frac{\pi}{2}\pi(1-\delta)(p+\delta\omega/2\omega_0),$$

$$\phi_e = \pi(p - p\delta + \delta\omega/2\omega_0) = p\pi + \delta\phi_e; \tag{P3.3.5}$$

$$\delta\phi_e = \pi(\delta\omega/2\omega_0 - p\delta).$$

We now perform a Taylor expansion of (P3.3.3) up to the second order in $\delta\phi_h$, and $\delta\phi_l$, which transforms (P3.3.3) into:

$$1 = (1 - \delta\phi_h^2/2)(1 - \delta\phi_l^2/2) - \xi\delta\phi_h\delta\phi_l. \tag{P3.3.6}$$

This equation can be simplified and then solved with respect to the position of the bandgap edge:

$$\delta\phi_h^2 + \delta\phi_l^2 + 2\xi\delta\phi_h\delta\phi_l = 0,$$

$$(\delta\omega/2\omega_0 + p\delta)^2 + (\delta\omega/2\omega_0 - p\delta)^2 + 2\xi(\delta\omega/2\omega_0 + p\delta)(\delta\omega/2\omega_0 - p\delta) = 0,$$

$$2(\delta\omega/2\omega_0)^2 + 2(p\delta)^2 + 2\xi((\delta\omega/2\omega_0) - (p\delta)^2) = 0,$$

$$\delta\omega = \pm 2p\omega_0\delta\sqrt{\frac{\xi-1}{\xi+1}}. \tag{P3.3.7}$$

Finally, from (P3.3.7) it follows that the size of a bandgap centered around $2p$ harmonic of ω_0 is:

$$\Delta\omega = 2|\delta\omega| = 4p\omega_0\delta\sqrt{\frac{\xi-1}{\xi+1}}. \tag{P3.3.8}$$

Solution of Problem 3.4

Part (a)

After substitution of (P3.4.3), (P3.4.4), and (P3.4.5) into (P3.4.2), matrix multiplication, and keeping only the terms up to the first order in $\delta\omega$ we get:

$$M_{\text{hlh}} = -\frac{1}{2r_0}\begin{pmatrix} 1+r_0^2 & 1-r_0^2 \\ 1-r_0^2 & 1+r_0^2 \end{pmatrix} +$$

$$\frac{\delta}{2r_0^2}\begin{pmatrix} C_r(1-r_0^2) - ir_0(C_h(1+r_0^2)+2C_1r_0) & C_r(1+r_0^2)+ir_0C_h(1-r_0^2) \\ C_r(1+r_0^2) - ir_0C_h(1-r_0^2) & C_r(1-r_0^2)+ir_0(C_h(1+r_0^2)+2C_1r_0) \end{pmatrix}.$$

$$\text{(P3.4.6)}$$

The eigenvalues and corresponding eigenvectors of (P3.4.6) can be found to be:

$$\lambda^{<1} = -r_0 - C_r\delta; \quad V_{\text{hlh}}^{<1} = \begin{pmatrix} -1 + 2i\delta\dfrac{C_h + C_1 r_0}{1-r_0^2} \\ 1 \end{pmatrix}$$

$$\lambda^{>1} = -\frac{1}{r_0} + \frac{C_r}{r_0^2}\delta; \quad V_{\text{hlh}}^{1} = \begin{pmatrix} 1 + 2ir_0\delta\dfrac{C_1 + C_h r_0}{1-r_0^2} \\ 1 \end{pmatrix}.$$

$$\text{(P3.4.7)}$$

Part (b)

$$M_{\text{dh}}M_{\text{hd}} = \begin{pmatrix} -i & 0 \\ 0 & i \end{pmatrix} + \delta\begin{pmatrix} C_h + C_0\dfrac{r_{d0}^2+1}{2r_{d0}} & C_0\dfrac{1-r_{d0}^2}{2r_{d0}} \\ C_0\dfrac{1-r_{d0}^2}{2r_{d0}} & C_h + C_0\dfrac{r_{d0}^2+1}{2r_{d0}} \end{pmatrix}. \quad \text{(P3.4.10)}$$

Part (c)

Substitution of (P3.4.7) and (P3.4.10) into (P3.4.1) leads to:

$$\begin{pmatrix} -i + \delta\left(C_h + C_0\dfrac{r_{d0}^2+1}{2r_{d0}}\right) & \delta C_0\dfrac{1-r_{d0}^2}{2r_{d0}} \\ \delta C_0\dfrac{1-r_{d0}^2}{2r_{d0}} & i + \delta\left(C_h + C_0\dfrac{r_{d0}^2+1}{2r_{d0}}\right) \end{pmatrix} \times \quad \text{(P3.4.11)}$$

$$\begin{pmatrix} 1 + 2ir_0\delta\dfrac{C_1 + C_h r_0}{1-r_0^2} \\ 1 \end{pmatrix} = \alpha\begin{pmatrix} -1 + 2i\delta\dfrac{C_h + C_1 r_0}{1-r_0^2} \\ 1 \end{pmatrix}.$$

Keeping only the terms up to the first order in $\delta\omega$, (P3.4.9) can be simplified:

$$\begin{cases} \left(-i + \delta\left(C_h + C_0\dfrac{r_{d0}^2+1}{2r_{d0}}\right)\right)\left(1 + 2ir_0\delta\dfrac{C_1 + C_h r_0}{1-r_0^2}\right) + \delta C_0\dfrac{1-r_{d0}^2}{2r_{d0}} \\ = \alpha\left(-1 + 2i\delta\dfrac{C_h + C_1 r_0}{1-r_0^2}\right) \\ \delta C_0\dfrac{1-r_{d0}^2}{2r_{d0}} + i + \delta\left(C_h + C_0\dfrac{r_{d0}^2+1}{2r_{d0}}\right) = \alpha, \end{cases}$$

$$\begin{cases} i - \delta \left(C_0 \dfrac{1}{r_{d0}} + \dfrac{4r_0 C_1 + (3 + r_0^2) C_h}{1 - r_0^2} \right) = \alpha \\ i + \delta \left(C_h + C_0 \dfrac{1}{r_{d0}} \right) = \alpha, \end{cases}$$

$$\frac{C_0}{2r_{d0}} = -\frac{r_0 C_1 + C_h}{1 - r_0^2}. \tag{P3.4.12}$$

Finally, substitution of (P3.4.3), (P3.4.4), (P3.4.5), (P3.4.8), and (P3.4.9) into (P3.4.12) gives:

$$\frac{\left(n_0^2 - n_{\mathrm{eff}} v_{\mathrm{g}}^{-1}\right)}{\left(n_0^2 - n_{\mathrm{eff}}^2\right)^{3/2}} \sqrt{n_{\mathrm{h}}^2 - n_{\mathrm{eff}}^2} \sqrt{n_1^2 - n_{\mathrm{eff}}^2}$$

$$= -\frac{\left(n_1^2 - n_{\mathrm{eff}} v_{\mathrm{g}}^{-1}\right) \sqrt{n_{\mathrm{h}}^2 - n_{\mathrm{eff}}^2} + \left(n_{\mathrm{h}}^2 - n_{\mathrm{eff}} v_{\mathrm{g}}^{-1}\right) \sqrt{n_1^2 - n_{\mathrm{eff}}^2}}{n_{\mathrm{h}}^2 - n_1^2}, \tag{P3.4.13}$$

which can be solved trivially for v_{g}. In the particular case of $n_0 = n_1$, expression (P3.4.13) can be further simplified:

$$v_{\mathrm{g}} = \frac{n_{\mathrm{eff}} \left(1 + r_0^3\right)}{n_{\mathrm{h}}^2 r_0^3 + n_1^2}. \tag{P3.4.14}$$

Finally, we demonstrate that in the case $n_0 = n_1$, at ω_0 the mode group velocity (P3.4.14) is always smaller than the mode phase velocity $v_{\mathrm{p}} = n_{\mathrm{eff}}^{-1}$ by showing that $v_{\mathrm{g}}^{-1} > v_{\mathrm{p}}^{-1}$. Indeed:

$$v_{\mathrm{g}}^{-1} - v_{\mathrm{p}}^{-1} = \frac{n_{\mathrm{h}}^2 r_0^3 + n_1^2}{n_{\mathrm{eff}} \left(1 + r_0^3\right)} - n_{\mathrm{eff}} = \frac{1}{n_{\mathrm{eff}}} \left(\frac{n_{\mathrm{h}}^2 r_0^3 + n_1^2}{1 + r_0^3} - n_{\mathrm{eff}}^2 \right)$$

$$= \frac{1}{n_{\mathrm{eff}}} \left(\frac{\left(n_{\mathrm{h}}^2 - n_{\mathrm{eff}}^2\right) r_0^3 + \left(n_1^2 - n_{\mathrm{eff}}^2\right)}{1 + r_0^3} \right) \underset{n_{\mathrm{eff}} < n_1 < n_{\mathrm{h}}}{>} 0. \tag{P3.4.15}$$

Chapter 5

Solution of Problem 5.1

Substitution of (P5.1.4) and (P5.1.5) into the coupler eigenequation (P5.1.3) leads to the following equation:

$$(\hat{H}_1 + \hat{H}_2 - \hat{U})(a_1 \,|\psi_1\rangle + a_2 \,|\psi_2\rangle) = (k_z^0 + \delta k_z)(a_1 \,|\psi_1\rangle + a_2 \,|\psi_2\rangle). \tag{P5.1.7}$$

Multiplying (P5.1.7) from the left by $\langle\psi_1|$ results in the following equation:

$$a_1 \,\langle\psi_1| \,\hat{H}_1 + \hat{H}_2 - \hat{U} \,|\psi_1\rangle + a_2 \,\langle\psi_1| \,\hat{H}_1 + \hat{H}_2 - \hat{U} \,|\psi_2\rangle$$

$$= (k_z^0 + \delta k_z)(a_1 \,\langle\psi_1 \mid \psi_1\rangle + a_2 \,\langle\psi_1 \mid \psi_2\rangle). \tag{P5.1.8}$$

Equation (P5.1.8) can be further simplified by collecting the terms of the first order and eliminating the terms of zeroth and second order:

$$a_1 \underbrace{\langle \psi_1 | \hat{H}_1 | \psi_1 \rangle}_{\text{0th order}} + a_1 \underbrace{\langle \psi_1 | \hat{H}_2 - \hat{U} | \psi_1 \rangle}_{\text{2nd order}} + a_2 \underbrace{\langle \psi_1 | \hat{H}_1 | \psi_2 \rangle}_{\text{1st order}} + a_2 \underbrace{\langle \psi_1 | \hat{H}_2 - \hat{U} | \psi_2 \rangle}_{\text{1st order}}$$

$$= a_1 \underbrace{k_z^0 \langle \psi_1 | \psi_1 \rangle}_{\text{0th order}} + a_2 \underbrace{k_z^0 \langle \psi_1 | \psi_2 \rangle}_{\text{1st order}} + a_1 \underbrace{\delta k_z \langle \psi_1 | \psi_1 \rangle}_{\text{1st order}} + \underbrace{a_2 \delta k_z \langle \psi_1 | \psi_2 \rangle}_{\text{2nd order}},$$

$$\text{(P5.1.9)}$$

finally leading to the following equation:

$$a_2 \langle \psi_1 | \hat{H}_2 - \hat{U} | \psi_2 \rangle = a_1 \delta k_z \langle \psi_1 | \psi_1 \rangle . \tag{P5.1.10}$$

In a similar manner, multiplying (P5.1.7) from the left by $\langle \psi_2 |$, and collecting the terms of the first order we arrive at the following equation:

$$a_1 \langle \psi_2 | \hat{H}_1 - \hat{U} | \psi_1 \rangle = a_2 \delta k_z \langle \psi_2 | \psi_2 \rangle . \tag{P5.1.11}$$

The system of (P5.1.10) and (P5.1.11) constitutes an eigenvalue problem with respect to the correction to the propagation constant δk_z. It can be trivially resolved to give:

$$\delta k_z = \pm \sqrt{\frac{\langle \psi_1 | \hat{H}_2 - \hat{U} | \psi_2 \rangle \langle \psi_2 | \hat{H}_1 - \hat{U} | \psi_1 \rangle}{\langle \psi_1 | \psi_1 \rangle \langle \psi_2 | \psi_2 \rangle}} = \pm \delta, \tag{P5.1.12}$$

with the corresponding supermode combinations:

$$\left| \psi_{c,\pm\delta} \right\rangle = | \psi_1 \rangle \pm | \psi_2 \rangle . \tag{P5.1.13}$$

Solution of Problem 5.2

From the expressions for the field components of a TE-polarized mode (3.1), (3.2) it follows that the transverse magnetic field is related to the transverse electric field in the following manner:

$$\mathbf{E}_{k_x}(z) = \mathbf{E}_{t,k_x}(z) = \hat{\mathbf{y}} E_{y,k_x}(z);$$

$$\mathbf{H}_{t,k_x}(z) = \hat{\mathbf{z}} H_{z,k_x}(z) = \hat{\mathbf{z}} \frac{k_x}{\omega} E_{y,k_x}(z). \tag{P5.2.1}$$

Therefore, the expression for the $\hat{\mathbf{x}}$ component of a modal flux can be written as:

$$\int_{-\infty}^{+\infty} dz \, \hat{\mathbf{x}} \left(\mathbf{E}_{t,k_x}^*(z) \times \mathbf{H}_{t,k_x}(z) + \mathbf{E}_{t,k_x}(z) \times \mathbf{H}_{t,k_x}^*(z) \right)$$

$$= 2 \frac{k_x}{\omega} \int_{-\infty}^{+\infty} dz \, |E_{y,k_x}(z)|^2 = 2 n_{\text{eff}} \left(\frac{d_c}{2} + (d_h + d_l) \frac{r_c^2}{1 - r_{TE}^2} \right). \tag{P5.2.2}$$

The perturbation to the dielectric profile with real refractive index n^r after addition of an imaginary contribution is $\delta \varepsilon = i 2 n^r n^i$. Substitution of $\delta \varepsilon$, as well as (P5.2.1) and

(P5.2.2), into the perturbation theory expression for the imaginary contribution to the propagation constant (5.44) leads to the following expression for the modal loss:

$$
\alpha_{abs}^{wg}[1/m] = 2\,|\beta_i| = 2\omega \frac{\left| \displaystyle\int_{-\infty}^{+\infty} dz \delta\varepsilon(z)\,\left|\mathbf{E}_{k_x}(z)\right|^2 \right|}{\displaystyle\int_{-\infty}^{+\infty} dz\,\hat{\mathbf{x}}\left(\mathbf{E}_{t,k_x}^*(z) \times \mathbf{H}_{t,k_x}(z) + \mathbf{E}_{t,k_x}(z) \times \mathbf{H}_{t,k_x}^*(z)\right)}
$$

$$
= 2\omega \frac{\left(n_c^r n_c^i \dfrac{d_c}{2} + (n_h^r n_h^i d_h + n_l^r n_l^i d_l)\dfrac{r_c^2}{1 - r_{TE}^2}\right)}{n_{eff}\left(\dfrac{d_c}{2} + (d_h + d_l)\dfrac{r_c^2}{1 - r_{TE}^2}\right)}.
\tag{P5.2.3}
$$

In the limit of large-core diameters:

$$
n_{eff} \underset{d_c \gg \lambda}{\simeq} n_c; \quad r_c^2 \underset{d_c \gg \lambda}{\simeq} \frac{1}{\varepsilon_h - \varepsilon_c}\left(\frac{\lambda}{2d_c}\right)^2; \quad r_{TE}^2 \underset{d_c \gg \lambda}{\simeq} \frac{\varepsilon_l - \varepsilon_c}{\varepsilon_h - \varepsilon_c},
\tag{P5.2.4}
$$

and (P5.2.3) can be further simplified:

$$
\alpha_{abs}^{wg}[1/m] = 2\omega\left(n_c^i + \left(\frac{n_h^r n_h^i}{n_c^r \sqrt{\varepsilon_h - \varepsilon_c}} + \frac{n_l^r n_l^i}{n_c^r \sqrt{\varepsilon_h - \varepsilon_c}}\right)\left(\frac{\lambda}{2d_c}\right)^3 \frac{1}{\varepsilon_h - \varepsilon_l}\right).
\tag{P5.2.5}
$$

Finally, from (P5.2.5) we conclude that the propagation loss of a core-guided mode becomes dominated by the bulk absorption loss of a gas filling the hollow core for the core diameters:

$$
d_c > \frac{\lambda}{2}\left(\left[\frac{n_h^i}{n_c^i}\frac{n_h^r}{n_c^r \sqrt{\varepsilon_h - \varepsilon_c}} + \frac{n_l^i}{n_c^i}\frac{n_l^r}{n_c^r \sqrt{\varepsilon_h - \varepsilon_c}}\right]\frac{1}{\varepsilon_h - \varepsilon_l}\right)^{1/3}.
\tag{P5.2.6}
$$

Chapter 6

Solution of Problem 6.1

For the case of TM polarization, the corresponding Maxwell's equations in the plane-wave formulation are:

$$
\text{TM}: \sum_{G',G\in G_{X-M}} \kappa(G - G')\left|k_t^{X-M} + G'\right|^2 E_{z,k_t^{X-M}}(G') = \omega_{k_t^{X-M}}^2 E_{z,k_t^{X-M}}(G), \tag{P6.1.3}
$$

where $G_{X-M} = [G_1 = (0,0),\, G_2 = (-2\pi/a, 0)]$. Equation (P6.1.3) can be expanded as:

$$\begin{cases} \kappa(0)\left|\mathbf{k}_t^{\text{X-M}} + \mathbf{G}_1\right|^2 E_{z,\mathbf{k}_t^{\text{X-M}}}(\mathbf{G}_1) + \kappa(\mathbf{G}_1 - \mathbf{G}_2)\left|\mathbf{k}_t^{\text{X-M}} + \mathbf{G}_2\right|^2 E_{z,\mathbf{k}_t^{\text{X-M}}}(\mathbf{G}_2) \\ = \omega_{\mathbf{k}_t^{\text{X-M}}}^2 E_{z,\mathbf{k}_t^{\text{X-M}}}(\mathbf{G}_1) \\ \kappa(\mathbf{G}_2 - \mathbf{G}_1)\left|\mathbf{k}_t^{\text{X-M}} + \mathbf{G}_1\right|^2 E_{z,\mathbf{k}_t^{\text{X-M}}}(\mathbf{G}_1) + \kappa(0)\left|\mathbf{k}_t^{\text{X-M}} + \mathbf{G}_2\right|^2 E_{z,\mathbf{k}_t^{\text{X-M}}}(\mathbf{G}_2) \\ = \omega_{\mathbf{k}_t^{\text{X-M}}}^2 E_{z,\mathbf{k}_t^{\text{X-M}}}(\mathbf{G}_2), \end{cases} \tag{P6.1.4}$$

which can be, furthermore, simplified by taking into account that $\kappa(\mathbf{G}_2 - \mathbf{G}_1) = \kappa(\mathbf{G}_1 - \mathbf{G}_2) = \kappa(2\pi/a)$. Thus, (P6.1.4) transforms into:

$$\left((\pi/a)^2 + k_y^2\right)\begin{pmatrix} \kappa(0) & \kappa(2\pi/a) \\ \kappa(2\pi/a) & \kappa(0) \end{pmatrix}\begin{pmatrix} E_{z,\mathbf{k}_t^{\text{X-M}}}(\mathbf{G}_1) \\ E_{z,\mathbf{k}_t^{\text{X-M}}}(\mathbf{G}_2) \end{pmatrix}$$

$$= \omega_{\mathbf{k}_t^{\text{X-M}}}^2 \begin{pmatrix} E_{z,\mathbf{k}_t^{\text{X-M}}}(\mathbf{G}_1) \\ E_{z,\mathbf{k}_t^{\text{X-M}}}(\mathbf{G}_2) \end{pmatrix}. \tag{P6.1.5}$$

The eigenproblem (P6.1.5) allows simple solution in the form:

$$\omega_1(k_y) = \sqrt{\left((\pi/a)^2 + k_y^2\right)(\kappa(0) + \kappa(2\pi/a))},$$

$$\left(E_{z,\mathbf{k}_t^{\text{X-M}}}(\mathbf{G}_1), E_{z,\mathbf{k}_t^{\text{X-M}}}(\mathbf{G}_2)\right) = \frac{1}{2}(1,1),$$

$$\omega_2(k_y) = \sqrt{\left((\pi/a)^2 + k_y^2\right)(\kappa(0) - \kappa(2\pi/a))},$$

$$\left(E_{z,\mathbf{k}_t^{\text{X-M}}}(\mathbf{G}_1), E_{z,\mathbf{k}_t^{\text{X-M}}}(\mathbf{G}_2)\right) = \frac{1}{2}(-1,1). \tag{P6.1.6}$$

In Fig. S6.1.1 solid curves plot dispersion relations of the perturbation split bands in (k_y, ω) coordinates for the nonvanishing refractive-index-contrast $\kappa(2\pi/a) \neq 0$.

To generate the plots we use $r = a/\sqrt{2\pi}$; $f = 0.5$; $\varepsilon_a = 2.56$; $\varepsilon_b = 1.96$, resulting in:

$$\kappa(0) = f\varepsilon_a^{-1} + (1-f)\varepsilon_b^{-1} = 0.4504$$

$$\kappa(2\pi/a) = 2f(\varepsilon_a^{-1} - \varepsilon_b^{-1})\frac{J_1(2\pi r/a)}{2\pi r/a} = -0.0236, \tag{P6.1.7}$$

and $\omega_1(k_y) < \omega_2(k_y)$. For comparison with an unperturbed case, the dotted curve shows the dispersion relation of the degenerate bands when the index contrast is zero $\omega = \sqrt{\left((\pi/a)^2 + k_y^2\right)\kappa(0)}$.

Using expressions for the field expansion coefficients (P6.1.6) we can now find the electric field intensities $|E_z|^2$ in the photonic-crystal eigenstates of the two perturbation split bands. Thus, substituting (P6.1.6) into (P6.1.2) we find, for the lower frequency state with $\omega_1(k_y)$:

$$E_{z,\omega_1}(x,y) = \left|E_{z,\omega_1}(\mathbf{G}_1)\exp(i\left(\mathbf{k}_t^{\text{X-M}} + \mathbf{G}_1\right)\mathbf{r}_t) + E_{z,\omega_1}(\mathbf{G}_2)\exp(i\left(\mathbf{k}_t^{\text{X-M}} + \mathbf{G}_2\right)\mathbf{r}_t)\right|^2$$

$$= 1/4\left|\exp(i(x\pi/a + yk_y)) + \exp(i(-x\pi/a + yk_y))\right|^2$$

$$= \cos^2(x\pi/a), \tag{P6.1.8}$$

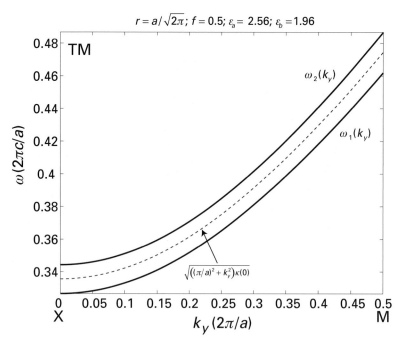

Figure S6.1.1 Dispersion relations of the perturbation split TM bands along the X–M direction $\mathbf{k}_t = (\pi/a, k_y)$ for the small-refractive-index-contrast square photonic-crystal lattice.

while for the higher frequency state with $\omega_2(k_y)$:

$$E_{z,\omega_2}(x, y) = \sin^2(x\pi/a). \tag{P6.1.9}$$

Field plots are presented in Fig. S6.1.2.

For the case of TE polarization, the corresponding Maxwell's equations in the plane-wave formulation are:

$$\text{TE}: \sum_{\mathbf{G}',\mathbf{G}\in\mathbf{G}_\omega} \kappa(\mathbf{G} - \mathbf{G}')\left(\mathbf{k}_t^{\text{X-M}} + \mathbf{G}'\right)\left(\mathbf{k}_t^{\text{X-M}} + \mathbf{G}\right) H_{z,\mathbf{k}_t^{\text{X-M}}}(\mathbf{G}') = \omega_{\mathbf{k}_t^{\text{X-M}}}^2 H_{z,\mathbf{k}_t^{\text{X-M}}}(\mathbf{G}),$$

$$\tag{P6.1.10}$$

where as before $\mathbf{G}_{\text{X-M}} = [\mathbf{G}_1 = (0, 0), \mathbf{G}_2 = (-2\pi/-2\pi a, 0)]$. Equation (P6.1.10) can be expanded as:

$$\begin{cases} \kappa(0)\left|\mathbf{k}_t^{\text{X-M}}+\mathbf{G}_1\right|^2 H_{z,\mathbf{k}_t^{\text{X-M}}}(\mathbf{G}_1)+\kappa(\mathbf{G}_1-\mathbf{G}_2)\left(\mathbf{k}_t^{\text{X-M}}+\mathbf{G}_1\right)\left(\mathbf{k}_t^{\text{X-M}}+\mathbf{G}_2\right) H_{z,\mathbf{k}_t^{\text{X-M}}}(\mathbf{G}_2) \\ = \omega_{\mathbf{k}_t^{\text{X-M}}}^2 H_{z,\mathbf{k}_t^{\text{X-M}}}(\mathbf{G}_1) \\ \\ \kappa(\mathbf{G}_2-\mathbf{G}_1)\left(\mathbf{k}_t^{\text{X-M}}+\mathbf{G}_2\right)\left(\mathbf{k}_t^{\text{X-M}}+\mathbf{G}_1\right) H_{z,\mathbf{k}_t^{\text{X-M}}}(\mathbf{G}_1)+\kappa(0)\left|\mathbf{k}_t^{\text{X-M}}+\mathbf{G}_2\right|^2 H_{z,\mathbf{k}_t^{\text{X-M}}}(\mathbf{G}_2) \\ = \omega_{\mathbf{k}_t^{\text{X-M}}}^2 H_{z,\mathbf{k}_t^{\text{X-M}}}(\mathbf{G}_2), \end{cases} \tag{P6.1.11}$$

which can, furthermore, be simplified by taking into account that $\kappa(\mathbf{G}_2 - \mathbf{G}_1) = \kappa(\mathbf{G}_1 - \mathbf{G}_2) = \kappa(2\pi/a)$. Thus, (P6.1.11) transforms into:

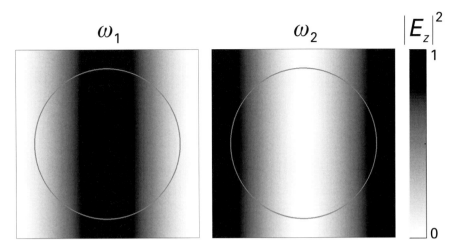

Figure S6.1.2 Electric field intensities $|E_z|^2$ of the TM-polarized photonic-crystal eigenstates in the two perturbation split bands. Field intensities are independent of the value of k_y. Intensities are shown in a single lattice cell of a photonic crystal.

$$
\begin{pmatrix}
\left((\pi/a)^2 + k_y^2\right)\kappa(0) & -\kappa(2\pi/a)\left((\pi/a)^2 - k_y^2\right) \\
-\kappa(2\pi/a)\left((\pi/a)^2 - k_y^2\right) & \left((\pi/a)^2 + k_y^2\right)\kappa(0)
\end{pmatrix}
\begin{pmatrix}
H_{z,\mathbf{k}_t^{\mathrm{X-M}}}(\mathbf{G}_1) \\
H_{z,\mathbf{k}_t^{\mathrm{X-M}}}(\mathbf{G}_2)
\end{pmatrix}
$$

$$
= \omega^2_{\mathbf{k}_t^{\mathrm{X-M}}}
\begin{pmatrix}
H_{z,\mathbf{k}_t^{\mathrm{X-M}}}(\mathbf{G}_1) \\
H_{z,\mathbf{k}_t^{\mathrm{X-M}}}(\mathbf{G}_2)
\end{pmatrix}. \tag{P6.1.12}
$$

The eigenproblem (P6.1.12) allows a simple solution in the form:

$$
\omega_1 = \sqrt{\kappa(0)\left((\pi/a)^2 + k_y^2\right) + \kappa(2\pi/a)\left((\pi/a)^2 - k_y^2\right)},
$$

$$
\left(H_{z,\mathbf{k}_t^{\mathrm{X-M}}}(\mathbf{G}_1), H_{z,\mathbf{k}_t^{\mathrm{X-M}}}(\mathbf{G}_2)\right) = \frac{1}{2}(-1, 1),
$$

$$
\omega_2 = \sqrt{\kappa(0)\left((\pi/a)^2 + k_y^2\right) - \kappa(2\pi/a)\left((\pi/a)^2 - k_y^2\right)},
$$

$$
\left(H_{z,\mathbf{k}_t^{\mathrm{X-M}}}(\mathbf{G}_1), H_{z,\mathbf{k}_t^{\mathrm{X-M}}}(\mathbf{G}_2)\right) = \frac{1}{2}(1, 1). \tag{P6.1.13}
$$

In Fig. S6.1.3, solid curves plot dispersion relations of the two perturbation split bands in (k_y, ω) coordinates for the nonvanishing refractive-index-contrast $\kappa(2\pi/a) \neq 0$.

Using the expressions for the field expansion coefficients (P6.1.13) we can now find the magnetic field intensities $|H_z|^2$ in the photonic-crystal eigenstates of the two perturbation split bands. Thus, substituting (P6.1.13) into (P6.1.2) we find, for the lower frequency state with $\omega_1(k_y)$:

$$
H_{z,\omega_1}(x, y) = \left|H_{z,\omega_1}(\mathbf{G}_1)\exp(i\left(\mathbf{k}_t^{\mathrm{X-M}} + \mathbf{G}_1\right)\mathbf{r}_t) + H_{z,\omega_1}(\mathbf{G}_2)\exp(i\left(\mathbf{k}_t^{\mathrm{X-M}} + \mathbf{G}_2\right)\mathbf{r}_t)\right|^2
$$

$$
= 1/4\left|\exp(i(x\pi/a + yk_y)) - \exp(i(-x\pi/a + yk_y))\right|^2 = \sin^2(x\pi/a), \tag{P6.1.14}
$$

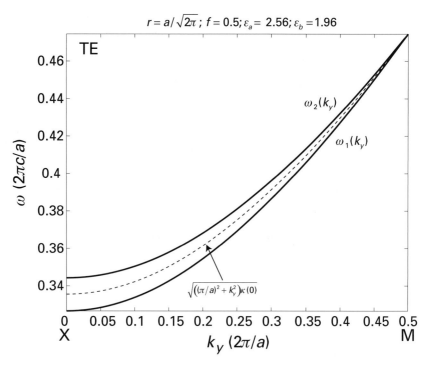

Figure S6.1.3 Dispersion relations of the perturbation split TE bands along the X–M direction $k_t = (\pi/a, k_y)$ for the small refractive-index-contrast square photonic-crystal lattice.

while for the higher frequency state with $\omega_2(k_y)$:

$$H_{z,\omega_2}(x, y) = \cos^2(x\pi/a). \qquad (P6.1.15)$$

Field plots for magnetic fields are the same as in Fig. S6.1.2 but reversed in their order. Thus, the magnetic field distribution in the lowest frequency mode $\omega_1(k_y)$ is that of an electric field distribution in Fig. S6.1.2 but for the higher frequency $\omega_2(k_y)$, and vice versa.

Solution of Problem 6.2

Using the discrete Fourier transform in the $\hat{x}$ direction and the continuous Fourier transform in the $\hat{y}$ direction, similarly to (6.53), the magnetic field of a TE-polarized guided mode can be rewritten in terms of the integral over the first Brillouin zone only:

$$H_{z,k_x}(\mathbf{r}) = \sum_{\mathbf{G}} \int_{-\pi/a}^{+\pi/a} dk_y \, H_z(\mathbf{k} + \mathbf{G}) \exp(i(\mathbf{k} + \mathbf{G})\mathbf{r})$$

$$\mathbf{G} = \bar{b}_x P_x + \bar{b}_y P_y; \quad \bar{b}_x = \frac{2\pi}{a}\hat{x}; \quad \bar{b}_y = \frac{2\pi}{a}\hat{y}; \left(P_x, P_y\right) \subset \text{integers}$$

$$\mathbf{k} = k_x\hat{x} + k_y\hat{y}; \quad k_x \text{ is a constant.} \qquad (P6.2.1)$$

The inverse dielectric function of a photonic-crystal waveguide is given by (6.54). Using the discrete and continuous Fourier transforms it can be furthermore presented as in

(6.55). Substituting (P6.2.1) and (6.55) into Maxwell's equation (6.10) written in terms of only the magnetic field $\omega^2 \mathbf{H} = \nabla \times (1/\varepsilon(\mathbf{r})\cdot\nabla \times \mathbf{H})$, we get:

$$\omega^2(k_x)\hat{z} \sum_{G'} \int_{-\pi/a}^{+\pi/a} dk'_y\, H_z(\mathbf{k}' + \mathbf{G}') \exp\left(i(\mathbf{k}' + \mathbf{G}')\mathbf{r}\right)$$

$$= \nabla \times \left(\begin{array}{l} \left[\dfrac{1}{\varepsilon_b} + \left(\dfrac{1}{\varepsilon_a} - \dfrac{1}{\varepsilon_b}\right)\sum_{G''} S_a(\mathbf{G}'')\exp\left(i\mathbf{G}''\mathbf{r}\right) + \left(\dfrac{1}{\varepsilon_d} - \dfrac{1}{\varepsilon_a}\right)\right. \\ \qquad \left.\times \sum_{G''} \dfrac{a}{2\pi} \int_{-\pi/a}^{+\pi/a} dk''_y\, S_d(k''_y\hat{y} + \mathbf{G}'')\exp\left(i(k''_y\hat{y} + \mathbf{G}'')\mathbf{r}\right)\right]\cdot \\ \nabla \times \left[\sum_{G'}\hat{z} \int_{-\pi/a}^{+\pi/a} dk'_y\, H_z(\mathbf{k}' + \mathbf{G}')\exp\left(i(\mathbf{k}' + \mathbf{G}')\mathbf{r}\right)\right] \end{array} \right) \text{, (P6.2.2)}$$

which can be further simplified to be:

$$\omega^2(k_x) \sum_{G'} \int_{-\pi/a}^{+\pi/a} dk'_y\, H_z(\mathbf{k}' + \mathbf{G}') \exp\left(i(\mathbf{k}' + \mathbf{G}')\mathbf{r}\right)$$

$$= \left[\dfrac{1}{\varepsilon_b} \sum_{G'} \int_{-\pi/a}^{+\pi/a} dk'_y\, H_z(\mathbf{k}' + \mathbf{G}')\left|\mathbf{k}' + \mathbf{G}'\right|^2 \exp\left(i(\mathbf{k}' + \mathbf{G}')\mathbf{r}\right)\right]$$

$$+ \left[\left(\dfrac{1}{\varepsilon_a} - \dfrac{1}{\varepsilon_b}\right)\sum_{G'}\sum_{G''} \int_{-\pi/a}^{+\pi/a} dk'_y\, H_z(\mathbf{k}' + \mathbf{G}')S_a(\mathbf{G}'')(\mathbf{k}' + \mathbf{G}' + \mathbf{G}'')\right.$$

$$\left.\times (\mathbf{k}' + \mathbf{G}')\exp(i(\mathbf{k}' + \mathbf{G}' + \mathbf{G}'')\mathbf{r})\right]$$

$$+ \left[\left(\dfrac{1}{\varepsilon_d} - \dfrac{1}{\varepsilon_a}\right)\dfrac{a}{2\pi}\sum_{G'}\sum_{G''} \int_{-\pi/a}^{+\pi/a} dk'_y \int_{-\pi/a}^{+\pi/a} dk''_y\, H_z(\mathbf{k}' + \mathbf{G}')S_d(k''_y\hat{y} + \mathbf{G}'')\right.$$

$$\left.\times (\mathbf{k}' + \mathbf{G}' + k''_y\hat{y} + \mathbf{G}'')(\mathbf{k}' + \mathbf{G}')\exp(i(\mathbf{k}' + \mathbf{G}' + k''_y\hat{y} + \mathbf{G}'')\mathbf{r})\right]. \qquad \text{(P6.2.3)}$$

Multiplying the left- and right-hand sides of (P6.2.3) by $\frac{1}{(2\pi)^2}\exp(-i(\mathbf{k}+\mathbf{G})\mathbf{r})$, integrating over the 2D vector $\mathbf{r}$, and using the orthogonality of the 2D plane waves in the form:

$$\dfrac{1}{(2\pi)^2}\int_{\infty} d\mathbf{r}\, \exp(i\mathbf{k}\mathbf{r}) = \delta(\mathbf{k}),$$

we finally get the governing equations for the plane-wave method applied to TE polarization:

$$\omega^2(k_x) H_z(\mathbf{k}+\mathbf{G}) = \frac{1}{\varepsilon_b} H_z(\mathbf{k}+\mathbf{G}) |\mathbf{k}+\mathbf{G}|^2$$

$$+ \left(\frac{1}{\varepsilon_a}-\frac{1}{\varepsilon_b}\right) \sum_{\mathbf{G}'} H_z(\mathbf{k}+\mathbf{G}') S_a(\mathbf{G}-\mathbf{G}')(\mathbf{k}+\mathbf{G})(\mathbf{k}+\mathbf{G}')$$

$$+ \left(\frac{1}{\varepsilon_d}-\frac{1}{\varepsilon_a}\right) \frac{a}{2\pi} \sum_{\mathbf{G}'} \int_{-\pi/a}^{+\pi/a} dk_y' H_z(\mathbf{k}'+\mathbf{G}') S_d(\mathbf{k}-\mathbf{k}'+\mathbf{G}-\mathbf{G}')(\mathbf{k}+\mathbf{G})(\mathbf{k}'+\mathbf{G}')$$

$$\mathbf{G} = \bar{b}_x P_x + \bar{b}_y P_y; \quad \mathbf{G}' = \bar{b}_x P_x' + \bar{b}_y P_y'; \quad \bar{b}_x = \frac{2\pi}{a}\hat{\mathbf{x}}; \quad \bar{b}_y = \frac{2\pi}{a}\hat{\mathbf{y}};$$

$$(P_x, P_y, P_x', P_y',) \subset \text{integers}$$

$$\mathbf{k} = k_x \hat{\mathbf{x}} + k_y \hat{\mathbf{y}}; \quad \mathbf{k}' = k_x \hat{\mathbf{x}} + k_y' \hat{\mathbf{y}}$$

$$k_x \text{ is a modal propagation constant}; (k_y, k_y') \subset \left(-\frac{\pi}{a}, \frac{\pi}{a}\right). \qquad \text{(P6.2.4)}$$

Solution of Problem 6.3

Using continuous Fourier transforms in the $\hat{\mathbf{x}}$ and $\hat{\mathbf{y}}$ directions, the magnetic field can be rewritten similarly to (6.79) in terms of the integral over the first Brillouin zone (FBZ):

$$H_{z,\omega_{\text{res}}}(\mathbf{r}) = \sum_{\mathbf{G}} \int_{\text{FBZ}} d\mathbf{k} H_z(\mathbf{k}+\mathbf{G}) \exp\left(i(\mathbf{k}+\mathbf{G})\mathbf{r}\right)$$

$$\mathbf{G} = \bar{b}_x P_x + \bar{b}_y P_y; \quad \bar{b}_x = \frac{2\pi}{a}\hat{\mathbf{x}}; \quad \bar{b}_y = \frac{2\pi}{a}\hat{\mathbf{y}}; (P_x, P_y) \subset \text{integers}$$

$$\mathbf{k} = k_x \hat{\mathbf{x}} + k_y \hat{\mathbf{y}}; \quad (k_x, k_y) \subset \left(-\frac{\pi}{a}, \frac{\pi}{a}\right). \qquad \text{(P6.3.1)}$$

The inverse dielectric function of a photonic-crystal waveguide is given by (6.80). Using the discrete and continuous Fourier transforms it can be furthermore presented as in (6.81). Substituting (P6.3.1) and (6.81) into Maxwell's equation (6.10) written in terms of only the magnetic field $\omega^2 \mathbf{H} = \nabla \times (1/\varepsilon(\mathbf{r}) \cdot \nabla \times \mathbf{H})$ we get:

$$\omega^2(k_x) \hat{\mathbf{z}} \sum_{\mathbf{G}'} \int_{\text{FBZ}} d\mathbf{k}' H_z(\mathbf{k}'+\mathbf{G}') \exp\left(i(\mathbf{k}'+\mathbf{G}')\mathbf{r}\right)$$

$$= \nabla \times \left(\begin{array}{c} \left[\frac{1}{\varepsilon_b} + \left(\frac{1}{\varepsilon_a}-\frac{1}{\varepsilon_b}\right) \sum_{\mathbf{G}''} S_a(\mathbf{G}'') \exp\left(i\mathbf{G}''\mathbf{r}\right) + \left(\frac{1}{\varepsilon_d}-\frac{1}{\varepsilon_a}\right) \right. \\ \left. \times \sum_{\mathbf{G}''} \left(\frac{a}{2\pi}\right)^2 \int_{\text{FBZ}} d\mathbf{k}'' S_d(\mathbf{k}''+\mathbf{G}'') \exp\left(i(\mathbf{k}''+\mathbf{G}'')\mathbf{r}\right) \right] \cdot \\ \nabla \times \left[\sum_{\mathbf{G}'} \hat{\mathbf{z}} \int_{\text{FBZ}} d\mathbf{k}' H_z(\mathbf{k}'+\mathbf{G}') \exp\left(i(\mathbf{k}'+\mathbf{G}')\mathbf{r}\right) \right] \end{array} \right), \qquad \text{(P6.3.2)}$$

which can be further simplified to be:

$$\omega_{res}^2 \sum_{G'} \int_{FBZ} d\mathbf{k}' H_z(\mathbf{k}' + \mathbf{G}') \exp\left(i(\mathbf{k}' + \mathbf{G}')\mathbf{r}\right)$$

$$= \left[\frac{1}{\varepsilon_b} \sum_{G'} \int_{FBZ} d\mathbf{k}' H_z(\mathbf{k}' + \mathbf{G}') |\mathbf{k}' + \mathbf{G}'|^2 \exp\left(i(\mathbf{k}' + \mathbf{G}')\mathbf{r}\right) \right]$$

$$+ \left[\left(\frac{1}{\varepsilon_a} - \frac{1}{\varepsilon_b} \right) \sum_{G'} \sum_{G''} \int_{FBZ} d\mathbf{k}' H_z(\mathbf{k}' + \mathbf{G}') S_a(\mathbf{G}'')(\mathbf{k}' + \mathbf{G}' + \mathbf{G}'') \right.$$

$$\left. \times \ (\mathbf{k}' + \mathbf{G}') \exp\left(i(\mathbf{k}' + \mathbf{G}' + \mathbf{G}'')\mathbf{r}\right) \right]$$

$$+ \left[\left(\frac{1}{\varepsilon_d} - \frac{1}{\varepsilon_a} \right) \left(\frac{a}{2\pi} \right)^2 \sum_{G'} \sum_{G''} \int_{FBZ} d\mathbf{k}' \int_{FBZ} d\mathbf{k}'' H_z(\mathbf{k}' + \mathbf{G}') S_d(\mathbf{k}'' + \mathbf{G}'') \right.$$

$$\left. \times \ (\mathbf{k}' + \mathbf{G}' + \mathbf{k}'' + \mathbf{G}'')(\mathbf{k}' + \mathbf{G}') \exp(i(\mathbf{k}' + \mathbf{G}' + \mathbf{k}'' + \mathbf{G}'')\mathbf{r}) \right]. \qquad \text{(P6.3.3)}$$

Multiplying the left- and right-hand sides of (P6.3.3) by $\frac{1}{(2\pi)^2} \exp(-i(\mathbf{k} + \mathbf{G})\mathbf{r})$, integrating over the 2D vector $\mathbf{r}$, and using the orthogonality of the 2D plane waves in the form:

$$\frac{1}{(2\pi)^2} \int_{\infty} d\mathbf{r} \, \exp(i\mathbf{k}\mathbf{r}) = \delta(\mathbf{k}),$$

we finally get the governing equations for the plane-wave method applied to TE polarization:

$$\omega_{res}^2 H_z(\mathbf{k} + \mathbf{G}) = \frac{1}{\varepsilon_b} H_z(\mathbf{k} + \mathbf{G}) |\mathbf{k} + \mathbf{G}|^2$$

$$+ \left(\frac{1}{\varepsilon_a} - \frac{1}{\varepsilon_b} \right) \sum_{G'} H_z(\mathbf{k} + \mathbf{G}') S_a(\mathbf{G} - \mathbf{G}')(\mathbf{k} + \mathbf{G})(\mathbf{k} + \mathbf{G}')$$

$$+ \left(\frac{1}{\varepsilon_d} - \frac{1}{\varepsilon_a} \right) \left(\frac{a}{2\pi} \right)^2 \sum_{G'} \int_{FBZ} d\mathbf{k}' H_z(\mathbf{k}' + \mathbf{G}') S_d(\mathbf{k} - \mathbf{k}' + \mathbf{G} - \mathbf{G}')(\mathbf{k} + \mathbf{G})(\mathbf{k}' + \mathbf{G}')$$

$$\mathbf{G} = \bar{b}_x P_x + \bar{b}_y P_y; \quad \mathbf{G}' = \bar{b}_x P_x' + \bar{b}_y P_y';$$

$$\bar{b}_x = \frac{2\pi}{a} \hat{\mathbf{x}}; \quad \bar{b}_y = \frac{2\pi}{a} \hat{\mathbf{y}}; \left(P_x, P_y, P_x', P_y', \right) \subset \text{integers}$$

$$\mathbf{k} = k_x \hat{\mathbf{x}} + k_y \hat{\mathbf{y}}; \quad (k_x, k_y) \subset \left(-\frac{\pi}{a}, \frac{\pi}{a} \right)$$

$$\mathbf{k}' = k_x' \hat{\mathbf{x}} + k_y' \hat{\mathbf{y}}; \quad (k_x', k_y') \subset \left(-\frac{\pi}{a}, \frac{\pi}{a} \right). \qquad \text{(P6.3.4)}$$

Solution of Problem 6.4

According to (6.67), the magnetic field of a guided mode can be expended into the linear combination of the modes of a perfectly periodic photonic crystal:

$$
\mathbf{H}^{\text{wg}}_{k_x} = \sum_{m=1}^{+\infty} \int_{-\pi/a}^{\pi/a} dk_y \, A_{m,k_y} \mathbf{H}^{\text{PC}}_{m,k_x,k_y}(\mathbf{r}) = \sum_{m=1}^{+\infty} \int_{-\pi/a}^{\pi/a} dk_y \, A_{m,k_y} \exp(i\mathbf{k}\mathbf{r}) \mathbf{U}^{\text{PC}}_{m,k_x,k_y}(\mathbf{r}),
$$

$$(P6.4.1)$$

where $\mathbf{U}^{\text{PC}}_{m,k_x,k_y}(\mathbf{r})$ is a periodic part of a Bloch solution. Expressing the expansion coefficients A_{m,k_y} in terms of the auxiliary functions ϕ_{m,k_y} using (6.73) we can rewrite (P6.4.1) as:

$$
\mathbf{H}^{\text{wg}}_{k_x} = \sum_{m=1}^{+\infty} \int_{-\pi/a}^{\pi/a} dk_y \, \frac{\phi_{m,k_y}}{\left(\omega^{\text{wg}}_{k_x}\right)^2 - \omega^2_{\text{PC},m}(k_x, k_y)} \exp(i\mathbf{k}\mathbf{r}) \mathbf{U}_{m,k_x,k_y}(\mathbf{r}). \qquad (P6.4.2)
$$

In what follows we consider the photonic-crystal-waveguide guided mode bifurcating from the lower edge of a fundamental bandgap, corresponding to the low index defect $\varepsilon_d < \varepsilon_a$. In the case of a weak defect, as established in Section 6.6.1, for a given value of a Bloch wavenumber k_x, the frequency of a guided mode will be very close to the frequency of a perfect photonic-crystal bulk state along the $X_1 - M$ direction:

$$
\omega^{\text{wg}}_{k_x} \underset{\Delta\omega \to 0}{=} \omega_{\text{PC},1}(k_x, \pi/a) + \Delta\omega(k_x). \qquad (P6.4.3)
$$

Assuming slow functional dependence of ϕ_{1,k_y} and $\mathbf{U}^{\text{PC}}_{1,k_x,k_y}(\mathbf{r})$ in the vicinity of $\mathbf{k} \approx \mathbf{k}_{\text{wg}} = (k_x, \pi/a)$, we can simplify (P6.4.2) as:

$$
\mathbf{H}^{\text{wg}}_{k_x} \underset{\mathbf{k}_{\text{wg}}=(k_x,\pi/a)}{\approx} \exp(i\mathbf{k}_{\text{wg}}\mathbf{r}) \mathbf{U}^{\text{PC}}_{1,k_x,\pi/a}(\mathbf{r}) \frac{\phi_{1,\pi/a}}{2\omega_{\text{PC},1}(k_x, \pi/a)}
$$

$$
\times \int_{-\pi/a}^{\pi/a} d(\Delta k_y) \frac{\exp(i\Delta k_y y)}{\Delta\omega(k_x) + |\gamma(k_x)| \delta k^2_y}. \qquad (P6.4.4)
$$

Performing coordinate transformation $\delta k_y = \xi/|y|$, assuming the distances from the core $|y| \gg a$, the integral in (P6.4.4) becomes:

$$
\int_{-\pi/a}^{\pi/a} d(\delta k_y) \frac{\exp(i\delta k_y y)}{\Delta\omega(k_x) + |\gamma(k_x)| \delta k^2_y} \underset{|y| \gg a}{\approx} \frac{|y|}{|\gamma(k_x)|} \int_{-\infty}^{+\infty} d\xi \frac{\exp(i\xi)}{\frac{\Delta\omega(k_x)|y|^2}{|\gamma(k_x)|} + \xi^2}
$$

$$
= \frac{\pi}{\sqrt{\Delta\omega(k_x)|\gamma(k_x)|}} \exp\left(-|y|\sqrt{\frac{\Delta\omega(k_x)}{|\gamma(k_x)|}}\right). \qquad (P6.4.5)
$$

Thus, defining the envelope function $\phi(\Delta\omega(k_x)|y|^2/|\gamma(k_x)|)$ as:

$$
\phi(\Delta\omega(k_x)|y|^2/|\gamma(k_x)|) = \exp\left(-\sqrt{\Delta\omega(k_x)|y|^2/|\gamma(k_x)|}\right), \qquad (P6.4.6)
$$

the magnetic field of a guided mode in the weak localization limit can be written up to a constant as:

$$\mathbf{H}^{\text{wg}}_{k_x}\underset{\mathbf{k}_{\text{wg}}=(k_x,\pi/a)}{\approx}\exp(i\mathbf{k}_{\text{wg}}\mathbf{r})\mathbf{U}^{\text{PC}}_{1,k_x,\pi/a}(\mathbf{r})\phi(\Delta\omega(k_x))\,|y|^2/|\gamma(k_x)|),\qquad(\text{P6.4.7})$$

thus exhibiting exponential decay outside a waveguide core region.

Note, that the form (P6.4.5) is only valid when $|y|\gg a$. From the functional dependence of an envelope function it follows that the characteristic size of a guided mode is $|y|\sim\sqrt{|\gamma(k_x)|/\Delta\omega(k_x)}$.

Solution of Problem 6.5

According to (6.90), the magnetic field of a defect state can be expended into the linear combination of the modes of a perfectly periodic photonic crystal:

$$\mathbf{H}^{\text{res}}_{\omega_{\text{res}}}=\sum_{m=1}^{+\infty}\int_{-\pi/a}^{\pi/a}dk_x\int_{-\pi/0a}^{\pi/a}dk_y\,A_{m,k_x,k_y}\mathbf{H}^{\text{PC}}_{m,k_x,k_y}(\mathbf{r})$$

$$=\sum_{m=1}^{+\infty}\int_{-\pi/a}^{\pi/a}dk_x\int_{-\pi/a}^{\pi/a}dk_y\,A_{m,k_x,k_y}\exp(i\mathbf{k}\mathbf{r})\mathbf{U}^{\text{PC}}_{m,k_x,k_y}(\mathbf{r}),\qquad(\text{P6.5.1})$$

where $\mathbf{U}^{\text{PC}}_{m,k_x,k_y}(\mathbf{r})$ is a periodic part of a Bloch solution. Expressing expansion coefficients A_{m,k_x,k_y} in terms of the auxiliary functions ϕ_{m,k_x,k_y} using (6.96) we can rewrite (P6.5.1) as:

$$\mathbf{H}^{\text{res}}_{\omega_{\text{res}}}=\sum_{m=1}^{+\infty}\int_{-\pi/a}^{\pi/a}dk_x\int_{-\pi/a}^{\pi/a}dk_y\frac{\phi_{m,k_x,k_y}}{\omega_{\text{res}}^2-\omega_{\text{PC},m}^2(k_x,k_y)}\exp(i\mathbf{k}\mathbf{r})\mathbf{U}_{m,k_x,k_y}(\mathbf{r}).\qquad(\text{P6.5.2})$$

In what follows, we consider a defect state bifurcating from the lower edge of a fundamental bandgap, corresponding to the lower index defect $\varepsilon_d<\varepsilon_a$. In the case of a weak defect, as established in Section 6.6.2 the resonator state frequency will be very close to the frequency of a perfect photonic-crystal bulk state at the M symmetry point:

$$\omega_{\text{res}}\underset{\Delta\omega\to0}{=}\omega_{\text{PC},1}(\pi/a,\pi/a)+\Delta\omega.\qquad(\text{P6.5.3})$$

Assuming slow functional dependence of ϕ_{1,k_x,k_y} and $\mathbf{U}^{\text{PC}}_{1,k_x,k_y}(\mathbf{r})$ in the vicinity of an M point $\mathbf{k}\approx\mathbf{k}_{\text{M}}=(\pi/a,\pi/a)$ we can simplify (P6.5.2) as:

$$\mathbf{H}^{\text{res}}_{\omega_{\text{res}}}\underset{\delta\mathbf{k}=\mathbf{k}-\mathbf{k}_{\text{M}}}{\approx}\exp(i\mathbf{k}_{\text{M}}\mathbf{r})\mathbf{U}^{\text{PC}}_{1,\pi/a,\pi/a}(\mathbf{r})\frac{\phi_{1,\pi/a,\pi/a}}{2\omega_{\text{PC},1}(\pi/a,\pi/a)}$$

$$\times\int_{-\pi/a}^{\pi/a}d(\delta k_x)\int_{-\pi/a}^{\pi/a}d(\delta k_y)\frac{\exp(i\delta\mathbf{k}\mathbf{r})}{\Delta\omega+|\gamma|\left(\delta k_x^2+\delta k_y^2\right)}.\qquad(\text{P6.5.4})$$

Performing the coordinate transformation $\delta\mathbf{k}=\boldsymbol{\xi}/|\mathbf{r}|$, assuming the distances $|\mathbf{r}|\gg a$,

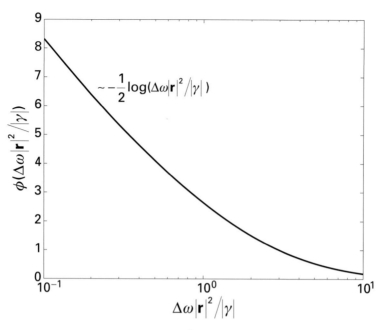

Figure S6.5.1 Envelope function $\phi(\delta\omega|\mathbf{r}|^2/|\gamma|)$ and its asymptotic for the small values of the argument. This form of an envelope function is valid for $|\mathbf{r}| \gg a$.

and assuming that $\mathbf{r}$ is directed along $\hat{\mathbf{x}}$, the integral in (P6.5.4) becomes:

$$
\int_{-\pi/a}^{\pi/a} \mathrm{d}(\delta k_x) \int_{-\pi/a}^{\pi/a} \mathrm{d}(\delta k_y) \frac{\exp(i\delta\mathbf{kr})}{\Delta\omega + |\gamma|\left(\delta k_x^2 + \delta k_y^2\right)}
$$

$$
\approx_{|\mathbf{r}|\gg a} \frac{1}{|\gamma|} \int_{-\infty}^{+\infty} \mathrm{d}\xi_x \int_{-\infty}^{+\infty} \mathrm{d}\xi_x \frac{\exp(i\xi_x)}{\dfrac{\Delta\omega\,|\mathbf{r}|^2}{|\gamma|} + |\xi|^2}. \tag{P6.5.5}
$$

This integral can be further simplified in cylindrical coordinates. Thus, defining the envelope function $\phi(\Delta\omega\,|\mathbf{r}|^2/|\gamma|)$ as:

$$
\phi(\Delta\omega\,|\mathbf{r}|^2/|\gamma|) = \int_{-\infty}^{+\infty} \mathrm{d}\xi_x \int_{-\infty}^{+\infty} \mathrm{d}\xi_x \frac{\exp(i\xi_x)}{\dfrac{\Delta\omega\,|\mathbf{r}|^2}{|\gamma|} + |\xi|^2}
$$

$$
= \int_{0}^{+\infty} \xi_\rho\,\mathrm{d}\xi_\rho \int_{0}^{2\pi} \mathrm{d}\xi_\theta \frac{\exp(i\xi_\rho\cos(\xi_\theta))}{\dfrac{\Delta\omega\,|\mathbf{r}|^2}{|\gamma|} + \xi_\rho^2} = 2\pi \int_{0}^{+\infty} \mathrm{d}\xi_\rho \frac{J_0(\xi_\rho)\xi_\rho}{\dfrac{\Delta\omega\,|\mathbf{r}|^2}{|\gamma|} + \xi_\rho^2}, \tag{P6.5.6}
$$

the magnetic field of a resonator state in the weak localization limit can be written up to a constant as:

$$\mathbf{H}^{res}_{\omega_{res}} \approx \exp(i\mathbf{k}_M\mathbf{r})\mathbf{U}^{PC}_{1,\pi/a,\pi/a}(\mathbf{r})\phi(\Delta\omega\,|\mathbf{r}|^2/|\gamma|). \qquad (P6.5.7)$$

Figure S6.5.1 shows the general form of an envelope function $\phi(\Delta\omega\,|\mathbf{r}|^2/|\gamma|)$ and indicates its asymptotics. Note that the form (P6.5.5) is only valid when $|\mathbf{r}| \gg a$. From the functional dependence of an envelope function, it follows that the characteristic size of a defect state is $|\mathbf{r}| \sim \sqrt{|\gamma|/\Delta\omega}$.

Chapter 7

Solution of Problem 7.1

Example of a MATLAB code for the truncation of (7.46) for the calculation of eigenvalues in (7.39) with $k_1 = k_2 = 0$, $E_0 = 15$, $I_0 = 6$.

```
I0 = 6;  E0 = 15;  Lx = pi;  Ly = pi;  N = 11;
dx = Lx/200;  dy = Ly/200;  kx = 2*pi/Lx;  ky = 2*pi/Ly;
x = 0:dx:Lx;  y = 0:dy:Ly;  [X Y] = meshgrid(x, y);
Pfunc = -E0./(1+I0*cos(X).^2.*cos(Y).^2);
k1 = 0;k2 = 0;
Pfourier = zeros(2*N+1);
for j = 1:1:2*N+1
    for l = 1:1:2*N+1
        integrand = Pfunc.*exp(-i*((j-N-1)*kx*X+(l-N-1)*ky*Y));
        Pfourier(j, l) = dx*dy*trapz(trapz(integrand))/(Lx*Ly);
    end
end
for k = 1:(2*N+1)^2
    for l = 1:(2*N+1)^2
        m  = -N+floor((k-1)/(2*N+1));  n = -N+mod(k-1, 2*N+1);
        ii = -N+floor((l-1)/(2*N+1));  jj = -N+mod(l-1, 2*N+1);
        if m-ii+N+1 < 1 | m-ii+N+1 > 2*N+1 | n-jj+N+1 < 1 |...
        n-jj+N+1 > 2*N+1
            L(k, l) = 0;
        else
            L(k, l) = Pfourier(m-ii+N+1, n-jj+N+1);
        end
    if k  = = 1
            L(k, l) = L(k, l)-(k1+m*kx)^2-(k2+n*ky)^2;
        end
        end
end
mu = sort(eig(-L));
mu(1:8)
```

Solution of Problem 7.2

(i) Substituting (7.49) into (7.50) we get:

$$\int_{-\infty}^{\infty}\int_{-\infty}^{\infty} B_m^*(x, y; k_1, k_2)B_n(x, y; \widehat{k}_1, \widehat{k}_2)dxdy$$

$$= \int_{-\infty}^{\infty}\int_{-\infty}^{\infty} e^{i(\widehat{k}_1-k_1)x}e^{i(\widehat{k}_2-k_2)y} G_m^*(x, y; k_1, k_2)G_n(x, y; \widehat{k}_1, \widehat{k}_2)dxdy$$

The integral above can be rewritten equivalently as:

$$\sum_{j_1=-\infty}^{+\infty}\sum_{j_2=-\infty}^{+\infty}\int_0^{\pi}\int_0^{\pi} e^{i(\widehat{k}_1-k_1)(x+j_1\pi)}e^{i(\widehat{k}_2-k_2)(y+j_2\pi)} \times$$

$$G_m^*(x + j_1\pi, y + j_2\pi; k_1, k_2)G_n(x, y; \widehat{k}_1, \widehat{k}_2)dxdy.$$

Owing to spatial periodicity of the G_n, G_m functions with a period π, the integral above can be finally rewritten as:

$$\left[\sum_{j_1=-\infty}^{+\infty} e^{i(\widehat{k}_1-k_1)j_1\pi}\right]\cdot\left[\sum_{j_2=-\infty}^{+\infty} e^{i(\widehat{k}_2-k_2)j_2\pi}\right] \qquad (P7.2.1)$$

$$\cdot\int_0^{\pi}\int_0^{\pi} e^{i(\widehat{k}_1-k_1)x}e^{i(\widehat{k}_2-k_2)y} G_m^*(x, y; k_1, k_2)G_n(x, y; \widehat{k}_1, \widehat{k}_2)dxdy.$$

(ii) Defining $u = e^{i(\widehat{k}-k)j\pi}$, from the well-known identity:

$$\sum_{j=-N}^{+N} u^j = \frac{u^{N+1} - u^{-N}}{u - 1} = \frac{u^{N+1/2} - u^{-N-1/2}}{u^{1/2} - u^{-1/2}},$$

it follows that:

$$\sum_{j_1=-N}^{+N} e^{i(\widehat{k}-k)j\pi} = \frac{e^{i(\widehat{k}-k)(N+1)\pi} - e^{-i(\widehat{k}-k)N\pi}}{e^{i(\widehat{k}-k)\pi} - 1} = \frac{\sin\left[(\widehat{k} - k)\pi(N + 1/2)\right]}{\sin\left[(\widehat{k} - k)\pi/2\right]}.$$

$$(P7.2.2)$$

(iii) Using the definition of the delta function in the form $\lim_{N\to\infty}$ $\sin(\alpha(N + 1/2))/\sin(\alpha/2) = 2\pi\,\delta(\alpha)$, we can now take a limit $N \to +\infty$ in the expression (P7.2.2):

$$\lim_{N\to+\infty}\sum_{j_1=-N}^{+N} e^{i(\widehat{k}-k)j\pi} = \lim_{N\to+\infty}\left(\frac{\sin\left[(\widehat{k} - k)\pi(N + 1/2)\right]}{\sin\left[(\widehat{k} - k)\pi/2\right]}\right)$$

$$= 2\pi\,\delta((\widehat{k} - k)\pi) = 2\delta(\widehat{k} - k). \qquad (P7.2.3)$$

(iv) The expression (P7.2.3) allows us to rewrite (P7.2.1) as:

$$\int_{-\infty}^{\infty}\int_{-\infty}^{\infty} B_m^*(x, y; k_1, k_2) B_n(x, y; \hat{k}_1, \hat{k}_2) dx dy$$

$$= (2\pi)^2 \, \delta(\hat{k}_1 - k_1)\delta(\hat{k}_2 - k_2) \left[\frac{1}{\pi^2} \int_0^\pi \int_0^\pi B_m^*(x, y; k_1, k_2) B_n(x, y; k_1, k_2) dx dy \right]$$

$$= (2\pi)^2 \, \delta(\hat{k}_1 - k_1)\delta(\hat{k}_2 - k_2)\delta_{m,n},$$

where we have used the normalization condition (7.51):

$$\frac{1}{\pi^2} \int_0^\pi \int_0^\pi B_m^*(x, y; k_1, k_2) B_n(x, y; k_1, k_2) dx dy$$

$$= \frac{1}{\pi^2} \int_0^\pi \int_0^\pi G_m^*(x, y; k_1, k_2) G_n(x, y; k_1, k_2) dx dy = \delta_{m,n}.$$

Note that the Kronecker delta in the last expression follows directly from the fact that two Bloch modes in question are solutions of a Hermitian eigenvalue problem (7.47) ($\varepsilon = 0$) with different eigenvalues, and, therefore, orthogonal to each other.

Solution of Problem 7.3

Example of a MATLAB code of the iteration method (7.69)–(7.71) for computing the defect mode in Fig. 7.8 (a). Other defect modes in (7.39) can be similarly computed.

```
Lx=10*pi; N=128; % mesh parameters
max_iteration=1e4; error_tolerance=1e-8;
x=-Lx/2:Lx/N:Lx/2-Lx/N; dx=Lx/N;
kx=[0:N/2-1 -N/2:-1]*2*pi/Lx;
y=x; dy=dx; ky=kx; [X, Y]=meshgrid(x, y);
[KX, KY]=meshgrid(kx, ky);
E0=15; I0=6; epsi=-0.6; c=4; DT=1.2;% scheme parameters
V=-E0./(1+I0*cos(X).^2.*cos(Y).^2.*(1+epsi*exp(-...
(X.^2+Y.^2).^4/128)));
U=exp(-(X.^2+Y.^2)/2.0); % initial conditions
for nn=1:max_iteration % iterations start
    Uold=U;
    LU=ifft2(-(KX.^2+KY.^2).*fft2(U))+V.*U;
    MinvLU=ifft2(fft2(LU)./(c+KX.^2+KY.^2));
    MinvU=ifft2(fft2(U)./(c+KX.^2+KY.^2));
    mu=-sum(sum(U.*MinvLU))/sum(sum(U.*MinvU));
    MinvLmuU=MinvLU+mu*MinvU;
    LmuMinvLmuU=ifft2(-(KX.^2+KY.^2).*fft2(MinvLmuU))+...
    (V+mu).*MinvLmuU;
    MinvLmuMinvLmuU=ifft2(fft2(LmuMinvLmuU)./(KX.^2+KY.^2+c));
    U=U-MinvLmuMinvLmuU*DT;
```

```
Uerror(nn)=sqrt(sum(sum(abs(U-Uold).^2))*dx*dy);Uerror(nn)
if Uerror(nn) < error_tolerance
break
end
end
imagesc(x, y, real(U))
```

Chapter 8

Solution of Problem 8.1

First, we show that in the domain of periodic functions with a period $2L$, the operator $\hat{H} = \partial^2/\partial x^2 + (\mu_0 - V(x))$ exhibits the following property:

$$\langle p|\,\hat{H}\,|u\rangle = \int_0^{2L} \left(p\partial^2 u/\partial x^2 + p\,(\mu_0 - V(x))\,u\right)dx$$

$$= p\partial u/\partial x|_0^{2L} - \int_0^{2L} (\partial p/\partial x \cdot \partial u/\partial x)dx + \int_0^{2L} (u\,(\mu_0 - V(x))\,p)dx$$

$$\underset{\substack{p(0)=p(2L)\\ \partial u/\partial x|_0 = \partial u/\partial x|_{2L}}}{=} \quad -u\partial p/\partial x|_0^{2L} + \int_0^{2L} \left(u\partial^2 p/\partial x^2 + u\,(\mu_0 - V(x))\,p\right)dx$$

$$\underset{\substack{u(0)=u(2L)\\ \partial p/\partial x|_0 = \partial p/\partial x|_{2L}}}{=} \quad \langle u|\,\hat{H}\,|p\rangle .$$

Now, multiplying on the left the nonhomogeneous equation $\hat{H}\,|u\rangle = f(x)$ by a solution $p(x)$ of a homogeneous equation $\hat{H}\,|p\rangle = 0$, and integrating over one period we get:

$$\int_0^{2L} p(x)f(x)dx = \langle p|\,\hat{H}\,|u\rangle = \langle u|\,\hat{H}\,|p\rangle = 0.$$

Solution of Problem 8.2

We first rewrite (8.10) as $p(x;\mu) = e^{i(k-k_0)x}q(x;\mu(k))$, where k_0 is the corresponding wavenumber of the band edge μ_0, and $q(x;\mu(k)) = e^{ik_0 x}\tilde{p}(x,\mu(k))$ is a periodic function with period $2L$. Note that $p(x;\mu_0) = q(x;\mu_0)$. Substituting the $p(x;\mu)$ function into (8.9), we have $q'' + 2i(k - k_0)q' - (k - k_0)^2 q - V(x)q + \omega q = 0$. Now we expand $\mu(k)$, $q(x,\mu(k))$ at the edge point $k = k_0$ as:

$$\mu = \mu_0 + D(k - k_0)^2 + O((k - k_0)^4)$$

$$q(x,\mu(k)) = p(x,\mu_0) + i(k - k_0)q^{(1)}(x) + (k - k_0)^2 q^{(2)}(x) + O((k - k_0)^3),$$

where D is given in (8.22). When these expansions are substituted into (8.9), to the order $O(k - k_0)$, we get $(q^{(1)})'' - V(x)q^{(1)} + \mu_0 q^{(1)} = -2p'(x; \mu_0)$ whose solution is $q^{(1)} = v(x)$ in view of (8.18). To the order $O((k - k_0)^2)$, we get:

$$(q^{(2)})'' - V(x)q^{(2)} + \mu_0 q^{(2)} = 2v'(x) + (1 - D)p(x; \mu_0).$$

For $q^{(2)}$ to have a periodic solution, it must satisfy the Fredholm condition:

$$\int_0^{2L} [2v'(x) + (1 - D)p(x; \mu_0)]p(x; \mu_0)dx = 0,$$

which is the identity (8.21).

Solution of Problem 8.3

First, we derive two constraints for the 2D envelope solutions. Multiplying (8.37) by $\bar{u}_x$ or $\bar{u}_y$, adding its conjugate equation, and integrating from $-\infty$ to $+\infty$, we get the following two constraints:

$$\int_{-\infty}^{\infty}\int_{-\infty}^{\infty} F'(x)\,|u(x, y)|^2\, dxdy = 0,$$

$$\int_{-\infty}^{\infty}\int_{-\infty}^{\infty} F'(y)\,|u(x, y)|^2\, dxdy = 0.$$

Substituting the perturbation expansion (8.43) of the solution $u(x, y)$ into the above equations, these constraints at the leading order become:

$$I_1(x_0, y_0) = \varepsilon^2 \int_{-\infty}^{\infty}\int_{-\infty}^{\infty} F'(x)\,|A_1 p_1(x)p_2(y) + A_2 p_2(x)p_1(y)|^2\, dxdy = 0, \quad \text{(P8.3.1)}$$

$$I_2(x_0, y_0) = \varepsilon^2 \int_{-\infty}^{\infty}\int_{-\infty}^{\infty} F'(y)\,|A_1 p_1(x)p_2(y) + A_2 p_2(x)p_1(y)|^2\, dxdy = 0. \quad \text{(P8.3.2)}$$

Here

$$A_k = A_k(X - X_0, Y - Y_0), k = 1, 2, \quad \text{(P8.3.3)}$$

and $(X_0, Y_0) = (\varepsilon x_0, \varepsilon y_0)$ is the center position of the envelope solution (A_1, A_2). The constraint (P8.3.1) can be rewritten as:

$$I_1(x_0, y_0) = \varepsilon^2 \int_{-\infty}^{\infty}\int_{-\infty}^{\infty} F'(x)[|A_1|^2\, p_1^2(x)p_2^2(y) + |A_2|^2\, p_2^2(x)p_1^2(y)$$
$$+ (A_1\bar{A}_2 + \bar{A}_1 A_2)p_1(x)p_2(x)p_1(y)p_2(y)]dxdy = 0. \quad \text{(P8.3.4)}$$

Since $F'(x)$ is antisymmetric, and $p_1^2(x)$, $p_2^2(x)$ are both symmetric in x, functions $F'(x)p_1^2(x)p_2^2(y)$, $F'(x)p_2^2(x)p_1^2(y)$, and $F'(x)p_1(x)p_2(x)p_1(y)p_2(y)$ have the following series expansions:

$$F'(x)p_1^2(x)p_2^2(y) = \sum_{m=1}^{\infty}\sum_{n=0}^{\infty} c_{m,n}^{(1)} \sin(2\pi mx/L)\cos(2\pi ny/L),$$

$$F'(x)p_2^2(x)p_1^2(y) = \sum_{m=1}^{\infty}\sum_{n=0}^{\infty} c_{m,n}^{(2)} \sin(2\pi mx/L)\cos(2\pi ny/L),$$

$$F'(x)p_1(x)p_2(x)p_1(y)p_2(y) = \sum_{m=1}^{\infty}\sum_{n=0}^{\infty} c_{m,n}^{(3)} \sin(2\pi mx/L)\cos(2\pi ny/L)$$

$$+ \sum_{m=0}^{\infty}\sum_{n=1}^{\infty} d_{m,n}^{(3)} \cos(2\pi mx/L)\sin(2\pi ny/L).$$

Here $d_{m,n}^{(3)} = 0$ or $c_{m,n}^{(3)} = 0$ (for all m, n) if $p_1(x)p_2(x)$ is even or odd respectively. Substituting the above Fourier expansions into (P8.3.4), to the leading order, we get:

$$I_1(x_0, y_0) = \varepsilon^2 \int_{-\infty}^{\infty}\int_{-\infty}^{\infty} \left\{ \left[c_{1,0}^{(1)}|A_1|^2 + c_{1,0}^{(2)}|A_2|^2 + c_{1,0}^{(3)}(A_1\overline{A}_2 + \overline{A}_1 A_2) \right] \sin(2\pi x/L) \right.$$

$$\left. + d_{1,0}^{(3)}(A_1\overline{A}_2 + \overline{A}_1 A_2)\sin(2\pi y/L) \right\} dxdy = 0.$$

Recalling (P8.3.3) and the symmetry assumptions in the exercise on envelope solutions, the above integral can be simplified to be:

$$I_1(x_0, y_0) = W_{1,1}\sin(2\pi x_0/L) + W_{1,2}\sin(2\pi y_0/L), \qquad (P8.3.5)$$

where

$$W_{1,1} = \varepsilon^2 \int_{-\infty}^{\infty}\int_{-\infty}^{\infty} \left\{ c_{1,0}^{(1)}|A_1(X, Y)|^2 + c_{1,0}^{(2)}|A_2(X, Y)|^2 \right.$$

$$\left. + c_{1,0}^{(3)}\left[A_1(X, Y)\overline{A}_2(X, Y) + \overline{A}_1(X, Y)A_2(X, Y) \right] \right\} \cos(2\pi x/L)dxdy,$$

$$W_{1,2} = \varepsilon^2 \int_{-\infty}^{\infty}\int_{-\infty}^{\infty} d_{1,0}^{(3)}\left[A_1(X, Y)\overline{A}_2(X, Y) + \overline{A}_1(X, Y)A_2(X, Y) \right] \cos(2\pi y/L)dxdy.$$

Repeating similar calculations for the integral of $I_2(x_0, y_0)$ in (P8.3.2), to the leading order, we can get:

$$I_2(x_0, y_0) = W_{2,1}\sin(2\pi x_0/L) + W_{2,2}\sin(2\pi y_0/L),$$

where expressions for $W_{2,1}$ and $W_{2,2}$ are similar to those for $W_{1,1}$ and $W_{1,2}$ above. Then, for the two constraints (S1) and (S2) to hold, we must have:

$$\sin(2\pi x_0/L) = \sin(2\pi y_0/L) = 0.$$

Thus, the envelope solution (A_1, A_2) can only be centered at four locations $(x_0, y_0) = (0, 0), (0, L/2), (L/2, 0), (L/2, L/2)$.

Index